Geminiviruses

R. Vinoth Kumar
Editor

Geminiviruses

Impact, Challenges and Approaches

Springer

Editor
R. Vinoth Kumar
School of Biological Sciences
University of East Anglia
Norwich, United Kingdom

ISBN 978-3-030-18250-2 ISBN 978-3-030-18248-9 (eBook)
https://doi.org/10.1007/978-3-030-18248-9

This Springer imprint is published by the registered company Springer Nature Switzerland AG.
The registered company address is: Gewerbestrasse 11, 6330 Cham, Switzerland

Preface

Viruses are the obligate parasites hindering the cultivation of all the economically important crops. Among them, the members of the family *Geminiviridae* cause significant crop loss by infecting monocots and dicots globally. Several researchers have made considerable efforts in studying the biology of geminiviruses and the diseases they cause. So, this book is intended to provide basic information about geminiviral diseases and their management. It introduces geminiviruses as one of the most important plant-infecting pathogens infecting several crops as well as weeds throughout the world. This book assembles the vast of knowledge on the distribution of geminiviruses and their associated satellites spread across various countries. It includes the genetic aspects of the viruses and diseases they cause among crop plants and several weeds. It analyses various evolutionary factors involved in the emergence of these geminiviral diseases. The combined knowledge of the type and distribution of geminiviruses helps in the designing of approaches to combat these pathogens. It also discusses the nomenclature and taxonomy of geminiviruses and the processes involved in its life cycle—replication and transcription. The important aspect of geminivirus biology is the involvement of the insect vectors in their cycle, and the details of this virus–vector relationship are included. In addition, a part of the book compiles the progress and advancements made in the molecular biology of the interactions and counter-interactions between the hosts and viruses. A chapter is dedicated to the advancement of trans-replication of satellite molecules and their effect on the geminiviral pathogenesis. Furthermore, the strategies and outcomes made to fight against these viruses through transgenics are just the tip of the iceberg. In the end, the approaches designed through integrated pest management to challenge these viruses in the field condition are detailed. Finally, this book assembles the recent and cutting-edge research on geminivirus—spread, pathogenesis and disease management. I believe that a book providing a broader knowledge of geminivirus biology is of great value to a wide range of readers from graduates to advanced researchers, breeders and plant pathologists who are familiar with plant virology.

Norwich, United Kingdom R. Vinoth Kumar

Contents

List of Contributors

Nicolas Bejerman Consejo Nacional de Investigaciones Científicas y Técnicas (CONICET), CABA, Argentina

Instituto de Patología Vegetal – Centro de Investigaciones Agropecuarias – Instituto Nacional de Tecnología Agropecuaria (IPAVE-CIAP-INTA), Córdoba, Argentina

Bhavin S. Bhatt Shree Ramkrishna Institute of Computer Education and Applied Sciences, Surat, India

Fenisha D. Chahwala School of Life Sciences, Central University of Gujarat, Gandhinagar, India

Omid Eini Department of Plant Protection, School of Agriculture, University of Zanjan, Zanjan, Iran

R. K. Gaur Department of Biosciences, Faculty of Arts, Science and Commerce, Mody University, Lakshmangarh, Sikar, Rajasthan, India

Vipin Hallan Academy of Scientific & Innovative Research (AcSIR), CSIR-Institute of Himalayan Bioresource Technology, Palampur, Himachal Pradesh, India

Plant Virology Lab, CSIR-IHBT, Palampur, Himachal Pradesh, India

Amir Hameed Department of Bioinformatics & Biotechnology, Government College University (GCU), Faisalabad, Pakistan

Zafar Iqbal Department of Plant Pathology, University of Florida, Gainesville, FL, USA

K. Kathiravan Molecular Plant Virology Laboratory, Department of Biotechnology, University of Madras, Guindy Campus, Chennai, Tamil Nadu, India

Jawaid A. Khan Plant virology Lab, Department of Biosciences, Jamia Millia Islamia, New Delhi, India

Aditya Kulshreshtha Academy of Scientific & Innovative Research (AcSIR), CSIR-Institute of Himalayan Bioresource Technology, Palampur, Himachal Pradesh, India

Plant Virology Lab, CSIR-IHBT, Palampur, Himachal Pradesh, India

Abhinav Kumar Research projects Lab, Department of Biotechnology, IILM College of Engineering and Technology, Greater Noida, Uttar Pradesh, India

R. Vinoth Kumar School of Biological Sciences, University of East Anglia, Norwich, UK

National Centre for Biological Sciences (NCBS-TIFR), Bengaluru, Karnataka, India

Rajesh Kumar Department of Biosciences, Faculty of Arts, Science and Commerce, Mody University, Lakshmangarh, Sikar, Rajasthan, India

Avinash Marwal Department of Biotechnology, Mohanlal Sukhadia University, Udaipur, Rajasthan, India

Megha Mishra Department of Biosciences, Faculty of Arts, Science and Commerce, Mody University, Lakshmangarh, Sikar, Rajasthan, India

S. U. Mohammed Riyaz Centre for Ocean Research, Sathyabama Institute of Science and Technology, Chennai, Tamil Nadu, India

Molecular Plant Virology Laboratory, Department of Biotechnology, University of Madras, Guindy Campus, Chennai, Tamil Nadu, India

Poonam Roshan Academy of Scientific & Innovative Research (AcSIR), CSIR-Institute of Himalayan Bioresource Technology, Palampur, Himachal Pradesh, India

Plant Virology Lab, CSIR-IHBT, Palampur, Himachal Pradesh, India

Rajrani Ruhel School of Life Sciences, Jawaharlal Nehru University, New Delhi, India

Sangeeta School of Life Sciences, Central University of Gujarat, Gandhinagar, India

Muhammad N. Sattar Department of Biotechnology, Muhammad Nawaz Sharif University of Agriculture, Multan, Pakistan

Department of Biotechnology, College of Agriculture and Food Science, King Faisal University, Al-Hasa, Kingdom of Saudi Arabia

Sara Shakir Boyce Thompson Institute, Ithaca, NY, USA

Nivedita Sharma School of Life Sciences, Jawaharlal Nehru University, New Delhi, India

Adel Ali Mohammed Al Shihi Department of Plant Protection, Ministry of Agriculture and Fisheries Wealth, Muscat, Oman

Achuit Kumar Singh Crop Improvement Division, ICAR – Indian Institute of Vegetable Research, Varanasi, India

Rakesh Kumar Verma Department of Biosciences, Faculty of Arts, Science and Commerce, Mody University, Lakshmangarh, Sikar, Rajasthan, India

B. K. Yadav School of Life Sciences, Central University of Gujarat, Gandhinagar, India

Syed Shan-e-Ali Zaidi National Institute for Biotechnology and Genetic Engineering (NIBGE), Faisalabad, Pakistan

Gembloux Agro-Bio Tech, University of Liège, Gembloux, Belgium

Classification, Taxonomy and Gene Function of Geminiviruses and Their Satellites

R. Vinoth Kumar

Abstract

The major constraint for the crop productivity throughout the world is due to the diseases caused by viruses. These plant-infecting viruses emerged as an unavoidable limiting factor and are responsible for severe crop losses in all major economically important plants. Among them, the members belonging to the family, *Geminiviridae*, are the most devastating pathogens that are transmitted by insect vectors. These geminiviruses cause diseases such as chlorotic, dwarf, leaf curl, mosaic, yellow mosaic and yellow vein in monocots and dicots across the tropical and sub-tropical countries. In addition, these geminiviruses use weeds as reservoir for the spread of diseases. Moreover, these viruses encode only a fewer proteins and rely majorly on the host factors for their replication, disease development and spread. This chapter introduces the readers to the classification and taxonomy of geminiviruses, genus/species demarcation thresholds and the nature of genomic component and satellites associated with geminiviral disease complexes. It also discusses the genome organization of viruses grouped into different genera, before giving a glimpse of the important functions of gene products it encode.

1 Family: *Geminiviridae*

Geminiviridae comprises of a group of plant-infecting insect-transmitted viruses containing non-enveloped circular ssDNA genome of ~2.8 kb in size. These viruses cause substantial crop losses in a large number of economically important vegetable

R. V. Kumar (✉)
School of Biological Sciences, University of East Anglia, Norwich, United Kingdom

National Centre for Biological Sciences (NCBS-TIFR), Bengaluru, Karnataka, India
e-mail: vinoth86_sls@jnu.ac.in

R. V. Kumar (ed.), *Geminiviruses*, https://doi.org/10.1007/978-3-030-18248-9_1

and food crops, ornamental plants and fibre crops worldwide (Navas-Castillo et al. 2011). In 1977, Harrison et al. coined the term '*Geminivirus*' based on their nature of twinned icosahedral particles. The geminate particles are packed in two incomplete icosahedra containing 22 pentameric capsomeres (Hesketh et al. 2018). Within one twinned particle, only one molecule of ssDNA can be encapsidated; thus for a bipartite genome, DNA-A and DNA-B are carried by two twinned particles (Jeske 2009). Virus replication occurs through rolling-circle and recombination-dependent mechanisms, but they do not encode any DNA polymerase (Jeske 2009). Hence they rely entirely on the infected cells for synthesizing their complementary strand in the nucleus by employing various host replication factors. For viral gene expression, the virus transcription occurs bi-directionally, and produces transcripts from both the complementary and virion sense strands leading to the generation of several overlapping viral transcripts (Brown et al. 2012).

1.1 Geminivirus Taxonomy

The usage of new molecular tools, such as rolling-circle amplification and high-throughput sequencing in the last decades have greatly helped the geminivirologists in identifying several novel geminivirus-like genomic components (Roossinck et al. 2015). Because of this near-global occurrence of several distinct geminiviruses, the *International Committee on Taxonomy of Viruses* (ICTV) has devised several guidelines for classifying the geminiviruses at genus level. Based on the insect vector, genome organization, genome-wide pairwise sequence identities and host range, nine genera, such as *Becurtovirus, Begomovirus, Capulavirus, Curtovirus, Eragrovirus, Grablovirus, Mastrevirus, Topocuvirus* and *Turncurtovirus* are included in *Geminiviridae* by geminivirus study group of the ICTV (Zerbini et al. 2017). Aphids, leafhoppers, treehoppers or whiteflies can transmit these geminiviruses.

1.1.1 Genus: *Becurtovirus*

The type species of the genus *Becurtovirus* is *Beet curly top Iran virus* (BCTIV) and two closely related groups, *Spinach curly top Arizona virus* (SCTAV) and *Exomis microphylla latent virus* are also included in this genus. Among these two species, BCTIV comprises four different strains (BCTIV-A, B, C, D). Unlike other geminiviruses, these members possess unique nona-nucleotides (TAAGATTCC) with a spliced replication-initiator protein (Rep) in the complementary sense strand (Fig. 1). The species and strain demarcation threshold value for this genus is fixed as 80% and 94%, respectively (Varsani et al. 2014a). The isolates belonging to BCTIV (A–D) has so far found infecting various dicot hosts, such as *Beta vulgaris*, *Vigna unguiculata*, *Solanum lycopersicum* and *Phaseolus vulgaris* in Iran (Yazdi et al. 2008; Soleimani et al. 2013). Moreover, an isolate of SCTAV was identified from *S. oleracea* plants in the Arizona region of USA (Hernandez-Zepeda et al. 2013). These dicot plants infecting becurtoviruses are reported to be transmitted by leafhoppers (Zerbini et al. 2017).

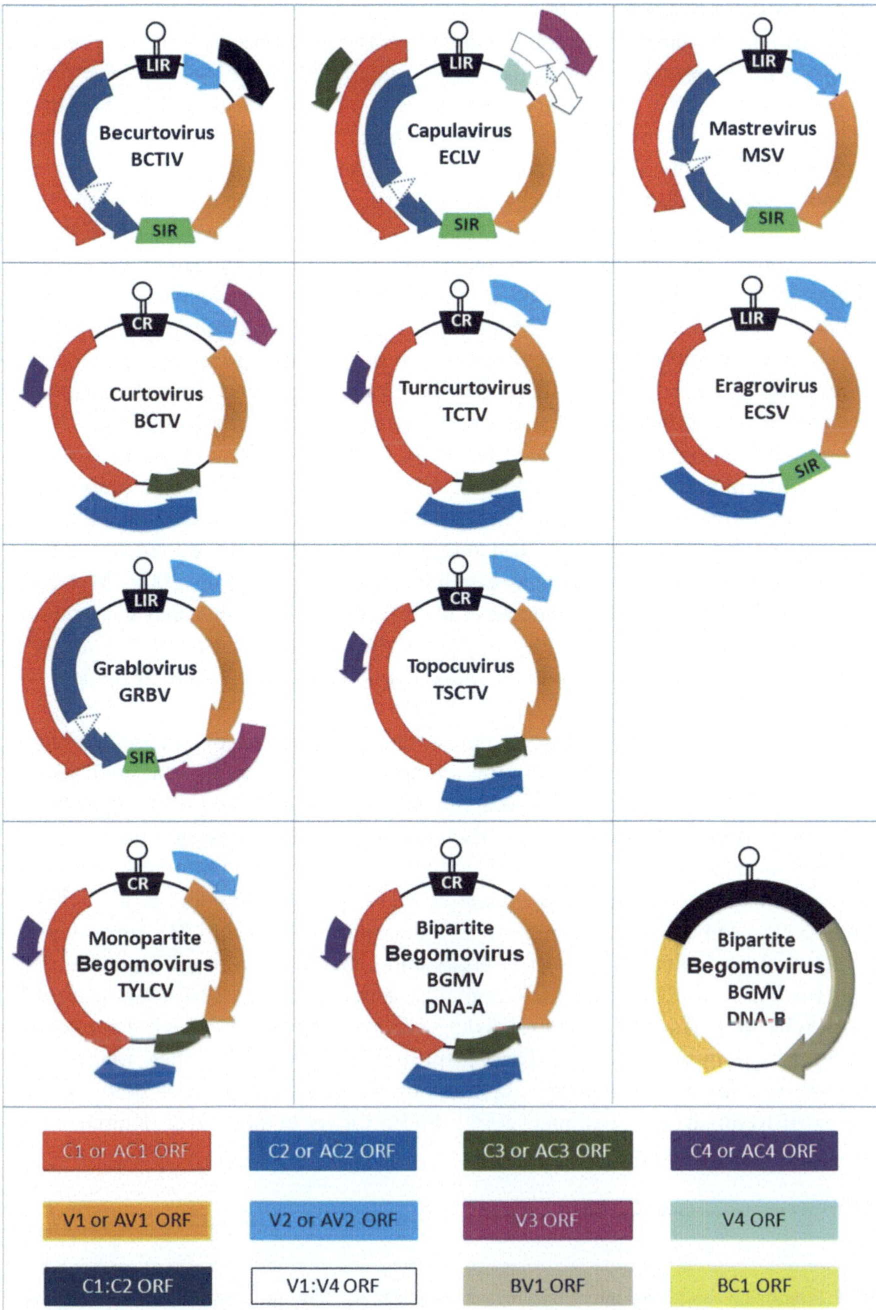

Fig. 1 *Genome organization of the representative geminiviruses belonging to different genera in the Geminiviridae family*. The virus-encoded ORFs are given in different colours and its description is provided at the bottom of the figure. The stem and loop structure indicates location of nona-nucleotides in the genome. The CR, SIR and LIR refer to common region, short intergenic region

1.1.2 Genus: *Begomovirus*

The genus name is derived from the name of its type member, *Bean golden mosaic virus* (BGMV). *Begomovirus* is the largest genus of the *Geminiviridae* family containing ~300 ICTV recognized virus species (Zerbini et al. 2017). The insect vectors, whiteflies (*Bemisia tabaci Genn.*) can transmit these viruses, and infect both monocots and dicots (Navas-Castillo et al. 2011; Brown et al. 2012). These begomoviruses are classified either as monopartite or bipartite (Brown et al. 2012) (Fig. 1). Monopartite begomoviruses that contain only DNA-A-like molecule are phloem limited and are not sap transmissible, whereas sap transmissible bipartite begomoviruses (with similar-sized DNA-A and DNA-B molecules) infect both phloem and non-phloem tissues (Melgarejo et al. 2013). According to the genome organization and phylogenetic segregation, these begomoviruses are divided into two regions: 'Old world (OW)' that includes Africa, Asia, Australia and Europe and 'New World (NW)' that includes America, Brazil and Mexico (Brown et al. 2012). However, the probable centre of its origin is found to be around South-east Asia (Nawaz-ul-Rehman and Fauquet 2009). Several factors such as recombination, pseudo-recombination, synergism, microsatellites, mutation, nucleotide diversity/ substitutions and vector transmission might be influencing the evolution of these begomoviruses and their associated satellites (Chakraborty et al. 2008; Melgarejo et al. 2013; Lima et al. 2017; Kumar et al. 2015b, 2017; Kumar and Chakraborty 2018). The strain demarcation threshold for this genus has been fixed as 94%, however the pairwise sequence identities of 91% have been proposed for demarcation of new species (Brown et al. 2015).

The DNA-A component have a conserved arrangement of six open reading frames (ORFs): two ORFs in the virion strand (AV1, AV2 in bipartite and V1, V2 in monopartite) and four in the complementary strand (AC1–AC4 in bipartite and C1–C4 in monopartite) (Fig. 1). The DNA-B component of bipartite viruses encode a movement protein (BC1) and a nuclear shuttle protein (BV1) to help in intra- and inter-cellular viral movements (Brown et al. 2012; Hanley-Bowdoin et al. 2013). A highly conserved non-coding region called common region (CR) separates the bidirectional transcription units of DNA-A and DNA-B (Lazarowitz and Shepherd 1992; Brown et al. 2012). The CR contains stem loop-structured nona-nucleotide (TAATATTAC) which acts as a cleavage site for Rep to initiate viral replication (Hanley-Bowdoin et al. 2013). Monopartite begomoviruses are often found along with satellite molecules namely, alphasatellites, betasatellites and deltasatellites (Nawaz-ul-Rehman and Fauquet 2009; Fiallo-Olive et al. 2012; Kumar et al. 2015a, 2017; Lozano et al. 2016).

Fig. 1 (continued) and long intergenic region, respectively. The full name of the virus isolates abbreviated here is *BCTIV Beet curly top Iran virus*, *ECLV Euphorbia caput-medusae latent virus*, *MSV Maize streak virus*, *BCTV Beet curly top virus*, *TCTV Turnip curly top virus*, *ECSV Eragrostis curvula streak virus*, *GRBV Grapevine red blotch virus*, *TSCTV Tomato pseudo curly top virus*, *TYLCV Tomato yellow leaf curl virus* and *BGMV Bean golden mosaic virus*

1.1.3 Genus: *Capulavirus*

The type species for the genus *Capulavirus* is *Euphorbia caput-medusae latent virus* (EcmLV). These EcmLV isolates were reported to infect *Euphorbia caput-medusae* plants in the South Africa. Roumagnac et al. (2015) observed the geminate particles by transmission electron microscopy from the purified preparations of EcmLV. Some of the virus species, such as *Alfalfa leaf curl virus* (ALCV), *French bean severe leaf curl virus* (FbSLCV) and *Plantago lanceolata latent virus* (PlLV) are also included in this genus. ALCV isolates were transmitted by aphids and are reported to infect *Medicago sativa* plants in France (Roumagnac et al. 2015). Similarly, in India, FbSLCV causes severe leaf curl disease in *Phaseolus vulgaris* and PlLV was identified in *Plantago lanceolata* from Finland (Roumagnac et al. 2015; Susi et al. 2017; Varsani et al. 2017).

The origin of replication sequence of this group of closely related viruses is found to be TAATATTAC. In common with the viruses belonging to the genus, *Becurtovirus* and *Mastrevirus*, the members of this genus also possess a long intergenic region (LIR) and a short intergenic region (SIR) along with a Rep protein which is expressed as a spliced complementary strand transcript (Varsani et al. 2017). The presence of a possible movement protein encoding ORF located in the 5′ direction from the coat protein differentiates them from the members of the other established genera of the *Geminiviridae* family (Fig. 1).

1.1.4 Genus: *Curtovirus*

The name *Curtovirus* was derived from the name of the type species of this genus: *Beet curly top virus* (BCTV) which is of the size of nearly 3 kb (Zerbini et al. 2017). Additionally, the *Spinach severe curly top virus* (SpSCTV) and *Horseradish curly top virus* (HrCTV) infecting *Spinacia oleracea* and *Armoracia rusticana* plants, respectively, are also grouped to this genus. They are transmitted by beet leafhopper (*Circulifer tenellus*) and infect several dicotyledonous plants, such as *Beta vulgaris*, *Capsicum annum*, *Phaseolus vulgaris* cv. Aluvori, *Solanum lycopersicum* and *Spinacia oleracea* (Stenger et al. 1990; Soto and Gilbertson 2003; Hernandez-Zepeda et al. 2013). Its geographical range was found around the Mediterranean region, the Middle East, the Indian subcontinent and North and Central America (Varsani et al. 2014b). Like other genera, they are monopartite in nature sharing less sequence homology with them. Its genome encodes for three ORFs in the virion strand and four ORFs in the complementary strand along with a SIR (Fig. 1). Virion sense strand encodes a coat protein, a regulator protein and a putative movement protein, whereas complementary sense strand encodes a Rep, replication enhancer (REn) protein, a silencing suppressor protein and a pathogenicity associated protein (Hormuzdi and Bisaro 1995).

1.1.5 Genus: *Eragrovirus*

Currently, *Eragrostis curvula streak virus* (ECSV) is the only species included in this genus. This virus codes for four ORFs (two ORFs each in the virion sense and the complementary sense strand) and the isolates of ECSV have nona-nucleotide motif as TAAGATTCC, at their presumed origin of viral DNA replication (Varsani et al. 2014a). The coat protein of ECSV resembles to the members of *Mastrevirus*, but the replication protein is quite similar to the viruses belonging to the genus, *Begomovirus*. The two strains of ECSV (A and B) were reported to be infecting a monocot species (*E. curvula*) in the South Africa (Varsani et al. 2009). The genome-wide per cent pairwise comparisons of 75% (for species) and 94% (for strain) are proposed as the demarcation threshold for the members of *Eragrovirus* (Varsani et al. 2014a).

1.1.6 Genus: *Grablovirus*

The type species of *Grablovirus* genus is *Grapevine red blotch virus* (GRBV) and *Prunus latent virus* and *Wild Vitis latent virus* are the other two virus species assigned to this genus. The isolates of GRBV are transmitted by the alfalfa treehoppers (Krenz et al. 2012; Bahder et al. 2016). The GRBV isolates infecting *Vitis vinifera* have been reported mainly from USA, followed by Canada and South Korea (Cieniewicz et al. 2018). These viruses possess TAATATTAC as the nona-nucleotide sequence for the initiation of virion sense strand synthesis (Varsani et al. 2017). The genome arrangement of its complementary sense strand appears similar to the members of *Capulavirus*, whereas it encodes three ORFs (V1, V2 and V3) in its virion sense strand (Fig. 1).

1.1.7 Genus: *Mastrevirus*

The type species for this genus is *Maize streak virus* (MSV), a monopartite virus that is transmitted by leafhoppers (Monjane et al. 2011; Zerbini et al. 2017). Mastreviruses are 2.7–2.8 kb in size and 37 distinct members are considered to be recognized species within this genus (Table 1). Eleven strains of *Maize streak virus* (MSV-A–K) and nine strains of *Panicum streak virus* (PanSV-A–I) are predominantly reported from the African countries (Muhire et al. 2013). The presence of the isolates of *Chickpea chlorosis virus*, *Chloris striate mosaic virus*, *Digitaria ciliaris striate mosaic virus*, *Digitaria didactyla striate mosaic virus*, *Paspalum striate mosaic virus* and *Sporobolus striate mosaic virus-1* and *2* are documented mainly from Australia (Kraberger et al. 2012, 2014). In general, these viruses are mostly found confined to the African countries and Australia, infecting both monocot and dicot plants, such as *Brachiaria* sp., *Cicer arietinum*, *Digitaria* sp., *Panicum maximum*, *Setaria* sp., *Urochloa* sp. and *Zea mays* (Kraberger et al. 2014). Moreover, mastreviruses are also identified from Germany, India, Japan and Pakistan (Muhire et al. 2013; Kraberger et al. 2014). Importantly, *Wheat dwarf virus* isolates infecting *Avena sativa*, *Hordeum vulgare* and *Triticum aestivum* plants are reported from China, Iran and several European countries (Kvarnheden et al. 2002; Ramsell et al. 2009).

The genome of the members of mastreviruses has nona-nucleotides (TAATATTAC) similar to other geminiviruses. Its genome encodes for two ORFs from the virion strand (capsid protein and movement protein) and the complementary strand encoded Rep protein expresses as a splicing product of C1 and C2 ORFs (Fig. 1). The virion sense strand encoded proteins are necessary for viral movement and encapsidation; whereas replication-associated proteins are encoded in the

Table 1 Type species, insect vector and species demarcation threshold of the genera in *Geminiviridae* family

Genera	Type species (abbreviated name)	ICTV recognized virus species	Insect vectors (common name)	Nona-nucleotides	Demarcation threshold at	
					Species level (%)	Strain level (%)
Becurtovirus	*Beet curly top Iran virus* (BCTIV)	3	*Circulifer haematoceps* (Leafhoppers)	TAAGATTCC	80	94
Begomovirus	*Bean golden mosaic virus* (BGMV)	388	*Bemisia tabaci* (Whiteflies)	TAATATTAC	91	94
Capulavirus	*Euphorbia caput-medusce latent virus* (EcmLV)	4	*Aphis craccivora* (Aphids)	TAATATTAC	78	–
Curtovirus	*Beet curly top virus* (BCTV)	3	*Circulifer tenellus* (Leafhoppers)	TAATATTAC	77	94
Mastrevirus	*Maize streak virus* (MSV)	37	*Cicadulina* sp., *Nesoclutha* sp., *Psammotettix alienus, Orosius* sp. (Leafhoppers)	TAATATTAC	78	94
Eragrovirus	*Eragrostis curvula streak virus* (ECSV)	1	Not identified	TAAGATTCC	75	94
Grablovirus	*Grapevein red blotch virus* (GRBV)	3	*Spissistilus festinus* (Treehoppers)	TAATATTAC	80	–
Topocuvirus	*Tomato severe curly top virus* (TSCTV)	1	*Micrutalis malleifera* (Treehoppers)	TTATATTAC	–	–
Turncurtovirus	*Turnip curly top virus* (TCTV)	2	*Circulifer haematoceps* (Leafhoppers)	TAATATTAC	75	95

complementary strand. Further, these ORFs are separated by LIR and SIR containing the origin of replication for the synthesis of virion and complementary strands, respectively (Kammann et al. 1991). They possess a unique characteristic of regulating their own gene expression through a post-transcriptional splicing event (Rojas et al. 2005).

1.1.8 Genus: *Topocuvirus*

A monopartite geminivirus, *Tomato pseudo curly top virus* of ~3 kb in size is the type member of this genus. This virus encodes two ORFs in the virion sense strand and four ORFs in the complementary sense strand. It is transmitted by treehoppers (*Micrutalis malleifera*) to dicot plants in the NW. Based on the genome organization, this virus species appears to be a recombinant between the genera, *Mastrevirus* and *Begomovirus* (Briddon et al. 1996).

1.1.9 Genus: *Turncurtovirus*

Only member in this genus includes *Turnip curly top virus* which is identified from *Brassica rapa* or *Raphanus sativus* plants (Briddon et al. 2010). Recently, *Turnip leaf roll virus* is also included in this genus. These phylogenetically distinct members are most closely resembled to the members of genus, *Curtovirus*. Its genome encodes six rather than seven proteins. It contains nona-nucleotides (TAATATTAC) similar to *Mastrevirus*, *Begomovirus*, *Curtovirus* and *Topocuvirus* (Table 1). A tentative strain demarcation threshold of 95% has been assigned and based on this criterion, four strains of *Turnip curly top virus* (TCTV-A–D) were proposed (Varsani et al. 2014a). These viruses are also identified from *Anchusa* sp., *Descurainia sophia*, *Hibiscus trionum* and *Solanum americanum* plants (Razavinejad and Heydarnejad 2013; Razavinejad et al. 2013).

1.2 Biological Functions of Geminivirus Components

1.2.1 DNA-A Component

The proteins encoded by DNA-A component are involved in the replication, gene expression, virus movement and encapsidation, and suppression of plant immunity.

The C1/AC1 protein is also called as replication-initiator protein (Rep) that is very much essential for the virus replication (Settlage et al. 1996; Hanley-Bowdoin et al. 2013). Though Rep proteins do not possess any similarity with known polymerases, it does share some similarity with the bacterial plasmid-encoded replication initiator proteins which undergo replication through rolling-circle mechanism (Ilyina and Koonin 1992). Also Rep proteins of begomovirus and a mastrevirus have been shown to be RNAi suppressors (Rodriguez-Negrete et al. 2013; Wang et al. 2014).

The transcription activator protein (TrAP) encoded by the C2/AC2 ORF is a multi-functional protein involved in virus replication, transactivation of late viral genes (which are needed for virus encapsidation) and several other host genes (Sunter and Bisaro 1992; Trinks et al. 2005; Caracuel et al. 2012). Several TrAPs also possess transcriptional gene silencing (TGS) and/or post-transcriptional gene silencing (PTGS) suppression activity (Dong et al. 2003; Buchmann et al. 2009; Hanley-Bowdoin et al. 2013; Kumar et al. 2015b).

REn protein encoded by C3/AC3 ORF can form homo-oligomers and it also heterodimerizes with Rep proteins (Pasumarthy et al. 2010). In addition, REn protein is also known to assist Rep protein in virus replication by interacting with various cell cycle regulators, such as pRBR and PCNA (Settlage et al. 1996; Castillo et al. 2003).

Like other viral proteins, C4/AC4-encoded proteins are shown to suppress host's gene silencing machinery, and these proteins also regulate host's brassinosteroid signalling, CLAVATA pathway and cell cycle regulation (Ismayil et al. 2018; Lai et al. 2009; Li et al. 2018; Mei et al. 2018).

The virion sense strand encoded coat protein (V1/AV1) and pre-coat protein (V2/AV2) are mainly involved in intra- and inter-cellular movements, virus encapsidation and insect transmission (Fondong 2013; Rojas et al. 2001; Ward and Lazarowitz 1999). In addition, pre-coat proteins of *Tomato yellow leaf curl virus* and BCTV possess suppression of TGS and PTGS activity, respectively (Wang et al. 2018; Luna et al. 2017). Also the interplay between host's RDR1 and pre-coat protein of tomato-infecting begomoviruses in symptom remission is demonstrated (Basu et al. 2018).

1.2.2 DNA-B Component

The DNA-A and DNA-B sequences are known to be divergent except a 100–200 nucleotide common region (Brown et al. 2012). This region is an important determinant for the replication of the viral genome by Rep proteins (Hanley-Bowdoin et al. 2013). In 1993, Evans and Jeske demonstrated that DNA-B component of *Abutilon mosaic virus* facilitates the spread of DNA-A component, but DNA-B is not found to be essential.

The DNA-B component encodes two ORFs, one each in virion and complementary strands which assist in virus movement (Hehnle et al. 2004; Lewis and Lazarowitz 2010). The BV1 is a movement protein (MP), which localizes in the plasma membrane, and helps in the intra- and inter-cellular movement of the viral molecules. The BC1, a nuclear shuttle protein (NSP) assists in the nucleo-cytoplasmic transport of viral components (Lazarowitz and Shepherd 1992). Hehnle et al. (2004) studied the interaction of NSP with plasmodesmata to increase the size exclusion limit of plasma membrane for the efficient transport of viral components between the host cells. The MPs are pathogenicity determinants and with NSP, they also play an important role in determining the host range (Pascal et al. 1993; Noueiry et al. 1994). In addition to their role in virus transportation, they also modulate host-mediated antiviral properties either by interfering mRNA decapping activity (by MP) or through translational suppression by NSP-interacting kinase 1 (by NSP) (Ye et al. 2015; Zorzatto et al. 2015).

1.2.3 Betasatellites

The monopartite begomoviruses are widely evolved by associating themselves with the satellite molecules called as betasatellites (Zhou 2013; Kumar and Chakraborty 2018). These satellite molecules are half the size (1.3 kb) of the helper component and encode a single ORF (βC1) in the complementary strand. These betasatellites require helper begomoviruses for their replication, encapsidation, insect transmission and systemic spread (Briddon and Stanley 2006). The selective replication of betasatellite by helper begomovirus-encoded Rep protein involves a novel DNA motif (Zhang et al. 2015). Furthermore, Ranjan et al. (2014) demonstrate the host specific role in the trans-replication or adaptation of betasatellites by distinct tomato-infecting begomoviruses. Several betasatellites have been reported to be pathogenicity determinant and help in symptom induction (Briddon and Stanley 2006; Gnanasekaran et al. 2019; Sivalingam and Varma 2012; Kumar et al. 2015a).

The betasatellite-encoded βC1 proteins have been found to be localized in the nucleus, cytoplasm and/or chloroplast (Cui et al. 2005; Bhattacharyya et al. 2015). In addition, it has been reported to possess both TGS and PTGS silencing suppression activity (Cui et al. 2005; Yang et al. 2011; Zhou 2013). The βC1 proteins are known to bind DNA, interact with SNF1-related kinase and S-adenosyl homocysteine hydrolase, and subvert host ubiquitination machinery to prevent its degradation (Yang et al. 2011; Shen et al. 2011, 2016; Jia et al. 2016).

1.2.4 Alphasatellites

Alphasatellites are circular ssDNA molecules of ~1.4 kb in size which are generally found with monopartite begomovirus-betasatellite complexes in the 'OW' (Zhou 2013; Siddiqui et al. 2016; Kumar et al. 2017). However, a few of them are also reported to be with the 'NW' begomoviruses (Paprotka et al. 2010; Romay et al. 2010). These alphasatellites can replicate independently in the infected cells, however, it depends on the helper viruses for insect transmission and systemic spread (Saunders and Stanley 1999; Kumar et al. 2017). A group of alphasatellites have been shown to ameliorate disease in the infected host by decreasing the accumulation of betasatellite molecules (Idris et al. 2011). However, *Chilli leaf curl alphasatellite* is found dispensable for symptom induction in the agro-inoculated *Nicotiana benthamiana* plants (Kumar et al. 2017). The capability of *Euphorbia yellow mosaic alphasatellite* in modulating symptoms, viral accumulation and whitefly transmission of the associated helper virus is recently reported (Mar et al. 2017). Also unusual combination of mastrevirus-satellite complexes have been identified from India and Puerto Rico in wheat and dragonflies, respectively (Rosario et al. 2013; Kumar et al. 2014). But the biological significance of these satellites in spreading the disease is not well studied.

1.2.5 Deltasatellites

Novel classes of helper begomovirus dependent satellite molecules called deltasatellites are identified from both 'OW' and 'NW'. These molecules are one-fourth of the size (600–750 nucleotides) of the helper begomoviruses and possess A-rich region (similar to betasatellites), a primary stem-loop sequence containing the nona-nucleotides, TAATATTAC. In addition, they also have a secondary stem-loop structure located between the SCR-like and A-rich regions. Unlike other geminivirus-associated satellites (alphasatellites and betasatellites), deltasatellites does not encode any ORFs, but it depends entirely on the helper geminiviruses. These molecules were identified from the geminivirus-infected plants, such as *Ipomeas* spp., *Malvastrum coromandelianum*, *Merremia dissecta* and *Sidastrum micranthum* (Fiallo-Olive et al. 2012; Lozano et al. 2016). Furthermore, deltasatellites were identified from *Bemisia tabaci* and are known to reduce the helper virus accumulation (Fiallo-olive et al. 2016). The effect of sweepovirus-associated deltasatellites on their helper viruses and its whitefly transmissibility is also ascertained (Hassan et al. 2016).

Acknowledgement The European Commission for granting Erasmus Mundus Action 2 post-doctorate scholarship through the BRAVE project (Grant: 2013-2536/001-001) is acknowledged.

References

Bahder BW, Zalom FG, Jayanth M, Sudarshana MR (2016) Phylogeny of geminivirus coat protein sequences and digital PCR aid in identifying *Spissistilus festinus* as a vector of grapevine red blotch-associated virus. Phytopathology 106:1223–1230

Basu S, Kushwaha NK, Singh AK, Sahu PP, Kumar RV, Chakraborty S (2018) Dynamics of a geminivirus encoded pre-coat protein and host RNA-dependent RNA polymerase 1 in regulating symptom recovery in tobacco. J Exp Bot 69(8):2085–2102

Bhattacharyya D, Prabu G, Kumar RK, Kushwaha NK et al (2015) A geminivirus betasatellite damages structural and functional integrity of chloroplasts leading to symptom formation and inhibition of photosynthesis. J Exp Bot 66(19):5881–5895

Briddon RW, Bedford ID, Tsai JH, Markham PG (1996) Analysis of the nucleotide sequence of the treehopper-transmitted geminivirus, tomato pseudo-curly top virus, suggests a recombinant origin. Virology 219(2):387–394

Briddon RW, Stanley J (2006) Subviral agents associated with plant single stranded DNA viruses. Virology 344:198–210

Briddon RW, Heydarnejad J, Khosrowfar F, Massumi H et al (2010) *Turnip curly top virus*, a highly divergent geminivirus infecting turnip in Iran. Virus Res 152:169–175

Brown JK, Fauquet CM, Briddon RW, Zerbini FM et al (2012) Family *Geminiviridae*. In: King AMQ, Adams MJ, Carstens EB, Lefkowitz EJ (eds) Virus taxonomy 9th report of the *international committee on taxonomy of viruses*. Elsevier Academic Press, London, pp 351–373

Brown JK, Zerbini FM, Navas-Castillo J, Moriones E et al (2015) Revision of *Begomovirus* taxonomy based on pairwise sequence comparisons. Arch Virol 160:1593–1619

Buchmann RC, Asad S, Wolf JN, Mohannath G, Bisaro DM (2009) Geminivirus AL2 and L2 proteins suppress transcriptional gene silencing and cause genome-wide reductions in cytosine methylation. J Virol 83:5005–5013

Caracuel Z, Lozano-Durán R, Huguet S, Arroyo-Mateos M, Rodríguez-Negrete EA, Bejarano ER (2012) C2 from *Beet curly top virus* promotes a cell environment suitable for efficient replication of geminiviruses, providing a novel mechanism of viral synergism. New Phytol 194 (3):846–858

Castillo AG, Collinet D, Deret S, Kashoggi A, Bejarano ER (2003) Dual interaction of plant PCNA with geminivirus replication accessory protein (REn) and viral replication protein (Rep). Virology 312:381–394

Chakraborty S, Vanitharani R, Chattopadhyay B, Fauquet CM (2008) More virulent recombination and asymmetric synergism between two distinct species of begomoviruses causing tomato leaf curl disease in India. J Gen Virol 89:818–828

Cieniewicz E, Thompson JR, McLane H, Perry KL, Dangl GS et al (2018) Prevalence and genetic diversity of Grabloviruses in free-living Vitis spp. Plant Dis 102(11):2308–2316.

Cui X, Li G, Wang D, Hu D, Zhou X (2005) A begomovirus DNA-β encoded protein binds DNA, functions as a suppressor of RNA silencing, and targets the cell nucleus. J Virol 79:10764–10775

Dong X, van Wezel R, Stanley J, Hong Y (2003) Functional characterization of the nuclear localization signal for a suppressor of posttranscriptional gene silencing. J Virol 77:7026–7033

Evans D, Jeske H (1993) DNA B facilitates, but is not essential for, the spread of *Abutilon mosaic virus* in agroinoculated *Nicotiana benthamiana*. Virology 194:752–757

Fiallo-Olive E, Martínez-Zubiaur Y, Moriones E, Navas-Castillo J (2012) A novel class of DNA satellites associated with New World begomoviruses. Virology 426(1):1–6

Fiallo-olive E, Tovar R, Navas-castillo J (2016) Deciphering the biology of deltasatellites from the New World: maintenance by New World begomoviruses and whitefly transmission. New Phytol 212(3):680–692

Fondong VN (2013) Geminivirus protein structure and function. Mol Plant Pathol 14:635–649

Gnanasekaran P, Kumar RK, Bhattacharyya D, Kumar RV, Chakraborty S (2019) Multifaceted role of geminivirus associated betasatellite in pathogenesis. Mol Plant Pathol 20(7):1019–1033

Hanley-Bowdoin L, Bejarano ER, Robertson D, Mansoor S (2013) Geminiviruses: masters at redirecting and reprogramming plant processes. Nat Rev Microbiol 11:777–788

Harrison BD, Barker H, Bock KR, Guthrie EJ, Meredith G, Atkinson M (1977) Plant viruses with circular single-stranded DNA. Nature 270:760–762

Hassan I, Orílio A, Fiallo-Olive E, Briddon RW, Navas-Castillo J (2016) Infectivity, effects on helper viruses and whitefly transmission of the deltasatellites associated with sweepoviruses (genus *Begomovirus*, family *Geminiviridae*). Sci Rep 6:30204

Hehnle S, Wege C, Jeske H (2004) Interaction of DNA with the movement proteins of geminiviruses revisited. J Virol 78:7698–7706

Hernandez-Zepeda C, Varsani A, Brown JK (2013) Intergeneric recombination between a new, spinach-infecting curtovirus and a new geminivirus belonging to the genus *Becurtovirus*: first New World exemplar. Arch Virol 158:2245–2254

Hesketh EL, Saunders K, Fisher C, Potze J, Stanley J, Lomonossoff GP, Ranson NA (2018) The 3.3 Å structure of a plant geminivirus using cryo-EM. Nat Commun 9(1):2369

Hormuzdi SG, Bisaro DM (1995) Genetic analysis of *Beet curly top virus:* examination of the roles of L2 and L3 genes in viral pathogenesis. Virology 206:1044–1054

Idris AM, Shahid MS, Briddon RW, Khan AJ, Zhu JK, Brown JK (2011) An unusual alphasatellite associated with monopartite begomoviruses attenuates symptoms and reduces betasatellite accumulation. J Gen Virol 92:706–717

Ilyina TV, Koonin EV (1992) Conserved sequence motifs in the initiator proteins for rolling circle DNA replication encoded by diverse replicons from eubacteria, eucaryotes and archaebacteria. Nucleic Acids Res 20:3279–3285

Ismayil A, Haxim Y, Wang Y, Li H, Qian L et al (2018) *Cotton leaf curl Multan virus* C4 protein suppresses both transcriptional and post-transcriptional gene silencing by interacting with SAM synthetase. PLoS Pathog 14(8):e1007282

Jeske H (2009) Geminiviruses. Curr Top Microbiol Immunol 331:185–226

Jia Q, Liu N, Xie K, Dai Y, Han S et al (2016) CLCuMuB βC1 subverts ubiquitination by interacting with NbSKP1s to enhance geminivirus infection in *Nicotiana benthamiana*. PLoS Pathog 12(6):e1005668

Kammann M, Schalk HJ, Matzeit V, Schaefer S, Schell J, Gronenborn B (1991) DNA replication of *Wheat dwarf virus*, a geminivirus, requires two cis-acting signals. Virology 184:786–790

Kraberger S, Thomas JE, Geering AD, Dayaram A, Stainton D et al (2012) Australian monocot-infecting mastrevirus diversity rivals that in Africa. Virus Res 169:127–136

Kraberger S, Harkins GW, Kumari SG, Thomas JE, Schwinghamer MW et al (2014) Evidence that dicot-infecting mastreviruses are particularly prone to inter-species recombination and have likely been circulating in Australia for longer than in Africa and the Middle East. Virology 444:282–291

Krenz B, Thompson JR, Fuchs M, Perry KL (2012) Complete genome sequence of a new circular DNA virus from grapevine. J Virol 86:7715

Kumar RV, Chakraborty S (2018) Evolution and emergence of Geminiviruses: reasons and consequences. In: Gaur RK, SMP K, Dorokhov YL (eds) Plant viruses: diversity, interaction and management. CRC Press, Boca Raton, pp 97–116

Kumar J, Kumar J, Singh SP, Tuli R (2014) Association of satellites with a mastrevirus in natural infection: complexity of Wheat dwarf India virus disease. J Virol 88:7093–7104

Kumar RV, Singh AK, Singh AK, Yadav T, Basu S et al (2015a) Complexity of begomovirus and betasatellite populations associated with chilli leaf curl disease in India. J Gen Virol 96:3157–3172

Kumar V, Mishra SK, Rahman J, Taneja J, Sundaresan G et al (2015b) *Mungbean yellow mosaic Indian virus* encoded AC2 protein suppresses RNA silencing by inhibiting *Arabidopsis* RDR6 and AGO1 activities. Virology 486:158–172

Kumar RV, Singh D, Singh AK, Chakraborty S (2017) Molecular diversity, recombination and population structure of alphasatellites associated with begomovirus disease complexes. Infect Genet Evol 49:39–47

Kvarnheden A, Lindblad M, Lindsten K, Valkonen JP (2002) Genetic diversity of *wheat dwarf virus*. Arch Virol 147:205–216

Lai J, Chen H, Teng K, Zhao Q, Zhang Z, Li Y et al (2009) RKP, a RING finger E3 ligase induced by BSCTV C4 protein, affects geminivirus infection by regulation of the plant cell cycle. Plant J 57:905–917

Lazarowitz SG, Shepherd RJ (1992) Geminiviruses: genome structure and gene function. Crit Rev Plant Sci 11:327–349

Lewis JD, Lazarowitz SG (2010) *Arabidopsis* synaptotagmin SYTA regulates endocytosis and virus movement protein cell-to-cell transport. Proc Natl Acad Sci USA 107:2491–2496

Li H, Zeng R, Chen Z, Liu X, Cao Z et al (2018) S-acylation of a geminivirus C4 protein is essential for regulating the CLAVATA pathway in symptom determination. J Exp Bot 9(18):4459–4468

Lima ATM, Silva JCF, Silva FN, Castillo-Urquiza GP, Silva FF et al (2017) The diversification of begomovirus populations is predominantly driven by mutational dynamics. Virus Evol 3(1): vex005

Lozano G, Trenado HP, Fiallo-Olive E, Chirinos D, Geraud-Pouey F et al (2016) Characterization of non-coding DNA satellites associated with sweepoviruses (genus *Begomovirus*, *Geminiviridae*)—definition of a distinct class of begomovirus-associated satellites. Front Microbiol 7:162

Luna AP, Rodríguez-Negrete EA, Morilla G, Wang L, Lozano-Duran R et al (2017) V2 from a curtovirus is a suppressor of post-transcriptional gene silencing. J Gen Virol 98:2607–2614

Mar TB, Mendes IR, Lau D, Fiallo-Olive E, Navas-Castillo J et al (2017) Interaction between the New World begomovirus *Euphorbia yellow mosaic virus* and its associated alphasatellite: effects on infection and transmission by the whitefly *Bemisia tabaci*. J Gen Virol 98:1552–1562

Mei Y, Yang X, Huang C, Zhang X, Zhou X (2018) *Tomato leaf curl Yunnan virus*-encoded C4 induces cell division through enhancing stability of Cyclin D 1.1 via impairing NbSK-η-mediated phosphorylation in *Nicotiana benthamiana*. PLoS Pathog 14(1):e1006789

Melgarejo TA, Kon T, Rojas MR, Paz-Carrasco L, Zerbini FM, Gilbertson RL (2013) Characterization of a New World monopartite begomovirus causing leaf curl disease of tomato in Ecuador and Peru reveals a new direction in geminivirus evolution. J Virol 87(10):5397

Monjane AL, Harkins GW, Martin DP, Lemey P, Lefeuvre P et al (2011) Reconstructing the history of maize streak virus strain a dispersal to reveal diversification hot spots and its origin in southern Africa. J Virol 85:9623–9636

Muhire B, Martin DP, Brown JK, Navas-Castillo J, Moriones E et al (2013) A genome-wide pairwise-identity-based proposal for the classification of viruses in the genus *Mastrevirus* (family *Geminiviridae*). Arch Virol 158:1411–1424

Navas-Castillo J, Fiallo-Olive E, Sanchez-Campos S (2011) Emerging virus diseases transmitted by whiteflies. Annu Rev Phytopathol 49:219–248

Nawaz-ul-Rehman MS, Fauquet CM (2009) Evolution of geminiviruses and their satellites. FEBS Lett 583:1825–1832

Noueiry AO, Lucas WJ, Gilbertson RL (1994) Two proteins of a plant DNA virus coordinate nuclear and plasmodesmal transport. Cell 76:925–932

Paprotka T, Metzler V, Jeske H (2010) The first DNA 1-like satellites in association with New World begomoviruses in natural infections. Virology 404:148–157

Pascal E, Goodlove PE, Wu LC, Lazarowitz SG (1993) Transgenic tobacco plants expressing the geminivirus BL1 protein exhibit symptoms of viral disease. Plant Cell 5:795–807

Pasumarthy KK, Choudhury NR, Mukherjee SK (2010) *Tomato leaf curl Kerala virus* (ToLCKeV) AC3 protein forms a higher order oligomer and enhances ATPase activity of replication initiator protein (Rep/AC1). Virol J 7:128

Ramsell JNE, Boulton MI, Martin DP, Valkonen JPT, Kvarnheden A (2009) Studies on the host range of the barley strain of *Wheat dwarf virus* using an agroinfectious viral clone. Plant Pathol 58:1161–1169

Ranjan P, Singh AK, Kumar RV, Basu S, Chakraborty S (2014) Host specific adaptation of diverse betasatellites associated with distinct Indian tomato-infecting begomoviruses. Virus Genes 48:334–342

Razavinejad S, Heydarnejad J (2013) Transmission and natural hosts of *Turnip curly top virus*. Iran J Plant Pathol 49:27–28

Razavinejad S, Heydarnejad J, Kamali M, Massumi H, Kraberger S, Varsani A (2013) Genetic diversity and host range studies of *turnip curly top virus*. Virus Genes 46:345–353

Rodriguez-Negrete E, Lozano-Duran R, Piedra-Aguilera A, Cruzado L, Bejarano ER, Castillo AG (2013) Geminivirus Rep protein interferes with the plant DNA methylation machinery and suppresses transcriptional gene silencing. New Phytologist 199:464–475

Rojas MR, Jiang H, Salati R, Xoconostle-Cázares B, Sudarshana MR, Lucas WJ, Gilbertson RL (2001) Functional analysis of proteins involved in movement of the monopartite begomovirus, tomato yellow leaf curl virus. Virology 291(1):110–125

Rojas MR, Hagen C, Lucas WJ, Gilbertson RL (2005) Exploiting chinks in the plant's armor: evolution and emergence of geminiviruses. Ann Rev Phytopathol 43:361–394

Romay G, Chirinos D, Geraud-Pouey F, Desvies C (2010) Association of an atypical alphasatellite with a bipartite New World begomovirus. Arch Virol 155:1843–1847

Roossinck MJ, Martin DP, Roumagnac P (2015) Plant virus metagenomics: advances in virus discovery. Phytopathology 105:716–727

Rosario K, Padilla-Rodriguez M, Kraberger S, Stainton D, Martin DP et al (2013) Discovery of a novel Mastrevirus and alphasatellite-like circular DNA in dragonflies (Epiprocta) from Puerto Rico. Virus Res 171:231–237

Roumagnac P, Granier M, Bernardo P, Deshoux M, Ferdinand R et al (2015) *Alfalfa leaf curl virus*: an aphid-transmitted geminivirus. J Virol 89:9683–9688

Saunders K, Stanley J (1999) A nanovirus-like DNA component associated with yellow vein disease of *Ageratum conyzoides*: evidence for interfamilial recombination between plant DNA viruses. Virology 264:142–152

Settlage SB, Miller AB, Hanley-Bowdoin L (1996) Interactions between geminivirus replication proteins. J Virol 70:6790–6795

Shen QT, Liu Z, Song FM, Xie Q, Hanley-Bowdoin L, Zhou XP (2011) Tomato SlSnRK1 protein interacts with and phosphorylates βC1, a pathogenesis protein encoded by a geminivirus β-satellite. Plant Physiol 157:1394–1406

Shen Q, Hu T, Bao M, Cao L, Zhang H et al (2016) Tobacco RING E3 ligase NtRFP1 mediates ubiquitination and proteasomal degradation of a Geminivirus-encoded βC1. Mol Plant 9 (6):911–925

Siddiqui K, Mansoor S, Briddon RW, Amin I (2016) Diversity of alphasatellites associated with cotton leaf curl disease in Pakistan. Virol Rep 6:41–52

Sivalingam PV, Varma A (2012) Role of betasatellite in the pathogenesis of a bipartite begomovirus affecting tomato in India. Arch Virol 157:1081–1092

Soleimani R, Matic S, Taheri H, Behjatnia SAA, Vecchiati M et al (2013) The unconventional geminivirus *Beet curly top Iran virus*: satisfying Koch's postulates and determining vector and host range. Ann Appl Biol 162:174–181

Soto MJ, Gilbertson RL (2003) Distribution and rate of movement of the curtovirus *Beet mild curly top virus* (Family *Geminiviridae*) in the beet leafhopper. Phytopathology 93:478–484

Stenger DC, Carbonaro D, Duffus JE (1990) Genomic characterization of phenotypic variants of *beet curly top virus*. J Gen Virol 71:2211–2215

Sunter G, Bisaro DM (1992) Transactivation of geminivirus AR1 and BR1 gene expression by the viral AL2 gene product occurs at the level of transcription. Plant Cell 4:1321–1331

Susi H, Laine A-L, Filloux D, Kraberger S, Farkas K et al (2017) Genome sequences of a capulavirus infecting *Plantago lanceolata* in the Aland archipelago of Finland. Arch Virol 162(7):2041–2045

Trinks D, Rajeswaran R, Shivaprasad PV, Akbergenov R, Oakeley EJ et al (2005) Suppression of RNA silencing by a geminivirus nuclear protein, AC2, correlates with transactivation of host genes. J Virol 79:2517–2527

Varsani A, Shepherd DN, Dent K, Monjane AL, Rybicki EP, Martin DP (2009) A highly divergent South African geminivirus species illuminates the ancient evolutionary history of this family. Virol J 6:36

Varsani A, Navas-Castillo J, Moriones E, Hernandez-Zepeda C, Idris A et al (2014a) Establishment of three new genera in the family *Geminiviridae*: *Becurtovirus*, *Eragrovirus* and *Turncurtovirus*. Arch Virol 159(8):2193–2203

Varsani A, Martin DP, Navas-Castillo J, Moriones E et al (2014b) Revisiting the classification of curtoviruses based on genome-wide pairwise identity. Arch Virol 159:1873–1882

Varsani A, Roumagnac P, Fuchs M, Navas-Castillo J, Moriones E et al (2017) *Capulavirus* and *Grablovirus*: two new genera in the family *Geminiviridae*. Arch Virol 162:1819–1831

Wang Y, Dang M, Hou H, Mei Y, Qian Y, Zhou X (2014) Identification of an RNA silencing suppressor encoded by a mastrevirus. J Gen Virol 95(9):2082–2088

Wang B, Yang X, Wang Y, Xie Y, Zhou X (2018) *Tomato yellow leaf curl virus* V2 interacts with host HDA6 to suppress methylation-mediated transcriptional gene silencing in plants. J Virol 92 (18):e00036-18

Ward BM, Lazarowitz SG (1999) Nuclear export in plants: use of geminivirus movement proteins for a cell-based export assay. Plant Cell 11(7):1267–1276

Yang X, Xie Y, Raja P, Li S, Wolf JN et al (2011) Suppression of methylation-mediated transcriptional gene silencing by βC1-SAHH protein interaction during geminivirus betasatellite infection. PLoS Pathog 7:e1002329

Yazdi HR, Heydarnejad J, Massumi H (2008) Genome characterization and genetic diversity of *beet curly top Iran virus*: a geminivirus with a novel nonanucleotide. Virus Genes 36:539–545

Ye J, Yang J, Sun Y, Zhao P, Gao S et al (2015) Geminivirus activates ASYMMETRIC LEAVES 2 to accelerate cytoplasmic DCP2-mediated mRNA turnover and weakens RNA silencing in *Arabidopsis*. PLoS Pathog 11(10):e1005196

Zerbini FM, Briddon RW, Idris A, Martin DP, Moriones E et al (2017) ICTV virus taxonomy profile: *Geminiviridae*. J Gen Virol 98:131–133

Zhang T, Xu X, Huang C, Qian Y, Li Z, Zhou X (2015) A novel DNA motif contributes to selective replication of geminivirus-associated betasatellite by a helper virus-encoded replication-related protein. J Virol 90(4):2077–2089

Zhou X (2013) Advances in understanding begomovirus satellites. Annu Rev Phytopathol 51:357–581

Zorzatto C, Machado JPB, Lopes KVG, Nascimento KJT, Pereira WA et al (2015) NIK1-mediated translation suppression functions as a plant antiviral immunity mechanism. Nature 52:679–682

Rolling Circle Replication and Transcription Processes in Geminiviruses

Nivedita Sharma and Rajrani Ruhel

Abstract

Geminivirus causes a tremendous loss in plant yield and the infection occurs in terminally differentiated plant cells wherein in order to complete its life cycle, the host gene expression is induced and also the host cell cycle machinery is modulated (Nagar et al. The Plant Cell 7:705–719, 1995; Hanley-Bowdoin et al. Nat Rev Microbiol 11:777–788, 2013). The small-sized (~2.7 kb) genome of geminivirus utilizes a bidirectional mode of transcription and has overlapping genes in different frames for its efficient usage. Functional studies on various mutants of the entire open reading frames of *Tomato golden mosaic virus* concluded that the Rep is the only viral protein absolutely necessary for its replication. In addition, it is also involved in the process of transcription and regulates the expression of certain viral genes. The main purpose of this chapter is to provide brief insights into the replication and transcription pathway of geminivirus.

1 Geminivirus Replication

The ssDNA (+) viral genome (~2.7 kb) of geminivirus is introduced into plant cell after whitefly feeding upon plants (Horns and Jeske 1991). The geminivirus replication takes place in the nuclei of the infected plant cell mainly through rolling circle replication mechanism (Saunders et al. 1991; Stenger et al. 1991) which proceeds through an intermediate dsDNA molecule which is also referred to as "replicative form" (Kammann et al. 1991; Saunders et al. 1992). AC1 (replication initiator protein, Rep) is the only viral open reading frame (ORF) that is indispensable for

N. Sharma · R. Ruhel (✉)
School of Life Sciences, Jawaharlal Nehru University, New Delhi, India

R. V. Kumar (ed.), *Geminiviruses*, https://doi.org/10.1007/978-3-030-18248-9_2

replication. Various host factors/proteins are involved in accomplishing the viral life cycle inside the plant. Geminivirus replicates through two different ways: rolling circle replication (RCR) and recombination-dependent replication (RDR).

1.1 Rolling Circle Replication

There are several extrachromosomal elements, such as a bacterial plasmid, ssDNA of bacteriophage, and RNA viroid which replicate their genomic material via RCR (Koepsel 1985; Gros et al. 1987; Branch et al. 1988). The RCR mode of replication is initiated after nick at a specific site called "Ori," origin of replication. This nicking produces a 3′-OH which serves as a primer during DNA synthesis. The elongation of DNA synthesis in phage produces multiple single-stranded linear copies of DNA which are called as concatemer. However, in replication of ssDNA phage genome and plasmids, after each round of replication, a unit length single-stranded molecule is formed which after cleavage is ligated to produce a circularized DNA molecule.

A number of evidence supported the idea of geminivirus adopting rolling circle mechanism for its replication. Various DNA forms detected upon *African cassava mosaic virus* (ACMV) infection clearly indicated rolling circle mechanism as the mode of geminivirus replication (Saunders et al. 1991). In addition to that, homology studies on the geminiviral Rep and the replication initiator proteins of the bacteriophages and eubacterial plasmid families had shown presence of the RCR motifs, namely motif I, motif II, and motif III in the amino-terminal half of Rep that aids in nicking and joining during the rolling circle replication (Ilyina and Koonin 1992).

The geminivirus replication via rolling circle mechanism is completed in three stages as shown in Fig. 1. In the first stage, viral ssDNA (+ strand) is converted to dsDNA intermediate replicative form with the help of host factors. As a result of feeding by the insect vectors, ssDNA genome is introduced into the plant cell. The viral genome enters the nucleus as part of ssDNA-CP (coat protein) complex. The entry to the nucleus is mediated by nuclear localization signal present in CP (Guerra-Peraza et al. 2005). Within the host nucleus, complementary strand synthesis occurs in which ssDNA molecule is converted into a covalently closed dsDNA replicative form (RF) by host DNA polymerase. This RF serves as a template for viral replication as well as its bidirectional transcription. In case of mastreviruses, an oligonucleotide complementary to the small intergenic region (SIR) is found packed within the virion which acts as a primer for DNA synthesis (Donson et al. 1984). Instead of begomoviruses, such a primer is generated within the large intergenic region by host RNA polymerase (Saunders et al. 1992). The dsDNA replicative form also interacts with cellular histone to form viral minichromosomes (Pilartz and Jeske 1992). Viral ssDNA genome is more vulnerable than dsDNA and therefore is more prone to damage. Recently, RAD51D, a paralog of RAD51, has been suggested to play a role during the complementary strand replication (CSR) of geminivirus to produce RF molecules (Richter et al. 2016). RAD51D is a key player in RAD51-independent single-strand annealing (SSA) recombination pathway in somatic cells of

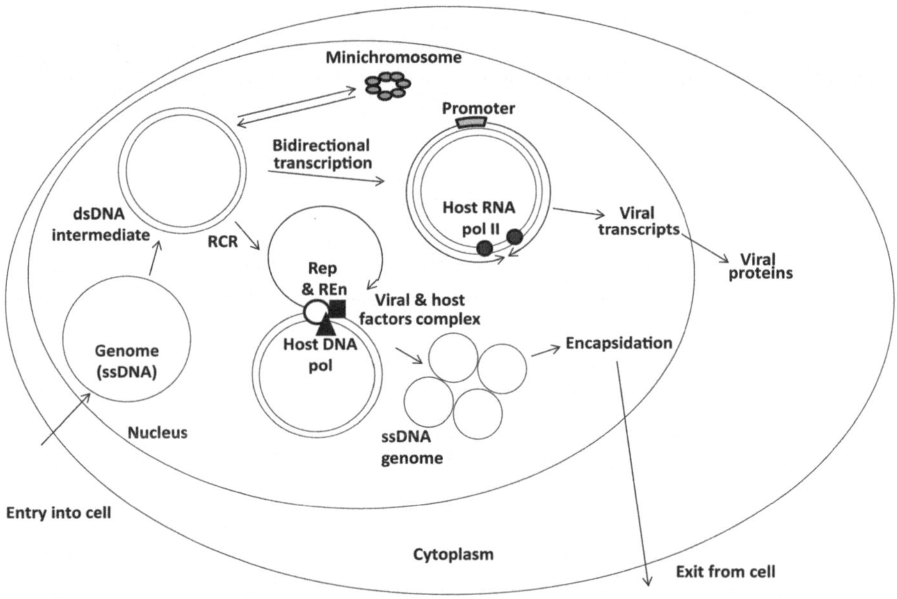

Fig. 1 Life cycle of geminivirus proceeds via a dsDNA intermediate which acts as template for rolling circle replication as well as bidirectional transcription

Arabidopsis thaliana, and the possible roles of SSA in complementary strand synthesis of the geminiviral replication have been suggested (Serra et al. 2013). In *A. thaliana*, four translesion synthesis (TLS) polymerases have been reported, namely Polη, Polζ, Polκ, and Rev1. Out of these, Polζ is believed to carry out CSR of geminiviral DNA (Richter et al. 2016). Supporting this, it has been found that TLS polymerases are constitutively expressed in differentiated plant cell where geminivirus replicates.

During the second stage of RCR, more RF DNAs are generated from RF. In this step, Rep (AC1) produces a nick on the (+) strand of the RF molecule. This nick is created at a conserved, specific nonamer sequence (TAATATT↓AC) present in the loop region of the stem-loop structure present in the common region of the circular DNA molecule (Laufs et al. 1995). Various cis elements are identified in the common region that affects the process of replication. Origin of replication on the TGMV (+)-strand possesses binding sites for two transcription factors, one at the TATA box and another at G-box. These sites are not required for viral replication. In contrast to this, other two elements, namely AG- and CA-motifs, are required for replication probably by binding to host factors. Following the cleavage, Rep protein recruits the host DNA polymerase which extends the 3′ end of the cleaved virion strand while Rep remains covalently bound to the 5′ terminus via a phosphotyrosine linkage. Rep is also known to interact with several host proteins during the course of replication. For example, it interacts with host RFC, RPA70, MCM, and PCNA (Rizvi et al. 2014). Rep acts as ATPase and is believed to act as a replicative helicase

(Choudhury et al. 2006; Clerot and Bernardi 2006). It belongs to the superfamily 3 (SF3) of helicases, and various amino acids of B′ motif have been shown to be essential for unwinding activity (George et al. 2014). Rep interacts with another viral protein REn (replication enhancer protein encoded by AC3) to enhance replication (Pasumarthy et al. 2010). Later during this step when the origin of replication is regenerated, another nick is introduced by Rep, and then Rep is transferred to the new 5′ terminus. Rep acts as ligase finally to produce a circular ssDNA molecule.

During the third stage of RCR, the newly synthesized ssDNA molecules are accumulated and encapsidated and new viral progenies are produced. During this final step, there is a shift from production of dsDNA molecules to the accumulation of ssDNA molecules. Various AC2 (TrAP) and AV2 mutants were found to result in a reduction in ssDNA levels and a corresponding increase in the level of dsDNA molecules (Hayes and Buck 1989; Hormuzdi and Bisaro 1993; Sunter et al. 1993). Thus, these viral factors are believed to control this transition step during replication. Both these viral proteins are implicated in the inhibition of minus strand synthesis and as a result of which ssDNA molecules produced are diverted from the replication pool toward encapsidation and viral assembly. Further, the host's transport machinery is utilized and exploited to spread the virus throughout the plant. Viral infection is established throughout the plant with the help of the nuclear shuttle protein (NSP encoded by BV1) and the movement protein (MP encoded by BC1) of the virus. Viral movement within the host occurs at two different levels: (a) short-distance cell-to-cell movement through plasmodesmata and (b) long-distance movement to the distal parts of the plant which occurs through the vascular system. NSP binds to the newly synthesized ssDNA viral genome and transports them from nucleus to cytoplasm (Pascal et al. 1994). NSP-interacting GTPase (NIG) interacts with NSP and helps in the release of the complex of NSP-DNA from the nuclear export machinery. Subsequently, cell-to-cell movement of viral DNA is carried out with the help of MP. In the case of monopartite viruses, movement of the viral DNA is carried out by coat protein (CP) and MP. CP can bind to viral DNA and contains the nuclear localization signal (NLS) (Guerra-Peraza et al. 2005). Movement of the viral genome in and out the nucleus is required for its replication and spread, respectively, which occurs via recognition by the nuclear transport machinery components. CP docks the viral genome at the nuclear pore and is believed to enter into the nucleus interacting with the importin alpha component of nuclear import machinery. Host's histone H3 interacts with both viral NSP and MP suggesting its possible participation in viral movement competent complex (Zhou et al. 2011). *Arabidopsis* calcium sensor protein synaptogmin A (SYTA) which also regulates endocytosis is found to be involved in MP-mediated cell-to-cell movement of the viral genome. It helps in targeting the MP-DNA complex to plasmodesmata via endocytosis (Lewis and Lazarowitz 2010) as well as by altering plasmodesmata permeability (Uchiyama et al. 2014).

1.1.1 Viral Proteins Involved in Replication Process

There are two viral proteins that are involved in the process of replication, namely AC1, replication initiator protein, and AC3, replication enhancer protein. ORF AC3

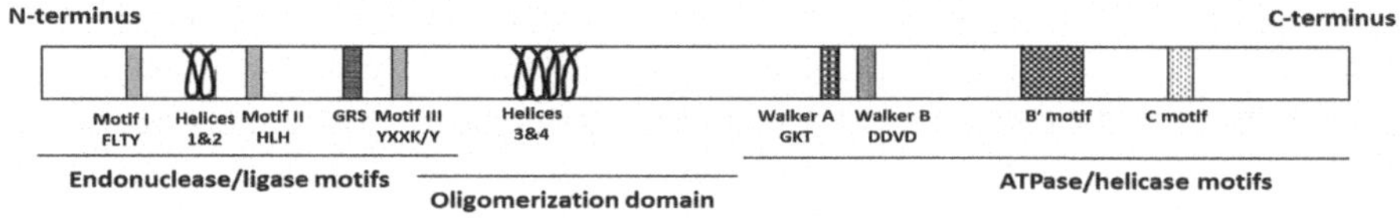

Fig. 2 Domain organization of replication initiator protein (Rep) showing endonuclease/ligase motifs, oligomerization domain, and ATPase/helicase domain present in the N-terminus, the central part, and the C-terminus of the Rep protein, respectively

encodes for a protein named as the replication enhancer (REn) as it increases viral replication (Sunter et al. 1990; Harrison and Robinson 1999; Hanley-Bowdoin et al. 2000). Mutation in AC3 ORF leads to reduced ssDNA and dsDNA accumulation (Garry Sunter et al. 1990). REn protein can oligomerize and bind to viral Rep protein. TGMV REn interacts with few host cell cycle proteins like maize retinoblastoma homolog (RBR1) and Arabidopsis PCNA (Settlage et al. 2001; Castillo et al. 2003; Hanley-Bowdoin et al. 2004). Replication initiator protein plays a central and indispensable role in viral replication. It is a multifunctional protein and is involved in initiation, elongation, and termination step during the course of replication process by virtue of its modular nature which has been depicted in Fig. 2.

The amino-terminal of Rep protein possesses endonuclease, ligase, and sequence-specific DNA-binding activity and ATP-dependent topoisomerase I activity (Fontes et al. 1992; Laufs et al. 1995; Orozco et al. 1997; Chatterji et al. 2000; Pant et al. 2001), while the ATPase and helicase activity is contributed by amino acid residues in the C-terminal (Desbiez et al. 1995; Choudhury et al. 2006; Clerot and Bernardi 2006; George et al. 2014). Amino acid residues present in the central region of Rep form the oligomerization domain which also is essential for interacting with many host factors, such as proliferating cell nuclear antigen (PCNA), retinoblastoma-related protein (RBR), and geminivirus Rep-interacting kinase (GRIK) (Rizvi et al. 2014). Rep functions as a site- and strand-specific endonuclease during the initiation of geminivirus replication. The catalytic site for endonuclease activity is formed by motif III (YXXKD/E) where the hydroxyl group of the Y (tyrosine) residue forms a covalent bond with the 5′-PO4 of the cleaved DNA strand. Motif I (FLTY) and motif II (IILII) are required for specific dsDNA binding and metal binding, respectively. Apart from RCR motifs (I–III), between motif I and motif II are present helix 1 and helix 2 which implicate the Rep protein's DNA-binding and endonuclease activity (Orozco and Hanley-Bowdoin 1998). DNA binding is oligomerization dependent as studies in Tomato golden mosaic virus (TGMV) Rep protein showed that DNA-binding domain spans from 1 to 130 amino acid region and overlaps with the oligomerization domain (120–180 amino acid residues). However, this oligomerization of Rep protein does not implicate its DNA cleavage and ligation activity. Geminivirus Rep sequence (GRS) is another motif present between motif II and motif III in Rep protein. This motif is found to have a significant degree of conservation and comprises uncharacterized sequence constituting of two clusters of amino acids (Nash et al. 2011). It has been found to be required for the replication initiation as the GRS mutations resulted in impaired DNA cleavage activity.

Rep protein belongs to AAA+ (ATPase associated with various cellular activities) family of ATPases and is grouped within the SF3of helicases (Gorbalenya and Koonin 1993). Four conserved SF3 motifs of Rep helicase are, namely A, B, B′, and C, all of which are located in approximately 100 amino acid residues stretch of C-terminal of Rep protein. AAA+ ATPase domain of the geminiviral Rep is found to be much reduced and lacks structural element such as conserved arginine finger (Clerot and Bernardi 2006). Motif A and Motif B serve as the nucleoside triphosphate (NTP)-binding pocket and metal ion coordination site, respectively, and thus are required for ATP hydrolysis. Motif C is essential for interacting with the PO4 at the gamma position of ATP and an "apical" water molecule. Motif B′ is involved in nonspecific ssDNA-binding during DNA unwinding process (George et al. 2014).

1.2 Recombination-Dependent Replication

In infected plant cells, a reservoir of geminiviral molecules results from not only RCR, but also recombination and repair pathways which are error-prone processes. Recombination and repair pathways are believed to be the major forces to overcome plant defense as well as for its evolution. An increased homologous recombination of transgenes upon geminivirus infection has been found, and it has been observed that this recombination occurs in tissue-specific manner (Richter et al. 2014). Recombination-dependent replication has been suggested as an important part of viral replication in case of incomplete replication due to DNA damage or low processivity of polymerases (Jeske et al. 2001; Preiss and Jeske 2003; Alberter et al. 2005; Ruschhaupt et al. 2013). Invasion of a homologous region of the circular dsDNA molecule by a viral ssDNA fragment occurs with the assistance of host recombination proteins. This marks the initiation of RDR. After this, the invaded ssDNA is extended on the viral template strand by the host DNA polymerase. RDR produces a heterogeneous pool of linear dsDNAs that accumulate at high levels upon geminivirus infection. The priming during RDR does not involve Rep activity. Instead, Rep helps in the release of the ssDNA genome fragment from the heterogeneous linear dsDNA molecule which can again enter the replication cycle. The long linear dsDNAs containing more than one origin of replication can be transcribed to generate viral mRNAs by host RNA Polymerase II. Rep protein that is produced after translation starts replication of the long linear dsDNA bearing more than one origin of replication (Pooggin 2013).

2 Interactions with Rep Protein and Its Implication on Viral DNA Accumulation

Geminivirus replication initiator protein interacts with various host factors and drives the environment within the host suitable for viral propagation.

2.1 Plant Retinoblastoma-Related Proteins (RBR)

Rb family of protein regulates the progression of cell cycle. Expression of a homolog of Rb (RBR) of maize results in reduced viral DNA replication in wheat cells. WDV RepA contains LXCXE motif which is required to physically interact with RBR protein (Xie et al. 1995). This interaction seems critical for replication as the mutants incapable of this interaction have impaired replication. Maize retinoblastoma-related proteins, RBR1 and RBR2, interact with both Rep and D-type cyclin (Ach et al. 1997). However, TGMV Rep lacks LXCXE motif and interacts with RBR protein with a distinct motif, namely helix4 motif (Arguello-Astorga et al. 2004). This helix 4 motif comprises charged amino acid residues which are involved in the interaction with RBR and also in viral replication. Rep was able to re-replicate in fission yeast. It was surprising as fission yeast lacks RB homolog. But the alternative interaction with cyclins through RXL motif has been proposed that regulates replication (Hipp et al. 2014).

2.2 Proliferating Cell Nuclear Antigen (PCNA)

Geminivirus infection induces the expression of PCNA in the mature infected cells (Nagar et al. 1995; Hanley-Bowdoin et al. 2013). Tomato PCNA has been shown to interact with TYLCSV Rep protein as well as with REn protein (Castillo et al. 2003). PCNA binds to IMYMV Rep protein. A 134–183 amino acid stretch of IMYMV Rep is required for the interaction; on the other hand, amino acid residues of PCNA involved in the interaction are dispersed throughout PCNA (Bagewadi et al. 2004). This interaction downregulates the endonuclease and ATPase function of the Rep protein.

2.3 Replication Factor C (RFC)

Replication factor C is a multimeric protein that loads the PCNA onto the DNA during the replication process. Wheat large subunit of replication factor C complex (TmRFC 1) binds to WDV Rep protein in the DNA/Rep/TmRFC-1 complexes which resemble the pre-initiation complex and thus assists in the further assembly of elongation complex for the viral replication (Luque et al. 2002).

2.4 Replication Protein A-32 (RPA-32)

RPA is a heterotrimeric protein that binds to ssDNA and, in addition to replication, is involved in repair and recombination. RPA32 subunit interacts with C-terminal of MYMIV Rep protein and downregulates the endonuclease activity while upregulating its ATPase activity (Singh et al. 2007). Thus, it has been believed that the interaction with RPA32 in addition might limit the replication initiation and drives to elongation phase of RCR.

2.5 Recombination Enzymes

RAD51 and RAD54 are two repair and recombination proteins which interact with Rep protein (Kaliappan et al. 2012; Suyal et al. 2013). They might play a critical role in case of replicational stress by stabilizing the replication fork. The N-terminal of RAD54 binds to the oligomerization domain of MYMIV-Rep and enhances its nicking, ATPase, as well as helicase activities (Kaliappan et al. 2012). However surprisingly, studies on rad54 mutant showed no effect on the CSR, RCR, or RDR, even though it physically interacts with Rep (Richter et al. 2015).

2.6 Sumoylation-Conjugating Enzyme-1

The N-terminal of Rep binds to sumoylation-conjugating enzyme (SCE-1) (Castillo et al. 2004). The K68 and K96 amino acid residues in the N-terminal of Rep are found to interact with SCE-1 and when mutated abolished the interaction and reduced the viral accumulation in infected plants (Sanchez-Duran et al. 2011).

2.7 Histones

Geminiviral genomic DNA has been shown to assemble as minichromosomes (Pilartz and Jeske 1992). TGMV Rep protein interacts with Histone-3 which suggests the probable implication of this interaction in replication and transcription process (Kong and Hanley-Bowdoin 2002). It has been hypothesized that Rep recruitment on the viral genome and its interaction with H3 may help in the removal of the nucleosomal block and thus helps in its efficient transcription and replication. Rep protein also interacts with a kinesin motor protein (GRIMP) that is involved in mitosis process. Apart from that it also interacts with a kinase, Geminivirus rep-interacting kinase (GRIK). These interactions might inhibit the cell from entry into the mitotic phase.

2.8 NAC Domain-Containing Proteins

GRAB1 and GRAB2 proteins belong to NAC domain-containing proteins family which are involved in plant development. They have a unique C-terminal which contains negatively charged residues, while the N-terminal is conserved, and interact with WDV Rep to inhibit the replication (Xie et al. 1999).

3 Geminivirus Transcription

Various studies done on geminiviruses mRNA have confirmed that the transcription in geminivirus genome occurs in a bidirectional manner (Townsend et al. 1985; Hanley-Bowdoin et al. 1989; Petty et al. 1988; Sunter and Bisaro 1989; Frischmuth et al. 1991; Mullineaux et al. 1993). The resulting viral transcript refers to both the virion and complementary sense ORF. The initiation of viral transcription occurs downstream of either initiator elements or consensus TATA box motifs and the viral transcripts are polyadenylated both suggesting that they are transcribed by RNA polymerase II.

Previously transient systems have also been developed to identify the geminiviral proteins that play an important role in viral transcription. In a study, the 5′ intergenic region which contained the promoter sequences was fused to β-glucuronidase or luciferase coding sequence, and the reporter activity was observed in the presence of mutant viral DNA component. This experiment suggested that AL2 enhances the reporter gene activity containing the AR1 or BR1 promoter region (Sunter and Bisaro 1991). The process of transcription is highly complex in geminivirus and frequently leads to the production of multiple overlapping RNAs. Moreover, the RNA-processing pattern is different for different subgroups of geminiviruses.

In case of bipartite subgroup III, geminiviruses transcription profile has been studied in detail. In *Tomato golden mosaic virus*, DNA-A component (TGMV) serves as a template for six RNAs, whereas DNA-B encodes four RNAs. From each genome, a single virion sense RNA is transcribed which further translates to either coat protein or BR1 (Hanley-Bowdoin et al. 1999). On the other hand, complementary sense transcription process is more complex comprising of multiple overlapping RNA with different 5′ end and a common 3′ end. Both the complementary and virion sense RNA overlap for polyadenylation site at the 3′ end (Hanley-Bowdoin et al. 1999). The complementary sense RNA from TGMV DNA-A (TGMV) possesses different coding capacity whereas complementary sense RNA from TGMV DNA-B encodes only BL1. The largest transcript (AL61) encodes for the entire left portion of TGMV DNA-A, and it is the only RNA which is translated to produce full-length Rep protein. There are two RNAs (AL2540 and AL2515) which are translated to AL4 and likewise two smallest RNAs known to code for AL2 and AL3 from their first open reading and second open reading, respectively. However, there is no RNA known to encode for AL3 as its first ORF showing that AL3 is encoded by polycistronic mRNA. The polycistronic nature and translational properties of many TGMV DNA-A complementary sense RNA have been studied by translation in vitro (Thommes and Buck 1994). The complementary sense transcription process is almost similar for other subgroups II and III geminiviruses (Frischmuth et al. 1991; Mullineaux et al. 1993). In the subgroup I geminiviruses, transcription occurs bidirectionally from multiple initiation sites and terminate at overlapping polyadenylation signals (Morris-Krsinich et al. 1985; Accotto et al. 1989; Dekker et al. 1991; Wright et al. 1997). Contrary to the subgroup I, RNA processing is a vital component of expression strategy (Wright et al. 1997). The complementary strand encodes two ORF C1 and C2 which together code for viral replication protein. Many studies on transcript mapping of *Wheat dwarf virus*

(WDV), *Maize streak virus* (MSV), *Digitaria streak virus* (DSV), and *Tobacco yellow dwarf virus*(TYDV) have revealed that a spliced mRNA is fused with C1 and C2 sequence (Schalk et al. 1989; Mullineaux et al. 1990; Dekker et al. 1991; Morris-Krsinich et al. 1992; Wright et al. 1997). Moreover, the spliced sequences are AT rich and contain a potential branch point with consensus splicing signal which is specific to plant intron. This indicates that RNA is processed by host machinery and the complementary strand is partially processed. The Rep A polypeptide may be encoded by C1 ORF of unspliced RNA, but till now there is no complete information suggesting the synthesis of Rep A during infection (Wright et al. 1997). On the other hand, a study on the replication of mutant virus clearly showed that spliced C1:C2 mRNA and its resulting Rep protein are sufficient and essential for replication (Schalk et al. 1989; Wright et al. 1997).

4 Transcript Mapping in Geminiviruses

Promoter mapping of different geminiviruses, such as mastreviruses MSV (Fenoll et al. 1988), DSV, WDV (Dekker et al. 1991) and bipartite begomoviruses, has revealed that promoters are located between 5′ end of the first complementary sense strand and the virion sense ORF. The complementary strand promoter in mastrevirus DSV regulates the synthesis of C1 and C protein (Accotto et al. 1989; Mullineaux et al. 1990; Dekker et al. 1991). In case of mastrevirus DSV, transcription of the viral sense strand is not experimentally shown, but in other mastreviruses WDV, it has been showed that a signal promoter element can regulate the synthesis of V1 and V2 transcripts (Hofer et al. 1992). In mastrevirus MSV, the virion sense gene expression is highly regulated involving differential start point (Wright et al. 1997). Two mRNAs are known to be transcribed to virion sense strand. Moreover, a shorter abundant mRNA translated to produce coat protein and less abundant transcript (AV1ORF) code for movement protein (Morris-Krsinich et al. 1985; Wright et al. 1997). The virion sense gene expression is highly regulated in MSV which involves differential transcription start site (Wright et al. 1997). In geminiviruses, cis-acting elements bring out transcription in both strands and are reported to be located in the long intergenic region (LIR) of mastrevirus (Heyraud-Nitschke et al. 1995). Earlier sequences analysis of LIR region showed a little sequence homology, but nevertheless, some conserved motifs have been found (Argüello-Astorga et al. 1994). Two TATA boxes and a GC-rich box (except in ACMV) are located on either side of LIR. The viral sense promoter activity was reported to be significantly reduced due to the deletion of GC-rich box of MSV LIR (Fenoll et al.1988; Willment et al. 2007). In bipartite begomovirus, mutation in the TGMV genome revealed the important feature of TATA box and G-box motif (CACGTG), which is needed to activate some of the complementary sense genes (Eagle and Hanley-Bowdoin 1997). Polyadenylated transcript has been mapped for both virion and complementary sense ORF in TGMV (Hanley-Bowdoin et al. 1989; Sunter and Bisaro 1989). Transcription starts downstream of either consensus TATA box motif (Breathnach and Chambon 1981). In TGMV DNA-A, a single transcript AR319 has been mapped and many overlapping RNAs differing in 5′ ends but having a singe 3′ end were mapped (Hanley-Bowdoin et al. 1988; Sunter and Bisaro 1989; Wyant et al. 2012). The study dictates that several promoters do exist in addition to promoter present in the IR region. Transcript producing a functional protein AC1 initiates at nucleotide position 62 (AL-62), and it is also responsible for coding AC2 and AC3 proteins in TGMV. Two smaller RNA initiating at nucleotide position 16, 29 and 19, 35 (AL1629 and AL1935) were also reported to code for both AC2 and AC3 coding region, respectively. In TGMV, AL1629 transcript was found to be more abundant in infected plant (Sunter and Bisaro 1989). No RNA coding for AC3 alone has been known and no splicing evidence has not been reported so far. A more recent study showed that monoubiquitination of NbUBC2 and NbHUB1 is promoted by *Chilli leaf curl virus* (ChiLCV) Rep protein which binds to viral genome and ultimately promotes the trimethylation of histone 3 at lysine 4 on ChiLCV minichromosome leading to increased transcription of viral genes (Kushwaha et al. 2017).

5 Geminiviruses Promoter

5.1 Early Promoter

The term promoter denotes the sequences from where transcription starts, and in geminiviruses, the early transcript represents the complementary sense genes which appear during the early stage of infection. For instance, early transcript in geminivirus TGMV is the five overlapping complementary sense strand with variable 5′ ends and a common 3′ end (Hanley-Bowdoin et al. 1989)· In TGMV, AL-62 RNA is known to be the largest complementary sense transcript and encodes all the genes on the left arm of TGMV DNA-A component. AL-62 RNA is the only RNA that includes the 5′ end of AC1-coding region and can further translate to produce full-length AC1 protein (Eagle and Hanley-Bowdoin 1997). AC1 protein of TGMV negatively regulates the expression of its own gene. The repression takes place by interaction of the protein with its cognate binding site lying between transcription start point and consensus TATAA sequence (Eagle et al. 1994; Sunter et al. 1993). By doing a series of mutation and transitions in the IR, DNA sequences that allow promoting the activity of TGMV complementary sense strand have been identified. It is found that negative regulation of AC62 promoter involves multiple cis elements. Moreover, a mutation in the TATA box leads to reduced transcription activity and repression mediated by AC1 (Eagle and Hanley-Bowdoin 1997). In cereal-infecting mastrevirus, the differential complementary sense transcript splicing is common (Dekker et al. 1991). In mastrevirus MSV, replication-associated protein Rep and Rep A are synthesized from complementary strand by differentially spliced mRNA, which is a typical characteristic for cereal-infecting mastreviruses (Wright et al. 1997). In the transgenic plant, the deletion of MSV complementary sense promoter showed that promoter activity is restricted to the meristematic region. The activity of complementary sense strand promoter was studied using a GFP fusion protein in

N. benthamiana, and it was found that promoter activity was active in a heterologous system (Kumar et al. 2017; Gopal et al. 2007). In a study on *cotton leaf curl Burewala virus* (CLCuBuV), a region of DNA-A, i.e., −2595 to +292 was recognized as a bidirectional promoter which controlled the expression of the complementary sense strand and virion strand genes. The AC1 promoter showed almost sixfold stronger expression than AV1 virion sense promoter and two times more than CaMV 35S promoter. In this study, many cis-regulatory elements (CREs) and transcription factor binding sites were identified in promoter sequences (Ashraf et al. 2014).

A study on curtovirus, *Beet curly top virus* (BCTV), showed that the IR region alone is not sufficient to promote the C1 expression in transgenic Arabidopsis plant. On the other hand when the sequences that were extended into coding region of C1 were tested, it was found that there was a strong expression of the reporter gene in the vascular tissue. This result indicated that transcriptional activator element lies in 5′ part of ORF (Hur et al. 2007).

The complementary sense transcription unit of TGMV DNA-A encodes several overlapping RNAs. The AL62 mRNA along with the complete complementary sense ORF (AC1, AC2, and AC3) and two smaller RNAs one at 1629 (encode AC2 and AC3) and another at 1935 (encode AC3) are transcribed. A minimal sequence between −129 and −213 was identified which is required for AL1629 mRNA (Shung et al. 2006). A sequence between −129 and −184 bound plant nuclear protein was found to be unable to activate expression of a heterologous promoter. The possible reason may be due to two elements that participate in the activation, one between −213 and −184 and another between −184 and −129 (Shung et al. 2006). In another study, similar regions in the geminivirus, *African Cassava Mosaic Virus* (ACMV) and *Mungbean Yellow Mosaic India Virus* (MYMIV), were found to regulate the expression of AC2 and AC3 in addition to AC1 (Zhan et al. 1991; Vanderschuren et al. 2007; Shivaprasad et al. 2005).

Two complementary sense transcripts in mastrevirus DSV appear to regulate the synthesis of 41Kd fusion of C1 and C2 ORF and a 30KD C1 product, respectively (Accotto et al. 1989; Mullineaux et al. 1990; Dekker et al. 1991). It is known that geminivirus DNA associates with host nucleosome as a viral minichromosome. In a study on bipartite begomovirus AbMV, one nucleosome-free gap was observed and interestingly this gap co-localized with the sequence related to bipartite begomovirus TGMV AL-1629 promoter. It was hypothesized that this gap could act as an interaction site for additional host factors for the AL1629 transcription (Pilartz and Jeske 1992; Pilartz and Jeske 2003). A nine base pair sequence bound to the nuclear protein was identified, and a two- to sixfold reduction in the accumulation of AL1629 mRNA was found upon the mutation of the binding site in Arabidopsis and tobacco. Moreover, viral sequence influenced the AC2 and AC3 expression and possessed some binding affinity to host soluble protein. This binding region was found to be conserved in several curtovirus and begomovirus showing a common expression mechanism (Tu and Sunter 2007).

Many monopartite begomoviruses possess an additional circular DNA molecule of approximately half the size of geminiviruses, i.e., 1.3 kb, known as betasatellite, which is reported to possess a single ORF directed in a complementary direction known as βc1 and encode for a protein responsible for pathogenicity. In monopartite geminivirus cotton leaf curl Multan virus (CLCuMV), a region within nucleotide position +1 to −989 of betasatellite was reported to bring Gus expression in a transient assay in tobacco whereas deletion in 5′ construct identified a 68 nt region (−139 to 207) to be important for promoter activity in βC1 transcription. Sequencing of the region identified a TATA box and G-box motif among several cis-regulatory elements in plants. The promoter activity was reduced to 40% during the mutation of G-box motif as compared to the 68 nucleotides βC1 promoter region (Eini et al. 2009).

5.2 Late Promoters

The promoter that regulates the transcription of virion sense strand usually becomes activated late during the infection cycle and so-called as late promoters. Studies were done on bipartite begomovirus ACMV, and TGMV indicates that late promoters are weaker than early promoters (Petty et al. 1988; Zhan et al. 1991; Eagle and Hanley-Bowdoin 1997). Studies revealed that in bipartite begomovirus TGMV, AV1 promoter confers kanamycin resistant to sixfold greater than the nopaline synthase promoter in *E. coli* (Petty et al. 1986). Further, a region of 158 bp was found to be responsible for this activity (Petty et al. 1986). Expression of Gus gene was enhanced during viral DNA replication from the above promoter revealing cross talk between replication and transcription from TGMV virion sense promoter. In mastrevirus MSV, the long intergenic sequence(LIR) and DNA sequence upstream V2 ORF (code for coat protein) were known to be only required for the temporal expression of a reporter gene in maize (Fenoll et al. 1988). Further, promoter activity was observed enhanced by a 122 bp upstream segment (UAS) that activates the core promoter in orientation and position-dependent manner. The UAS activated the CaMV 35S core promoter and bound specifically to proteins in nuclear extract of maize (Fenoll et al. 1988). The histochemical study revealed that MSV V2 promoter was active in callus cells, but in transgenic plants, its expression occurred in vascular tissue but was absent in root meristem. In this study, promoter showed a development-dependent expression pattern (Mazithulela et al. 2000). In another study, the same promoter activity was found to be highest in the early G2 phase of cell cycle (Nikovics et al. 2001).

In monopartite begomovirus CLCuV, the virion sense putative promoter activity was found to be low in transgenic tobacco plant (Yingqiu et al. 2001). In the related bipartite begomovirus *Cotton leaf curl Multan virus* (CLCuMV), it was found that the virion sense promoter needs the viral AC2 product (the transactivator protein TrAP) for optimal activity. The average strength of the promoter was found to be one-tenth of the CaMV 35S promoter through quantitative analysis (Xie et al. 2003). Usharani et al. studied the virion sense promoter of bipartite begomovirus MYMIV through transient agroinfiltration into the leaves of *N. benthamiana* and found that the promoter activity was independent of AC2 and the binding site for transcription factor was in its upstream region. However, the activity of the promoter was not compared with any known promoters (Usharani et al. 2006). The virion sense promoter of two curtoviruses (BCTV and beet severe curly top virus BSCTV) that drive the expression of Gus reporter gene in transgenic *A. thaliana* plants was used to study their promoter activities. Interestingly, it was found that the promoter of less virulent BCTV displayed a high expression level than the more virulent BSCTV. It was also found that the reporter gene expression was low in mature plants and promoter activity mainly occurred in dividing tissue. A conserved late element (CLE) motif of 30 nucleotides was reported to be present thrice in tandem in BCTV and a single copy in BSCTV promoter. The deletion of these elements in BCTV caused low expression of the promoter. These studies clearly indicate that the expression of gene relies on the developmental stage of plant and number of CLEs present (Hur et al. 2008; Hur et al. 2007; Borah et al. 2016).

The intergenic region (IR) of curtovirus *Spinach curly top virus* (SCTV) was studied for analyzing the presence of control element that influences the promoter activity. A sequence of 135–173 (within 43 base pair of translation site of CP gene) was identified to be important for the activation of virion sense promoter of SCTV. Apart from this, the same IR region could regulate the expression of the two other virion sense genes (V2 at 252 nt and V3 at 292 nt) of the curtovirus. An 84 bp fragment (−476 to −392) of viral DNA was found to act as a positive regulatory element which enhanced the expression of V2 and V3 genes up to 25-fold when placed upstream of CaMV 35S core promoter. However, the semi-quantitative RT-PCR results showed that the accumulation of the V2 transcript was quite low (45%) as compared to the V3 transcript (55%) in Arabidopsis plant. The SCTV virion sense promoter activity was specific to phloem tissue in transgenic *Nicotiana benthamiana* plant expressing Gus reporter gene. The expression by this promoter was observed to be independent of C2 activation from both homologous and heterologous origin (Rao and Sunter 2012).

Apart from the IR as discussed above, the putative promoter has also been located within the coding sequence of geminivirus DNA. For instance, a single abundant virion sense transcript was found to control the synthesis of V and V2 ORF in mastrevirus WDV (Dekker et al. 1991). Recently in some monopartite begomoviruses, a cryptic promoter referred to as AV3 has been discovered (Wang et al. 2013). This promoter had a prokaryotic ribosome binding site. Likewise in the monopartite begomovirus, *Ageratum yellow vein virus* NT (AYVV-NT), the sequence of 613–887 was found to drive expression of GFP reporter gene in complementary sense orientation in *E. coli.* Fine mapping revealed that nucleotide position 762–869 is sufficient for promoter activity. Similar results were observed for AV3 promoter from three geminiviruses, i.e., bipartite begomovirus ToLCV, *Squash leaf curl virus* (SqLCV), and monopartite begomovirus gonostegia mosaic virus TC isolate which displayed promoter activity (Wang et al. 2014).

6 Transactivator C2/AC2 Regulation

In begomoviral AC2/C2 proteins, three conserved domains have been discovered, and these are basic domains with nuclear localization signal (NLS) at N-terminus, a central DNA binding domain with zinc finger motif and an activator domain at C-terminus (Hartitz et al. 1999). Studies showed that in monopartite begomoviruses, namely Tomato yellow leaf curl virus (TYLCV) and bipartite begomovirus TGMV and ACMV, AV1 ORF expression is mediated by C2/AC2 (Sunter et al. 1990; Groning et al. 1994). Similarly, in another bipartite begomovirus cabbage leaf curl virus (CabLCuV), AC2 activates the AV1 promoter in mesophyll and vascular tissue (Xie et al. 2003). Another study reported that TGMV AC1 promoter repression and CabLCuV AV1 promoter activation mediated by sequence bind to a different nuclear factor of plants (Xie et al. 2003). Similar studies have reported that monopartite begomovirus tomato yellow leaf curl virus (TY LCV) and bipartite begomovirus mung bean yellow mosaic virus (MYMV) bind to common factors for regulating AV promoter (Lacatus and Sunter 2008). In TGMV, chromatin immunoprecipitation revealed that AC2 interacts with activator and repressor independently, thereby showing that cellular proteins interact with AC2 and bind to viral DNA element for repression and activation (Lacatus and Sunter 2008). Studies have shown that AC2-mediated activation occurs at the level of transcription (Sunter and Bisaro 1992). Moreover, a study on transgenic plants promoter revealed that AC2 regulates AV1 expression in all plants in two ways. Firstly, AV1 Promoter might be activated in mesophyll cells by an element positioned between −125 and −60 (Sunter and Bisaro 2003), and then the promoter is derepressed in phloem tissue by the sequences located between 1.2 and 1.5 upstream of transcript start site (Sunter and Bisaro 1997).

In a study, the role of virion sense and complementary sense promoter of the bipartite begomoviruses ACMV DNA-A was studied using two reporter genes. The DNA-A genome replication and AC2 expression were observed to lower the expression of complementary sense promoter in both protoplast and transgenic plants. However, the virion sense expression was highly activated by AC2 in protoplast system but not in transgenic tobacco (Frey et al. 2001). Other study showed that in bipartite begomovirus BGMV the viral sense transcript was transactivated by both AC2 protein and other by bipartite begomovirus TGMV AC2 protein in *N. benthamiana* protoplast (Hung and Petty 2001). But C2 protein from a curtovirus could not complement begomoviral AC2 protein function (Saunders and Stanley 1995; Sung and Coutts 1995). These findings suggest that transactivation mediated by AC2/C2 is conserved phenomena (Sunter and Bisaro 1997). There are many reports that indicate that binding of AC2 to double-stranded DNA is weak and sequence nonspecific (Noris et al. 1996; Sung and Coutts 1996; Hartitz et al. 1999). Since AC2 is found to accumulate within both nucleus and protoplast, there might be a probability that AC2 functions by protein–protein interaction (Borah et al. 2016). With regard to AC2, a minimal sequence was found to be sufficient for bipartite begomovirus TGMV AV1 promoter to be transactivated by AC2 in *Nicotiana benthamiana* protoplast. Two elements were found, first between −125 and −107 and second between −96 and −60 from transcription start site. The second element was found to interact with a repressor because its deletion enabled basal promoter activity even in the absence of AC2 (Sunter and Bisaro 2003). In a study, it was found that AC2 interacts with itself, which requires a zinc finger-like motif (CCHC) and cysteine residue for AV1 promoter activation. Bimolecular fluorescence complementation showed that AC2 dimerizes and accumulates in the nucleus (Yang et al. 2007). The probable reason may be due to the presence of nuclear localization signal (Trinks et al. 2005). This study concluded that AC2 self-interaction is correlated with nuclear localization and transcription activation (Yang et al. 2007).

The AC2 protein of bipartite begomovirus MYMV also suppresses the RNAi-mediated antiviral defense response in plants. When a mutation was done in the AC2 NLS region or in zinc finger domain, both the transactivation function and RNAi suppression were abolished. This result suggests that suppression of silencing by AC2 possibly needs transactivation of host suppressors. This result was in agreement with the study where several promoter Arabidopsis were strongly induced by MYMV and ACMV AC2 proteins (Trinks et al. 2005).

7 Concluding Remarks

The transcription process in geminivirus is a highly complex process, and many aspects are still not well understood. The viral sense promoters require transactivation by complementary sense viral gene product AC2, whereas the complementary sense promoter is suppressed by their own gene product (C1/AC1 and C4/AC4). This process reflects the inbuilt mechanism of infection by geminivirus in plants. The sequences of the promoter regulating AC1/C1 genes are highly variable for both viral sense genes and complementary sense genes. The result depicts the fact that promoters may contain few functional domains common to all geminiviruses. The C2 of monopartite viruses and AC2 of the bipartite virus are positionally analogous and share similar transcriptional activation activity, but other functions are not common. Therefore, it is important to understand the regulatory pathway during transcription of each virus separately. This is essential for promoters of DNA-B and betasatellites because of lack of sufficient information.

References

Accotto GP, Donson J, Mullineaux PM (1989) Mapping of Digitaria streak virus transcripts reveals different RNA species from the same transcription unit. EMBO J 8(4):1033–1039

Ach RA, Durfee T, Miller AB et al (1997) RRB1 and RRB2 encode maize retinoblastoma-related proteins that interact with a plant D-type Cyclin and Geminivirus replication protein. Mol Cell Biol 17(9):5077–5086

Alberter B, Rezaian MA, Jeske H (2005) Replicative intermediates of tomato leaf curl virus and its satellite DNAs. Virology 331(2):441–448

Argüello-Astorga G, Herrera-Estrella GL, Bustamante RR (1994) Experimental and theoretical definition of Geminivirus origin of replication. Plant Mol Biol 26:553–556

Arguello-Astorga G, Lopez-Ochoa GL, Kong LJ et al (2004) A novel motif in Geminivirus replication proteins interacts with the plant retinoblastoma-related protein. J Virol 78 (9):4817–4826

Ashraf MA, Shahid AA, Rao AQ, Bajwa KS et al (2014) Functional characterization of a bidirectional plant promoter from cotton leaf curl Burewala virus using an agrobacterium-mediated transient assay. Viruses 6(1):223–242

Bagewadi B, Shoajiang C, Sunil KL et al (2004) PCNA interacts with Indian Mung bean yellow mosaic virus rep and Downregulates rep activity. J Virol 78(21):11890–11903

Borah BK, Zarreen F, Baruah G, Dasgupta I (2016) Insights into the control of Geminiviral promoters. Virology 495:101–111

Branch AD, Benenfeld BJ, Robertson HD (1988) Evidence for a single rolling circle in the replication of potato spindle tuber Viroid. Proc Natl Acad Sci U S A 85(23):9128–9132

Breathnach R, Chambon P (1981) Organization and expression of Eucaryotic Split genes coding for proteins. Annu Rev Biochem 50:349–386

Castillo AG, Colline D, Deret S et al (2003) Dual interaction of plant PCNA with Geminivirus replication accessory protein (REn) and viral replication protein (rep). Virology 312(2):381–394

Castillo AG, Kong LJ, Hanley-Bowdoin L, Bejarano ER (2004) Interaction between a Geminivirus replication protein and the plant Sumoylation system. J Virol 78(6):2758–2769

Chatterji A, Chatterji U, Beachy RN, Fauquet CM (2000) Sequence parameters that determine specificity of binding of the replication-associated protein to its cognate site in two strains of tomato leaf curl virus-New Delhi. Virology 273(2):341–350

Choudhury NR, Malik PS, Singh DK et al (2006) The Oligomeric rep protein of Mungbean yellow mosaic India virus (MYMIV) is a likely Replicative Helicase. Nucleic Acids Res 34 (21):6362–6377

Clerot D, Bernardi F (2006) DNA Helicase activity is associated with the replication initiator protein rep of tomato yellow leaf curl Geminivirus. J Virol 80(22):11322–11330

Dekker EL, Woolston CJ, Xue Y et al (1991) Transcript mapping reveals different expression strategies for the Bicistronic RNAs of the Geminivirus wheat dwarf virus. Nucleic Acids Res 19 (15):4075–4081

Desbiez C, David C, Mettouchi A, Laufs J, Gronenborn B (1995) Rep protein of tomato yellow leaf curl Geminivirus has an ATPase activity required for viral DNA replication. Proc Natl Acad Sci U S A 92(12):5640–5644

Donson J, Morris-Krsinich BA, Mullineaux PM et al (1984) A putative primer for second-Strand DNA synthesis of maize streak virus is Virion-associated. EMBO J 3(13):3069–3073

Eagle PA, Hanley-Bowdoin L (1997) Cis elements that contribute to Geminivirus transcriptional regulation and the efficiency of DNA replication. J Virol 71(9):6947–6955

Eagle PA, Orozco BM, Hanley-Bowdoin L (1994) A DNA sequence required for Geminivirus replication also mediates transcriptional regulation. Plant Cell 6(8):1157–1170

Eini O, Behjatnia SAA, Dogra S et al (2009) Identification of sequence elements regulating promoter activity and replication of a Monopartite Begomovirus-associated DNA β satellite. J Gen Virol 90(1):253–260

Fenoll C, Black DM, Howell SH (1988) The Intergenic region of maize streak virus contains promoter elements involved in rightward transcription of the viral genome. EMBO J 7 (6):1589–1596

Fontes EP, Luckow VA, Hanley-Bowdoin L (1992) A Geminivirus replication protein is a sequence-specific DNA binding protein. Plant Cell 4(5):597–608

Frey PM, Scharer-Hernandez NG, Futterer J et al (2001) Simultaneous analysis of the bidirectional African cassava mosaic virus pro- Moter activity using two different Luciferase genes. Virus Genes 185(22):596–604

Frischmuth S, Frischmuth T, Jeske H (1991) Transcript mapping of Abutilon mosaic virus, a Geminivirus. Virology 185(2):596–604

George B, Ruhel R, Mazumder M et al (2014) Mutational analysis of the Helicase domain of a replication initiator protein reveals critical roles of Lys 272 of the B' motif and Lys 289 of the β-hairpin loop in Geminivirus replication. J Gen Virol 95(PART 7):1591–1602

Gopal P, Kumar PP, Sinilal B et al (2007) Differential roles of C4 and C1 in mediating suppression of post-transcriptional gene silencing: evidence for Transactivation by the C2 of Bhendi yellow vein mosaic virus, a Monopartite Begomovirus. Virus Res 123(1):9–18

Gorbalenya AE, Koonin EV (1993) Helicases: amino acid sequence comparisons and structure-function relationships. Curr Opin Struct Biol 3(3):419–429

Groning BR, Hayes RJ, Buck KW (1994) Simultaneous regulation of tomato Golden mosaic virus coat protein and AL1 gene expression: expression of the AL4 gene may contribute to suppression of the AL1 gene. J Gen Virol 75(4):721–726

Gros MF, Riele HT, Ehrlich SD (1987) Rolling circle replication of single-stranded DNA plasmid pC194. EMBO J 6(12):3863–3869

Guerra-Peraza O, Kirk D, Seltzer V et al (2005) Coat proteins of Rice Tungro bacilliform virus and Mungbean yellow mosaic virus contain multiple nuclear-localization signals and interact with Importin α. J Gen Virol 86(6):1815–1826

Hanley-Bowdoin L, Elmer JS, Rogers SG (1988) Transient expression of Heterologous RNAs using tomato Golden mosaic virus. Nucleic Acids Res 16(22):10511–10528

Hanley-Bowdoin L, Elmer LJS, Rogers SG (1989) Functional expression of the leftward open Reading frames of the a component of tomato Golden mosaic virus in transgenic tobacco plants. Plant Cell 1(11):1057–1067

Hanley-Bowdoin L, Settlage SB, Orozco BM et al (1999) Geminiviruses: models for plant DNA replication, transcription, and cell cycle regulation. Crit Rev Plant Sci 18(1):71–106

Hanley-Bowdoin L, Settlage SB, Orozco BM et al (2000) Geminiviruses: models for plant DNA replication, transcription, and cell cycle regulation. Crit Rev Biochem Mol Biol 35(2):105–140

Hanley-Bowdoin L, Settlage SB, Robertson D (2004) Reprogramming plant gene expression: a prerequisite to Geminivirus DNA replication. Mol Plant Pathol 5(2):149–156

Hanley-Bowdoin L, Bejarano ER, Robertson D et al (2013) Geminiviruses: masters at redirecting and reprogramming plant processes. Nat Rev Microbiol 11(11):777–788

Harrison BD, Robinson DJ (1999) Natural genomic and antigenic variation in whitefly-transmitted Geminiviruses (Begomoviruses). Annu Rev Phytopathol 37:369–398

Hartitz MD, Sunter G, Bisaro DM (1999) The tomato Golden mosaic virus Transactivator (TrAP) is a single-stranded DNA and zinc-binding Phosphoprotein with an acidic activation domain. Virology 263(1):1–14

Hayes RJ, Buck KW (1989) Replication of tomato Golden mosaic virus DNA B in transgenic plants expressing open Reading frames (ORFs) of DNA a: requirement of ORF AL2 for production of single-stranded DNA. Nucleic Acids Res 17(24):10213–10222

Heyraud-Nitschke F, Schumacher S, Laufs J et al (1995) Determination of the origin cleavage and joining domain of Geminivirus rep proteins. Nucleic Acids Res 23(6):910–916

Hipp K, Rau P, Schäfer B et al (2014) The RXL motif of the African cassava mosaic virus rep protein is necessary for Rereplication of yeast DNA and viral infection in plants. Virology 462–463(1):189–198

Hofer JM, Dekker EL, Reynold HV et al (1992) Coordinate regulation of replication and Virion sense gene expression in wheat dwarf virus. Plant Cell 4(2):213–223

Hormuzdi SG, Bisaro DM (1993) Genetic analysis of beet curly top virus: evidence for three Virion sense genes involved in movement and regulation of single- and double-stranded Dna levels. Virology 193(2):900–909

Horns T, Jeske H (1991) Localization of Abutilon mosaic virus (AbMV) DNA within leaf tissue by in situ hybridization. Virology 181(2):580–588

Hung HC, Petty ITD (2001) Functional equivalence of late gene promoters in bean Golden mosaic virus with those in tomato Golden mosaic virus. J Gen Virol 82(3):667–672

Hur J, Buckley KJ, Lee S, Davis KR (2007) Transcriptional activator elements for Curtovirus C1 expression reside in the 3 â€™ coding region of ORF C1. Mol Cells 23(1):80–87

Hur J, Choi YE, Buckley KJ, Lee S et al (2008) Identification of a promoter motif involved in Curtovirus sense-gene expression in transgenic Arabidopsis. Mol Cells 26(2):131–139

Ilyina TV, Koonin EV (1992) Conserved sequence motifs in the initiator proteins for rolling circle DNA replication encoded by diverse Replicons from Eubacteria, Eucaryotes and Archaebacteria. Nucleic Acids Res 20(13):3279–3285

Jeske H, Lütgemeier M, Preib W (2001) DNA forms indicate rolling circle and recombination-dependent replication of Abutilon mosaic virus. EMBO J 20(21):6158–6167

Kaliappan K, Choudhury NR, Suyal G, Mukherjee SK (2012) A novel role for RAD54: this host protein modulates Geminiviral DNA replication. FASEB J 26(3):1142–1160

Kammann M, Schalk HJ, Matzeit V, chaefer S et al (1991) DNA replication of wheat dwarf virus, a Geminivirus, requires two Cis-acting signals. Virology 184(2):786–790

Koepsel RR (1985) The replication initiator protein of plasmid pT181 has sequence-specific Endonuclease and Topoisomerase-like activities. Proc Natl Acad Sci 82(20):6845–6849

Kong LJ, Hanley-Bowdoin L (2002) A Geminivirus replication protein interacts with a protein Kinase and a motor protein that display different expression patterns during plant development and infection. Plant Cell 14(8):1817–1832

Kumar RV, Prasanna HC, Singh AK, Ragunathan D et al (2017) Molecular genetic analysis and evolution of Begomoviruses and Betasatellites causing yellow mosaic disease of Bhendi. Virus Genes 53(2):275–285

Kushwaha NK, Bhardwaj M, Chakraborty S (2017) The replication initiator protein of a Geminivirus interacts with host Monoubiquitination machinery and stimulates transcription of the viral genome. PLoS Pathog 13(8):e1006587. https://doi.org/10.1371/journal.ppat.1006587

Lacatus G, Sunter G (2008) Functional analysis of bipartite Begomovirus coat protein promoter sequences. Virology 376(1):79–89

Laufs J, Traut W, Heyraud F, Matzeit F et al (1995) In vitro cleavage and joining at the viral origin of replication by the replication initiator protein of tomato yellow leaf curl virus. Proc Natl Acad Sci U S A 92(9):3879–3883

Lewis JD, Lazarowitz SG (2010) Arabidopsis Synaptotagmin SYTA regulates Endocytosis and virus movement protein cell-to-cell transport. Proc Natl Acad Sci 107(6):2491–2496

Luque A, Andrés P, Burgos S, Ramirez-Parra E et al (2002) Interaction of Geminivirus rep protein with replication factor C and its potential role during Geminivirus DNA replication. Virology 302(1):83–94

Mazithulela G, Sudhakar D, Heckel T, Mehlo L et al (2000) The maize streak virus coat protein transcription unit exhibits tissue-specific expression in transgenic Rice. Plant Sci 155:21–29

Morris-Krsinich BAM, Mullineaux PM, Donson J, Boulton MI et al (1985) Bidirectional transcription of maize streak virus DNA and identification of the coat protein gene. Nucleic Acids Res 13:7237–7256

Morris-Krsinich BAM, Richardson KA, Haley A, Zhan XC et al (1992) The nucleotide sequence of the infectious cloned Dna component of tobacco yellow dwarf virus reveals features of Geminiviruses infecting monocotyledonous plants. Virology 187(2):633–642

Mullineaux PM, Guerineau F, Accotto GP (1990) Processing of complementary sense RNAs of Digitaria streak virus in its host and in transgenic tobacco. Nucleic Acids Res 18(24):7259–7265

Mullineaux PM, Rigden JE, Dry IB, Krake LR et al (1993) Mapping of the Polycistronic RNAs of tomato leaf curl Geminivirus. Virology 193(1):414–423

Nagar S, Pedersen TJ, Carrick KM, Hanley-Bowdoin L et al (1995) A geminivirus induces expression of a host DNA synthesis protein in terminally differentiated plant cells. Plant Cell 7(June):705–719

Nash TE, Dallas MB, Reyes MI, Buhrman GK et al (2011) Functional analysis of a novel motif conserved across Geminivirus rep proteins. J Virol 85(3):1182–1192

Nikovics K, Simidjieva J, Peres A, Ayaydin F et al (2001) Cell-cycle, phase-specific activation of maize streak virus promoters. Mol Plant-Microbe Interact 14(5):609–617

Noris E, Jupin I, Accotto GP, Gronenborn B (1996) DNA-binding activity of the C2 protein of tomato yellow leaf curl Geminivirus. Virology 217(2):607–612

Orozco BM, Hanley-Bowdoin L (1998) Conserved sequence and structural motifs contribute to the DNA binding and cleavage activities of a Geminivirus replication protein. J Biol Chem 273 (38):24448–24456

Orozco BM, Miller AB, Settlage SB, Hanley-Bowdoin L (1997) Functional domains of a Geminivirus replication protein. J Biol Chem 272(15):9840–9846

Pant V, Gupta D, Choudhury NR, Malathi VG et al (2001) Molecular characterization of the rep protein of the Blackgram isolate of Indian Mungbean yellow mosaic virus. J Gen Virol 82 (10):2559–2567

Pascal E, Sanderfoot AA, Ward BM, Medville R et al (1994) The Geminivirus BR1 movement protein binds single-stranded DNA and localizes to the cell nucleus. Plant Cell 6(7):995–1006

Pasumarthy KK, Choudhury NR, Mukherjee SK (2010) Tomato leaf curl Kerala virus (ToLCKeV) AC3 protein forms a higher order Oligomer and enhances ATPase activity of replication initiator protein (rep/AC1). Virol J 7:128

Petty ITD, Coutts RHA, Buck KW (1986) Geminivirus coat protein gene promoter sequences can function in Escherichia Coli. Nucleic Acids Res 14(12):5113

Petty ITD, Coutts RHA, Buck KW (1988) Transcriptional mapping of the coat protein gene of tomato Golden mosaic virus. J Gen Virol 69(1988):1359–1365

Pilartz M, Jeske H (1992) Abutilon mosaic Geminivirus double-stranded DNA is packed into Minichromosomes. Virology 189(2):800–802

Pilartz M, Jeske H (2003) Mapping of Abutilon mosaic Geminivirus Minichromosomes. J Virol 77 (20):10808–10818

Pooggin MM (2013) How can plant DNA viruses evade siRNA-directed DNA Methylation and silencing? Int J Mol Sci 14(8):15233–15259

Preiss W, Jeske H (2003) Multitasking in replication is common among Geminiviruses. J Virol 77 (5):2972–2980

Rao K, Sunter G (2012) Sequences within the spinach curly top virus Virion sense promoter are necessary for vascular-specific expression of Virion sense genes. Virology 432(1):10–19

Richter KS, Kleinow T, Jeske H (2014) Somatic homologous recombination in plants is promoted by a Geminivirus in a tissue-selective manner. Virology 452–453:287–296

Richter KS, Ende L, Jeske H (2015) Rad54 is not essential for any Geminiviral replication mode in Planta. Plant Mol Biol 87(1–2):193–202

Richter KS, Serra H, White CI, Jeske H (2016) The recombination mediator RAD51D promotes Geminiviral infection. Virology 493:113–127

Rizvi I, Choudhury NR, Tuteja N (2014) Insights into the functional characteristics of Geminivirus rolling-circle replication initiator protein and its interaction with host factors affecting viral DNA replication. Arch Virol 160(2):375–387

Ruschhaupt M, Darren PM, Lakay F, Bezuidenhout M et al (2013) Replication modes of maize streak virus mutants lacking RepA or the RepA-pRBR interaction motif. Virology 442 (2):173–179

Sanchez-Duran MA, Dallas MB, Ascencio-Ibanez JT, Reyes MI et al (2011) Interaction between Geminivirus replication protein and the SUMO-conjugating enzyme is required for viral infection. J Virol 85(19):9789–9800

Saunders K, Stanley J (1995) Complementation of African cassava mosaic virus AC2 gene function in a mixed bipartite Geminivirus infection. J Gen Virol 76(9):2287–2292

Saunders K, Lucy A, Stanley J (1991) DNA forms of the Geminivirus African cassava mosaic virus consistent with a rolling circle mechanism of replication. Nucleic Acids Res 19(9):2325–2330

Saunders K, Lucy A, Stanley J (1992) RNA-primed complementary-sense DNA synthesis of the Geminivirus African cassava mosaic virus. Nucleic Acids Res 20(23):6311–6315

Schalk HJ, Matzeit V, Schell J, Gronenborn B (1989) Wheat dwarf virus, a Geminivirus of Graminaceous plants needs splicing for replication. EMBO J 8(2):359–364

Serra H, Ines OD, Degroote F, Gallego ME et al (2013) Roles of XRCC2, RAD51B and RAD51D in RAD51-independent SSA recombination. PLoS Genet 9(11):e1003971

Settlage SB, Miller AB, Gruissem W, Hanley-Bowdoin L (2001) Dual interaction of a Geminivirus replication accessory factor with a viral replication protein and a plant cell cycle regulator. Virology 279(2):570–576

Shivaprasad PV, Akbergenov R, Trinks D, Veluthambi K et al (2005) Promoters, transcripts, and regulatory proteins of Mungbean yellow mosaic Geminivirus promoters, transcripts, and regulatory proteins of Mungbean Yellow Mosaic Geminivirus. J Virol 79(13):8149–8163

Shung CY, Sunter J, Sirasanagandla SS, Sunter G (2006) Distinct viral sequence elements are necessary for expression of tomato Golden mosaic virus complementary sense transcripts that direct AL2 and AL3 gene expression. Mol Plant-Microbe Interact 19(12):1394–1405

Singh DK, Mohammad NI, Choudhury NR, Karjee S et al (2007) The 32 kDa subunit of replication protein a (RPA) participates in the DNA replication of Mung bean yellow mosaic India virus (MYMIV) by interacting with the viral rep protein. Nucleic Acids Res 35(3):755–770

Stenger DC, Revington GN, Stevenson MC, Bisaro DM (1991) Replicational release of Geminivirus genomes from Tandemly repeated copies: evidence for rolling-circle replication of a plant viral DNA. Proc Natl Acad Sci 88(18):8029–8033

Sung YK, Coutts RHA (1995) Pseudorecombination and complementation between potato yellow mosaic Geminivirus and tomato Golden mosaic Geminivirus. J Gen Virol 76(11):2809–2815

Sung YK, Coutts RHA (1996) Potato yellow mosaic Geminivirus AC2 protein is a sequence non-specific DNA binding protein. FEBS Lett 383(1–2):51–54

Sunter G, Bisaro DM (1989) Transcription map of the B genome component of tomato Golden mosaic virus and comparison with a component transcripts. Virology 173(2):647–655

Sunter G, Bisaro DM (1991) Transactivation in a Geminivirus: AL2 gene product is needed for coat protein expression. Virology 180(1):416–419

Sunter G, Bisaro DM (1992) Transactivation of Geminivirus AR1 and BR1 gene expression by the viral AL2 gene product occurs at the level of transcription. Plant Cell 4(10):1321–1331

Sunter G, Bisaro DM (1997) Regulation of a Geminivirus coat protein promoter by AL2 protein (TrAP): evidence for activation and Derepression mechanisms. Virology 232(2):269–280

Sunter G, Bisaro DM (2003) Identification of a minimal sequence required for activation of the tomato Golden mosaic virus coat protein promoter in protoplasts. Virology 305(2):452–462

Sunter G, Hartitz MD, Hormuzdi SG, Brough CL et al (1990) Genetic analysis of tomato Golden mosaic virus: ORF AL2 is required for coat protein accumulation while ORF AL3 is necessary for efficient DNA replication. Virology 179(1):69–77

Sunter G, Hartitz MD, Bisaro DM (1993) Tomato Golden mosaic virus leftward gene expression: autoregulation of Geminivirus replication protein. Virology 195(1):275–280

Suyal G, Mukherjee SK, Choudhury NR (2013) The host factor RAD51 is involved in Mungbean yellow mosaic India virus (MYMIV) DNA replication. Arch Virol 158(9):1931–1941

Thommes PA, Buck KW (1994) Synthesis of the tomato Golden mosaic virus AL1, AL2, AL3 and AL4 proteins in vitro. J Gen Virol 75(8):1827–1834

Townsend R, Stanley J, Curson SJ, Short MN (1985) Major Polyadenylated transcripts of cassava latent virus and location of the gene encoding coat protein. EMBO J 4(1):33–37

Trinks D, Rajeswaran R, Shivaprasad PV, Akbergenov R et al (2005) Suppression of RNA silencing by a Geminivirus nuclear protein, AC2, correlates with Transactivation of host genes. J Virol 79(4):2517–2527

Tu J, Sunter G (2007) A conserved binding site within the tomato Golden mosaic virus AL-1629 promoter is necessary for expression of viral genes important for pathogenesis. Virology 367 (1):117–125

Uchiyama A, Shimada-Beltran H, Levy A, Zheng JY et al (2014) The Arabidopsis Synaptotagmin SYTA regulates the cell-to-cell movement of diverse plant viruses. Front Plant Sci 5:584

Usharani KS, Periasamy M, Malathi VG (2006) Studies on the activity of a bidirectional promoter of Mungbean yellow mosaic India virus by Agroinfiltration. Virus Res 119(2):154–162

Vanderschuren H, Akbergenov R, Pooggin MM, Hohn T et al (2007) Transgenic cassava resistance to African cassava mosaic virus is enhanced by viral DNA-A bidirectional promoter-derived siRNAs. Plant Mol Biol 64(5):549–557

Wang WC, Hsu YH, Lin NS, Wu CY et al (2013) A novel prokaryotic promoter identified in the genome of some Monopartite Begomoviruses. PLoS One 8(7):e70037

Wang WC, Wu CY, Lai YC, Lin NS et al (2014) Characterizationof the cryptic AV3 promoter of Ageratum yellow vein virus in prokaryotic and eukaryotic systems. PLoS One 9:e108608

Willment JA, Martin DP, Palmer KE, Schnippenkoetter WH et al (2007) Identification of long Intergenic region sequences involved in maize streak virus replication. *J Gen Virol* 88. (Pt 6:1831–1841

Wright EA, Heckel T, Groenendijk J, Davies JW et al (1997) Splicing features in maize streak virus Virion- and complementary-sense gene expression. Plant J 12(6):1285–1297

Wyant PS, Kober S, Schwierzok A, Kocher C et al (2012) Cloned tomato Golden mosaic virus Back in tomatoes. Virus Res 167(2):397–403

Xie Q, Suárez-López P, Gutiérrez C (1995) Identification and analysis of a retinoblastoma binding motif in the replication protein of a plant DNA virus: requirement for efficient viral DNA replication. EMBO J 14(16):4073–4082

Xie Q, Sanz-Burgos AP, Guo H, García JA et al (1999) GRAB proteins, novel members of the NAC domain family, isolated by their interaction with a Geminivirus protein. Plant Mol Biol 39 (4):647–656

Xie Y, Liu Y, Meng M, Chen L et al (2003) Isolation and identification of a super strong plant promoter from cotton leaf curl Multan virus. Plant Mol Biol 53(1–2):1–14

Yang X, Baliji S, Buchmann RC, Wang H et al (2007) Functional modulation of the Geminivirus AL2 transcription factor and silencing suppressor by self-interaction. J Virol 81 (21):11972–11981

Yingqiu X, Yule L, Zhen Z (2001) Expressing activity of promoter elements of large Intergenic region from cotton leaf curl virus in host plant. Sci China Ser C Life Sci 44(1):8–17

Zhan X, Haley A, Richardson K, Morris B (1991) Analysis of the potential promoter sequences of African cassava mosaic virus by transient expression of the ??-Glucuronidase gene. J Gen Virol 72(11):2849–2852

Zhou Y, Rojas MR, Park MR, Seo YS et al (2011) Histone H3 interacts and Colocalizes with the nuclear shuttle protein and the movement protein of a Geminivirus. J Virol 85(22):11821–11832

Distribution of Geminivirus in the Indian Subcontinent

Bhavin S. Bhatt, Fenisha D. Chahwala, Sangeeta, B. K. Yadav, B. Singh, and Achuit K. Singh

Abstract

Viral diseases cause havoc on crop yield, both qualitatively and quantitatively. *Geminiviridae* is the largest family of plant viruses and constitutes an important group of plant pathogens with genomes of ssDNA. Geminiviruses are characterized by particle morphology of twinned incomplete icosahedra. Geminiviruses derived their name from unique structure and geometry of virus particles, where two icosahedrals are joined together. Family *Geminiviridae* is further classified into nine different genera on the basis of nature of genome, host plant infection, and vector requirements for disease transmission. These viruses cause significant yield loss to economically important plants. Disease outbreaks on cotton, cassava, tomato, and other important horticultural plants were reported to have major crop loss due to virus infection. Further, molecular interactions and presence of satellite molecules enable virus particles to break innate immunity of plants and revoke disease outbreaks. Also introduction of exotic species, transfer of plant materials across continents, and vector migration are also important factors which contribute to widespread distribution of geminiviruses. India and the Indian subcontinent have experienced and are experiencing major loss due to infection by geminiviruses. Novel recombinant viruses, host switching, and newer satellite molecules continue to be reported from the Indian subcontinent. Tropical humid atmosphere and crop diversity are major factor for vector multiplication and hence virus transmission too. This chapter reviews the major geminiviral crop infections in the Indian subcontinent.

B. S. Bhatt
Shree Ramkrishna Institute of Computer Education and Applied Sciences, Surat, India

F. D. Chahwala · Sangeeta · B. K. Yadav
School of Life Sciences, Central University of Gujarat, Sector-30, Gandhinagar, India

B. Singh · A. K. Singh (✉)
Crop Improvement Division, ICAR—Indian Institute of Vegetable Research, Varanasi, India

R. V. Kumar (ed.), *Geminiviruses*, https://doi.org/10.1007/978-3-030-18248-9_3

1 Introduction

Plants growing in the natural environment are sessile in nature and can be attacked continuously by omnipresent microorganisms, namely viruses, bacteria, and fungi. Plant viruses are intracellular parasites, which need the vector for movement from one plant to another plant. It directly affects the economy by reducing the production of plant products. Such losses due to viral diseases impact heavily on crop production and are one of the major thrust for the detailed study of plant viruses (Teng 1985). Chemically, viruses are nucleoprotein in nature, where viruses' genome is encapsidated in protein shell. Viral genome (DNA or RNA) encoded proteins modify host cellular machinery for their replication, movement and transmission efficiently. Small genome size and ability to multiply within host cell have made virus particles extremely dynamic and diverse. On the other hand, possessing a small genome and restricted host range made virus particles a model organism to understand the concepts and principles of molecular biology.

Structurally, plant viruses resemble other viruses. The viral nucleic acid is encapsidated in the closed shell or tube-like structure, made of protein, termed as the capsid. Plant viruses are intracellular parasites, which need the vector for movement from one plant to another plant. It directly affects the economy by reducing production of plant products, so in these cases, we can say it is the most dangerous guest of the host plant. Viruses are responsible for causing around the US $30 billion loss in the yield every year (Valérie Nicaise 2014). According to *International Committee for Taxonomy of Viruses* (ICTV), there are around 49 families and 79 genera of plant viruses which have been discovered and reported to date.

2 Family *Geminiviridae*

Geminivirus constitutes an important group of plant pathogens with the genome of ssDNA. Geminivirus genome comprises a closed circular ssDNA of 2.6 kb–3.0 kb size with an intergenic region which has geminivirus signature nonanucleotide sequence that is recognition and initiation site for viral DNA replication. Geminivirus derived its name from unique structure and geometry of virus particles that look like small balls stuck together. A single molecule of covalently closed circular single-stranded viral sense DNA is encapsidated in each paired particle. This family is the most devastating to plants and is responsible for causing significant loss, especially in the tropical and subtropical regions.

Family *Geminiviridae* constitutes the largest number of viruses. Biological, genome partition and vector requirement for transmission have made the base for the partition of Geminivirus into nine different genera (Varsani et al. 2014, 2017).

3 Geminiviruses in Indian Subcontinent

The *Geminiviridae* family has been divided on basis of genome organization, host range and insect vector into nine genera: *Mastrevirus*, *Curtovirus*, *Topocuvirus*, *Begomovirus*, *Becurtovirus*, *Eragrovirus*, *Turncurtovirus*, *Grablovirus* and *Capulavirus* (Brown et. al. 2012) (Table 1).

Among nine genera of family *Geminiviridae*, prevalence of three genera *viz.* Capulavirus, Mastrevirus and Begomovirus is widespread in the Indian subcontinent. Due to tropical humid climate, vector population of these viruses is widespread in this area and hence reports of disease occurrence and outbreaks are available from the different parts of the Indian subcontinent. Amongst these three genera, occurrence of begomovirus is wide spread in the Indian subcontinent.

Table 1 Characteristics of genera of family *Geminiviridae*

Genus	Type Species	Acronym	Genome	Host range	Insect vector
Becurtovirus	*Beet curly top Iran virus*	BCTIV	Monopartite	Dicotyledonous plants	Leafhopper
Begomovirus	Bean golden mosaic virus	BGMV	Monopartite and Bipartite	Dicotyledonous plants	Whitefly
Capulavirus	caput-medusae latent virus	CMLV	Monopartite	Dicotyledonous plants	Aphid
Curtovirus	*Beet curly top virus*	BCTV	Monopartite	Dicotyledonous plants	Leafhopper
Eragrovirus	*Eragrostis curvula streak virus*	ECSV	Monopartite	Monocotyledonous plants	–
Grablovirus	Grapevine red-blotch associated virus	GRBV	Monopartite	Dicotyledonous plants	Treehopper
Mastrevirus	Maize streak virus	MSV	Monopartite	Mostly monocotyledonous plants (except for tobacco yellow dwarf virus and bean yellow dwarf virus which infect dicots).	Leafhopper
Topocuvirus	Tomato pseudo-curly top virus	TPCTV	Monopartite	Dicotyledonous plants	Treehopper
Turncurtovirus	*Turnip curly top virus*	TCTV	Monopartite	Dicotyledonous plants	Leafhopper

A large number of begomoviruses are continuously reported from different geographical regions. These begomoviruses are either have newer more virulent features or they may have an infection on newer host species. Thus, careful examination on occurrence and spreading of diseases on continuous basis is vital need to design timely and efficient measures to manage viral infections.

The subsequent segments of this chapter highlight our current knowledge of occurrence of geminiviruses in the Indian subcontinent, their interaction and virulent strategies and major symptoms of the virus infection.

4 Genus *Capulavirus* (Type Species: Caput-Medusae Latent Virus, CMLV)

4.1 Introduction

Nowadays, high-throughput technologies for nucleotide sequencing methods are used to discover previously unknown viruses. *Capulavirus* genus name was derived from the virus *caput-medusae latent virus*. The viruses belonging to *Capulavirus* genus are transmitted by aphids. Four species are present in this genus: Alfalfa leaf curl virus, Euphorbia caput-medusae latent virus, French bean severe leaf curl virus (FbSLCV), and Plantago lanceolata latent virus (Varsani et al. 2017).

4.2 Important Capulavirus in Indian Subcontinent

4.2.1 French Bean Severe Leaf Curl Virus

Out of four virus species reported for genus *Capulavirus*, only one virus, French bean Severe Leaf Curl Virus (FbSLCV) reported from India was associated with severe leaf curl disease of French bean. Furthermore, Bernardo et al. (2016) reported virus *Caput-medusae latent virus* from South Africa which shares the maximum identity of 78% with this FbSLCV isolate.

5 Genus *Mastrevirus* (Type Species: Maize Streak Virus, MSV)

5.1 Introduction

Maize streak disease was first observed on maize in 1901 in South Africa in Hawaii region. Mastreviruses are leafhopper-transmitted monopartite viruses infecting monocots. Well-characterized subgroup I pathogens include maize streak virus (MSV) and wheat dwarf virus (WDV). Two other members of this genus, TYDV (Tomato yellow dwarf virus) and BeYDV (Bean yellow dwarf virus), also infect dicotyledonous species (Kraberger et al. 2013). They are generally monopartite in nature where sole DNA-A component is responsible for causing diseases. Recently,

Kumar et al. (2012a, b, c) for the first time proved association of two alphasatellite species, a Cotton leaf curl Multan alphasatellite (CLCuMA) and a Guar leaf curl alphasatellite (GLCuA), with *wheat dwarf India virus* (WDIV). Mastrevirus infection was first confirmed in 1992 in India (Horn et al. 1993). Mastrevirus infection was shown on sugarcane, wheat, and chickpea from various parts of the Indian subcontinent (Horn et al. 1993, 1994; Kumar et al. 2012b; Haider et al. 2011) (Table 2). However, the presence of Mastrevirus was shown on Bajra, that is, Bajra streak virus in 1972, but it is not yet confirmed and is designated as unassigned species.

5.2 Important Mastreviruses of the Indian Subcontinent

5.2.1 Wheat Dwarf India Virus

A very first time maize streak disease was observed in 1972 in India (Seth et al. 1972). During 2010 and 2011, wheat plants were found affected by dwarf diseases across the country. Plants were showing symptoms such as sterile spikes, yellowing of leaves and dwarfism. The presence of *Psammotettix* sp. (Leafhopper) in affected fields was suspected to be that of geminivirus. On the basis of PCR and RT-PCR studies, it was confirmed that there was absence of BYDV-MAV and BYDV-PAV in infected samples. However, association of a new species of mastrevirus named wheat dwarf India virus was identified (Kumar et al. 2012b). Agro-inoculation of wheat seedlings by infectious clones of virus results in the dwarfism of wheat plants, while mock inoculated control wheat seedlings were healthy and tall. Yet, typical streak phenotype was not observed in any of the inoculated wheat plants. Wheat dwarf India virus was also reported from Bihar, Maharashtra, Uttar Pradesh, Rajasthan, and Madhya Pradesh, which results in significant reduction in production of wheat (Kumar et al. 2012a, b, c, Kumar et al. 2014a, b).

There are reports that the association of begomovirus satellite molecules with mastrevirus increases severity in plants. During 2014, there was a report on the association of two alphasatellites and one betasatellite molecule with *wheat dwarf India virus* (Kumar et al. 2014a, b). This was the first report on the association of satellite molecule with WDV. Guar leaf curl alphasatellite and Cotton leaf curl Multan betasatellite were associated with WDV. Ageratum yellow leaf curl betasatellite is also associated with WDV. The satellite molecule tends to increase WDV accumulation in plant and suppresses the small RNAs' accumulation related to diseases (Kumar et al. 2014a, b). One study showed the co-infection of mastrevirus and begomovirus on cotton and *Xanthium strumarium*. In this way, WDV in association with alpha and betasatellites tends to cause severe symptoms in wheat.

5.2.2 Chickpea Chlorotic Dwarf Virus

Mastrevirus infection on dicot plant has been distributed in Asia, Africa, and Australia (Nahid et al. 2008). It infects important dicots (Kraberger et al. 2013). Horn et al. (1993) first reported the chickpea chlorotic dwarf virus (CpCDV) in

Table 2 List of major geminiviruses associated with various crops and weeds in the Indian subcontinent

Genus	Species	Abbreviation	Host	Country of origin
Mastrevirus	*Chickpea chlorotic dwarf virus*	CpCDV-C[PK-Fai6–06]	Chickpea	Pakistan
		CpCDV-D[PK-BGR-08]	Chickpea	Pakistan
	Wheat dwarf India virus	WDIV-[IN-10]	Wheat	India
Capulavirus	*French bean severe leaf curl virus*	FbSLSV-[IN-10]	French bean	India
Begomovirus	*Ageratum enation virus*	AEV-IN[IN-Kan-08]	Ageratum	India
		AEV-NP[NP-99]	Ageratum	Nepal
		AEV-UP [IN-UP-Ag10–10]	Ageratum	India
		AYVSLV-[LK-99]	Ageratum	Sri Lanka
	Bhendi yellow vein Bhubaneswar virus	BYVBhV-[IN-Ori-03]	Bhendi	India
	Bhendi yellow vein Haryana virus	BYVMV-Har[IN-Har-07]	Bhendi	India
	Bhendi yellow vein mosaic virus	BYVMV-IN[IN-mad]	Bhendi	India
		BYVMV-[IN-Mah-NOL751]	Bhendi	India
		BYVMV-[PK-Fai201–95]	Bhendi	Pakistan
		BYVMV-TN[IN-Coi4–04]	Bhendi	India
		BYVMV-Tha[IN-Tha-05]	Bhendi	India
	Catharanthus yellow mosaic virus	CaYMV-[PK-Isl-DR151]	Catharanthus	Pakistan
	Chilli leaf curl India virus	ChiLCINV-[IN-08]	Chilli	India
	Chilli leaf curl Kanpur virus	ChiLCKaV-[IN-Kam-08]	Chilli	India
	Chilli leaf curl Vellanad virus	ChiLCVV-[IN-Vel-08]	Chilli	India
	Chilli leaf curl virus	ChiLCV-PK[PK-Mul-98]	Chilli	Pakistan
		ChiLCV-IN[IN-Amr-Pap-09]	Papaya	India
		ChiLCV-chi[IN-chi-05]	Chilli	India
		ChiLCV[BD-Gaz]	Chilli	India
		ChiLCV-JO[IN-Pon-Hib-07]	Hibiscus	India
		ChiLCV-Kha[PK-Kha-04]	Chilli	Pakistan

	Clerodendron yellow mosaic virus	ClYMV-[IN-Iari-06]		India
	Corchorus golden mosaic virus	CoGMV-IN[IN-Bah-08]	Corchorus	India
	Corchorus yellow vein mosaic virus	CoYV-[IN-Mah-CEA8–11]	Corchorus	India
	Cotton leaf curl Allahabad virus	CLCuAlV-Al[PK-Ala804a-96]	Cotton	Pakistan
		CLCuAlV-ha[IN-Kar-OY77-Okr-05]	Okra	India
		CLCuAlV-Ka[IN-Kar-OY81B-Okr-05]	Okra	India
		CLCuAlV-lo[PK-Mul-Lob-06]	Cotton	Pakistan
		CLCuAlV-mu[PK-Mul-Pun-06]	Cotton	Pakistan
	Cotton leaf curl Bangalore virus	CLCuBaV-[IN-Ban-04]	Cotton	India
	Cotton leaf curl Kokhran virus	CLCuKoV-Ko[PK-Man806b-96]	Cotton	Pakistan
		CLCuKoV-Bu[PK-Veh-06]	Cotton	Pakistan
		CLCuKoV-La[PK-Lay-11]	Cotton	Pakistan
		CLCuKoV-Lu[IN-Luc-Ct-Bea10]	Cyamopsis	India
		CLCuKoV-Sha[PK-Sha-05]	Cotton	Pakistan
	Cotton leaf curl Multan virus	CLCuMuV-Dar[PK-Mul-Dar1–06]	Cotton	Pakistan
		CLCuMuV-Fai[PK-Yaz62–95]	Cotton	Pakistan
		CLCuMuV-Hib[IN-Hib1–11]	Hibiscus	India
		CLCuMuV-his[PK-Mul-H65–1-97]	Hibiscus	Pakistan
		CLCuMuV-PK[PK-Mul-06]	Cotton	Pakistan
		CLCuMuV-Ra[IN-Sri-94]	Cotton	India
	Croton yellow vein mosaic virus	CroYVMV-[IN]	Croton	India
	Dolichos yellow mosaic virus	DoYMV-[BD-Gaz]	Dolichos	Bangladesh
	French bean leaf curl virus	FbLCV-[IN-Kan-11]	French bean	India-
	Hemidesmus yellow mosaic virus	HemYMV-[IN-Tir-H1–12]	Hemidesmus	India
	Hollyhock leaf curl virus	HoLCV-[PK-Fai-20-4-06]	Hollyhock	Pakistan
	Horsegram yellow mosaic virus	HgYMV-[IN-Coi]	Horsegram	India
	Indian cassava mosaic virus	ICMV-Jat[IN-Dha-08]	Cassava	India
		ICMV-Ker[IN-Ker2–02]	Cassava	India

(continued)

Table 2 (continued)

Genus	Species	Abbreviation	Host	Country of origin
	Jatropha leaf curl virus	JLCuV-ND[IN-ND-07]	Jatropha	India
		JLCuV-Gu[IN-Guj-09]	Jatropha	India
	Jatropha mosaic India virus	JMINV-[IN-Luc-09]	Jatropha	India
	Jatropha yellow mosaic virus	JYMV-[IN-Kat-08]	Jatropha	India
	Mesta yellow vein mosaic Bahraich virus	MeYVMBaV-[IN-Bah-07]	Mesta	India
	Mesta yellow vein mosaic virus	MeYVMV-and[IN-Ama27–08]	Mesta	India
		MeYVMV-[PK-CM-09]	Mesta	Pakistan
		MeYVMV-ben[IN-Bar-06]	Mesta	India
	Mungbean yellow mosaic India virus	MYMIV-[IN-ND-Bg3–91]	Mungbean	India
		MYMV-[TH-Mg2]	Mungbean	Thailand
	Okra enation leaf curl virus	OELCuV-[IN-SorEL10–06]	Okra	India
	Papaya leaf crumple virus	PaLCrV-[IN-Pan-08]	Papaya	India
		PaLCuV-Cir[IN-PaND13–12]	Papaya	India
		PaLCuV-Ast[PK-Luc-as-11]	Papaya	Pakistan
		PaLCuV-A[IN-WB-Cr-Cro-08]	Croton	India
		PaLCuV-Ama[PK-Luc-am-11]	Amaranthus	Pakistan
		PaLCuV-Luc[IN-Luc]	Papaya	India
		PaLCuV-Sik[in-Sik-Cal-10]	Calotropis	India
		PaLCuV-Rh[PK-Mia-Rc-07]	Rhynchosia	Pakistan
		PaLCuV-soy[IN-Luc-Soy-11]	Soybean	India
		PaLCuV-PK[PK-Cot-02]	Cotton	Pakistan
		PaLCuV-Pun[PK-Pun-Cro-06]	Croton	Pakistan
		PaLCuV-IN[IN-pat-Rad-09]	Radish	India
		PaLCuV-Lah[PK-Lah-HYDNA-Alc-06]	Hollyhock	Pakistan
		PaLCuV-Tob[IN-Luc-Nic-10]	Nicotiana	India
		PaLCuV-tom[IN-CTM-tom-06]	Tomato	

	Pedilanthus leaf curl virus	PeLCV-[PK-Mul-06]	Pedilanthus	Pakistan
		PeLCV-Eu[PK-RYK1-To-04]	Tomato	Pakistan
		PeLCV-Sb[PK-NS-Sb-09]	Soybean	Pakistan
	Pepper leaf curl Bangladesh virus	PepLCBV-BD[BD-Bog-99]	Pepper	Bangladesh
		PepLCBV-PK[PK-Kha-04]	Pepper	Pakistan
		PepLCBV-IN[IN-Coi-08]	Pepper	India
		PepLCBV-[PK-Lah-04]	Pepper	Pakistan
	Pepper leaf curl Lahore virus	PepLCLaV-[IN-Luc-11]	Pepper	India
	Radish leaf curl virus	RaLCuV-[IN-Var-03]	Radish	India
		RaLCuV-to[PK-Bih-To-09]	Tobacco	Pakistan
	Rhynchosia yellow mosaic India virus	RhYMIV-[IN-Thi-JRH1–09]	Rhynchosia	India
		RhYMV-[PK-Lah33–07]	Rhynchosia	Pakistan
	Rose leaf curl virus	RoLCuV-[IN-Raj-SikAS24–14]	Rose	India
	Spinach yellow vein virus	SpiYVV-[IN-Sik-AS22]	Spinach	India
	Sri Lankan cassava mosaic virus	SLCMV-LK[LK-Col-98]	Cassava	Sri Lanka
		SLCMV-IN[IN-Adi-03]	Cassava	India
	Sunn hemp leaf distortion virus	SHLDV-[IN-Bar-08]	Sunn hemp	India
	Tobacco leaf curl Pusa virus	TbLCPuV-to[IN-Pus-09]	Tobacco	India
	Tomato leaf curl Bangalore virus	ToLCBaV-A[IN-Ban1]	Tomato	India
		ToLCBaV-[IN-Hes-TC265–10]	Tomato	India
		ToLCBaV-B[IN-Ban5]	Tomato	India
		ToLCBaV-D[IN-KerII-05]	Tomato	India
		ToLCBaV-C[IN-Ban4–97]	Tomato	India
	Tomato leaf curl Bangladesh virus	ToLCBV-[BD-BD2]	Tomato	Bangladesh
	Tomato leaf curl Gujarat virus	ToLCGuV-[IN-Var-01]	Tomato	India
	Tomato leaf curl Joydebpur virus	ToLCJV-[IN-Var-Caa-10]	Chilli	India
	Tomato leaf curl Karnataka virus	ToLCKaV-ban[IN-Ban-93]	Tomato	India

(continued)

Table 2 (continued)

Genus	Species	Abbreviation	Host	Country of origin
	Tomato leaf curl Kerala virus	ToLCKeV-[IN-Ker3–07]	Tomato	India
	Tomato leaf curl New Delhi virus	ToLCNDV-[IN-ND-Svr-92]	Tomato	India
	Tomato leaf curl New Delhi virus 2	ToLCNDV2-[IN-IANDS1–11]	Tomato	India
	Tomato leaf curl New Delhi virus 4	ToLCNDV4-[IN-Jun-TC306–11]	Tomato	India
	Tomato leaf curl Palampur virus	ToLCPalV-[IN-pal-047]	Tomato	India
	Tomato leaf curl Patna virus	ToLCPatV-[IN-Pat-08]	Tomato	India
	Tomato leaf curl Pune virus	ToLCPuV-[IN-Pun-05]	Tomato	India
	Tomato leaf curl Rajasthan virus	ToLCRaV-[IN-Raj-05]	Tomato	India
	Tomato leaf curl Sri Lanka virus	ToLCLKV-[LK-Ban-97]	Tomato	Sri Lanka
	Velvet bean severe mosaic virus	VBSMV-[IN-Luc-08]	Velvet bean	India
	Vernonia yellow vein virus	VeYVV-[IN-Mad-05]	Vernonia	India

Pakistan on Kabuli type. This disease was later reported from Haryana, Punjab, Gujarat, Andhra Pradesh, and Madhya Pradesh regions in India. It was responsible for causing significant loss up to 75–90% in the field (Kanakala et al. 2013a, b). They also reported that CpCDV is responsible for stunt diseases in chickpea. CpCDV is responsible for causing symptoms in Kabuli- as well as Desi-type chickpea. The infected plants displayed symptoms like stunting, phloem browning, internode shortening, and leaf reddening in Desi-type, whereas leaf yellowing happens in Kabuli type. During field observation, different symptoms appeared at different times in plants like the initiation of reddening followed by discoloration and small leave phenotype after 45 days resulting in drying rot-like symptoms at the final stage (Kanakala et al. 2013a, b).

Orosius orientalis (leafhopper) is responsible for transmitting CpCDV in different families like *Solanaceae, Chenopodiaceae,* and *Leguminosae*. More recently, it also infected the *Capsicum annum*. An isolate of CpCDV from India shares a maximum nucleotide sequence identity with an CpCDV isolate from Pakistan. Initially, the agroinfected plants display symptoms such as small leaves, yellowing of terminal leaf and stunting of plants and later, they died before flowering. Constructed clones also caused symptoms on *N. benthamiana, N. glutinosa, N. tabacum, sesame,* soybean, black gram, mustard, French bean, and tomato (Kanakala et al. 2013a). There is no association of any alphasatellite or betasatellite molecule with CpCDV.

Recently, chickpea chlorotic dwarf virus was reported from spinach in natural field condition from Pakistan along with alpha- and betasatellites. Spinach is a very common vegetable crop. Presence of virus in symptomatic suspected leaves was confirmed by PCR amplification, and virus amplification was done by rolling circle replication (RCA) method. Sequencing analysis confirmed the presence of chickpea chlorotic virus in spinach. Apart from spinach, CpCDV infect many other dicot species, e.g., pepper (Akhtar et al. 2014), tomato (Zia-Ur-Rehman et al. 2015), cucumber (Hameed et al. 2017), cotton (Manzoor et al. 2014), and okra (Zia-Ur-Rehman et al. 2017).

5.2.3 Sugarcane Streak Virus

Sugarcane, the most important cash crop in Pakistan, was affected with geminivirus during 2012. PCR amplification studies revealed that Sugarcane maize streak virus was responsible for causing a significant loss in the field. Coat protein of sugarcane maize streak virus showed maximum identity with Mauritius isolate, Reunion isolate, and Zimbabwe SSV isolate. So, it was remarked as the same variants of the virus. This was the first report of mastrevirus infection sugarcane in Pakistan.

6 Genus *Begomovirus* (Type Species: Bean Golden Mosaic Virus, BGMV)

6.1 Introduction

Begomovirus is the largest genus in the family, *Geminiviridae*. Monopartite begomoviruses carry one genomic component, termed as DNA-A, while bipartite geminivirus possesses two genomic components, DNA-A and DNA-B. DNA-A component encodes for major proteins for virus replication and multiplication inside the host cell, while DNA-B cares for intra- and intercellular movement of virus particles (Brown et al. 2012; Hanley-Bowdoin et al. 2013). In the case of the bipartite genome, both genome components are essential for efficient disease transmission and systemic infection (Evans and Jeske 1993). Some of the monopartite geminiviruses are also associated with additional circular ssDNA molecules, such as betasatellite or alphasatellite, which are nearly half the size of DNA-A (Mansoor et al. 1999; Kumar et al. 2015, 2017).

6.2 Important Crops Affected by Begomovirus in the Indian Subcontinent

6.2.1 Tomato

Similar to other begomoviruses, tomato-infecting begomoviruses are also transmitted by an insect vector, white fly (*Bemisia tabaci* (*Gennadius*)). Lack of thick cuticle layer on tomato leaves, soft epidermal layer, fine hairs on the epidermis, and nutritionally rich leaf sap make tomato best-suited host for whitefly. Regarding the Indian subcontinent, ToLCV infection in tomato was reported by Vasudeva and Sam Raj for the first time in 1948 from the southern part of India (Vasudeva and Samraj 1948). Further, the disease was prevalent in tomato during the summer season in South India and autumn in North India (Saikia and Muniyappa 1986). So far, in the Indian subcontinent, 13 begomovirus species infecting tomato have been characterized, namely *Tomato leaf curl Bangalore virus* (ToLCBV), *Tomato leaf curl Gujarat virus* (ToLCGV), *Tomato leaf curl Karnataka virus* (ToLCKV), *Tomato leaf curl Kerala virus* (ToLCKeV), *Tomato leaf curl New Delhi virus* (ToLCNDV), *Tomato leaf curl Palampur virus* (ToLCPMV), *Tomato leaf curl Patna virus* (ToLCPaV), *Tomato leaf curl Pune virus* (ToLCPV), *Tomato leaf curl Ranchi virus* (ToLCRnV), *Tomato leaf curl Rajasthan virus* (ToLCRajV), *Tomato leaf curl Pakistan virus* (ToLCPkV), *Tomato leaf curl Sri Lanka virus* (ToLCSLV), and *Tomato leaf curl Bangladesh virus* (ToLCBDV) (Chatchawankanphanich and Maxwell 2002; Chakraborty et al. 2003; Kumar et al. 2008, 2016; Kumari et al. 2010; Pasumarthy et al. 2010). All these begomoviruses are monopartite except for ToLCNDV and ToLCPMV, whereas ToLCGV exists as both monopartite and bipartite nature. The primary host for ToLCV is tomato (*Solanum lycopersicum*). However, these begomoviruses are also recognized and infect more than 43 other plant species of families, such as *Cucurbitaceae*, *Solanaceae*, *Euphorbiaceae*, *Malvaceae*, and *Fabaceae* (Chigurupati et al. 2012).

6.2.2 Okra

Bhendi yellow vein mosaic virus (BYVMV) is one of the earliest reported begomovirus infecting okra; hence, most studies have been carried out on the BYVMV. BYVMV has been reported from different parts of the world but regarding the Indian subcontinent, the first report of BYVMV infection has been reported in 1924 from Mumbai, India (Kulkarni 1924), suggesting that India might be the origin of BYVMV. Later on, from a different part of India, BYVMV infection has been reported, but the incidence of the disease is frequently occurring in the southern part of India (Uppal et al. 1940; Verma 1955). BYVMV-infected okra plants showed vein enation, vein clearing, yellowing of mid veins, and typical mosaic symptoms of begomovirus infection. Reduced leaf size, fruit, and twisted fruit resulted in the significant loss of crop yield and in severe conditions crop yield loss is reported up to 96% (Pun and Doraiswamy 1999).

Molecular biology of BYVMV revealed the typical monopartite nature of begomovirus having a single component of circular single-stranded DNA of nearly 2.7 kb genome (Jose and Usha 2000). DNA-β of nearly 1.3 kb which encodes single protein βC1 is responsible for infectivity and symptom severity of the disease. In order to evaluate the role of DNA-β for disease severity, okra plants were agroinnoculated with BYVMV alone and BYVMV with its associated betasatellite molecule (BYVMB). Okra plants agroinnoculated with only BYVMV showed mild leaf curling symptoms. Whereas, okra plants agroinnoculated with BYVMV and BYVMB produced typical BYVMD symptoms of yellowing of veins (Jose and Usha 2003). These results clearly suggest an indispensable role of betasatellite in disease onset, progression, and severity. Although BYVMV is a typical monopartite whitefly-transmitted begomovirus having single genome component and associated betasatellite molecule but the association of DNA-B molecule with yellow vein mosaic disease of okra has been also reported from India (Venkataravanappa et al. 2013). Begomoviruses are very prone to undergo recombination and have a high rate of mutation. It replicates via rolling circle mode of replication with error-prone low fidelity DNA polymerase enzyme (Duffy et al. 2008; Duffy and Holmes 2009). The evolutionary analysis on BYVMV and associated betasatellite has revealed the ancestral relationship of BYVMV with cotton-infecting begomovirus. In mutation analysis study, the very high rate of nucleotide substitution in BYVMV (V1) and associated betasatellite (βC1) was observed indicating the mutation of BYVMV for host adaptation. Since cotton and okra belong to the same family of dicotyledonous plant group, there might be host adaptation of CLCuV for the evolution of BYVMV.

Okra enation leaf curl virus (OELCV) is another emerging monopartite begomovirus affecting okra production in India (Singh 1996). Almost in all parts of India, OELCV infection has been reported either from okra or other crops such as cotton and tomato. In Pakistan, okra enation leaf curl was reported in 1998 and was found to be one of the variants of begomovirus to cause cotton leaf curl epidemic during the 1990s (Zhou et al. 1998). The typical symptoms of OELCV infection are vein enation, curling of leaf blade and petiole, and stunted plant growth. In India, a geographical survey of begomovirus causing diseases in okra revealed the association of okra enation leaf curl betasatellite with okra enation leaf curl disease

(Krishnareddy et al. 2010). Intra-host infection enables OELCV with broad host range for evolutionary adaptation. Furthermore, infection of OELCV to intra-host cotton in Pakistan has been reported along with Cotton leaf curl Multan betasatellite and Cotton leaf curl Multan alphasatellite (Hameed et al. 2014). A study in Pakistan has shown the recombination between okra- and cotton (both crops belonging to *Malvaceae* family)-infecting begomoviruses resulting into the evolution of OELCV as a new species of virus (Serfraz et al. 2015). Bhendi yellow mosaic virus is the major parent of OELCV, which was not reported from Pakistan previously, and Cotton leaf curl Multan virus is the distant parent of OELCV.

Okra leaf curl virus (OLCV) is another monopartite begomovirus infecting okra. In the Indian subcontinent, OLCV, a potential pathogen of okra leaf curl disease, has been reported from Pakistan in 2001, and associated betasatellite was found to be involved in disease severity of okra (Mansoor et al. 2001). Alphasatellite is also found to be associated with okra leaf curl disease in Pakistan (Mansoor et al. 2003, 2006).

6.2.3 Legumes

Yellow mosaic diseases are a big constraint in crop productivity in the Indian subcontinent. *Fabaceae*, *Verbenaceae*, and *Malvaceae* families are the favorite host for yellow mosaic diseases. Mungbean yellow mosaic virus (MYMV), Horsegram yellow mosaic virus (HGYMV), and associated strains are causal agents for mosaic diseases. *Mungbean yellow mosaic virus, mungbean yellow mosaic India virus, Dolichos mosaic virus, Horsegram mosaic virus,* and *Rhynchosia yellow mosaic virus* are severely infecting agents.

Mungbean yellow mosaic virus and mungbean yellow mosaic India virus are the two major viruses infecting legume crops. Both viruses are isolated from India, Pakistan, and Sri Lanka. Interestingly, these viruses are restricted to the Indian subcontinent. Mungbean yellow mosaic virus "Indian" strain was first observed and reported in the late 1950s by Nariani (1960). It produces typical mosaic symptoms on leaves of infected plants and naturally transmitted by whitefly (Nene 1973). In addition to India, the virus is widely prevalent in the Indian subcontinent, Sri Lanka, Bangladesh, and Pakistan (Honda 1986). An epidemic of yellow mosaic disease of mungbean was also identified in Thailand in the 1980s (Honda et al. 1983). MYMIV infection is confined to Northern India, Pakistan, Nepal, Bangladesh, and Indonesia. Both viruses are transmitted by whitefly and mostly they are non-sap transmissible. Female whiteflies are a good transmitter of viruses compared to male whiteflies since females can retain virus up to 10 days compared to 3 days for male whiteflies. Disease occurrence through MYMIV infection is reported from Northern India, Pakistan, Nepal, Bangladesh, and Indonesia. While, MYMV infection is mostly restricted to Thailand, Vietnam, and Eastern Ghats and Deccan plateau of India (Islam et al. 2012; Tsai et al. 2013). MYMIV is important economically as it infects five major leguminous species, blackgram, mungbean, French bean, pigeonpea, and soybean, causing yield loss of about $300 million annually (Varma et. al. 1992). Natural infection of MYMV has been reported in Dolichos (Williams et al. 1968), urdbean (Ahmad and Harwood 1973), moth bean (Ahmad and Harwood 1973), mung bean (Nariani 1960), black gram (Vanitharani et al.

1996), French bean (Singh 1979), lima bean (Shahid et al. 2012), Horsegram, and pigeon pea (Biswas and Varma 2000).

MYMV and MYMIV produce yellow bright mosaic to golden bright mosaic symptoms on infected leaves. They produce a poor quality of seeds and fever flowers. In French bean, it produces mosaic and downward leaf curling symptoms associated with stunted growth. Seed-borne nature of MYMV on black gram was first proved by Kothandaraman et al. (2016).

HgYMV was first reported by Williams et al. (1968) in India. HgYMV was found as the causal agent of yellow mosaic disease (YMD). The incidence of disease ranged from 60 to 100% in summer and early rainy season. YMD is characterized by yellow mosaic patches on leaves, reduced leaf size, and dwarfism in severely affected plants (Muniyappa et al. 1987). The occurrence of HgYMV was found limited to Southern India (Borah and Dasgupta 2012; Varma and Malathi 2003). HgYMV is reported to infect 15 plant species of 9 genera of *Fabaceae* family. This includes *Arachis hypogea* (Muniyappa and Veeresh 1984), *Cajanus cajan* (Muniyappa and Veeresh 1984), *Glycine max* (Muniyappa and Reddy 1976), *Dolichos biflorus* (Williams et al. 1968), *Phaseolus aconitifolius*, *Phaseolus aureus*, *Phaseolus mungo*, *Phaseolus vulgaris* and *Phaseolus lunatus* (Muniyappa and Reddy 1976), and *Phaseolus limensis* (Muniyappa and Veeresh 1984).

6.2.4 Chilli

Chilli leaf curl virus (ChLCV) is the most devastating agent for chilli production in the Indian subcontinent. India, Pakistan, and Bangladesh are majorly affected by ChLCV. ChLCV and associated strains, namely Chilli leaf curl Bangladesh virus and Pepper leaf curl Sri Lanka virus, are spread throughout the Indian subcontinent. Leaf curl disease on chilli was first recorded from Sri Lanka in 1939 and from India in 1930 (Senanayake et al. 2012; Husain 1932). But the authenticated first report was noted in 2007 from India (Senanayake et al. 2007). Chilli leaf curl virus showed symptoms of leaf curling, rolling of leaf, leaf curling, vein enation, stunting of leaf, and lower production and quality of fruits (Dhanraj and Seth 1968; Mishra et al. 1963). In this decade, chilli leaf curl virus is prevalent in central to south India. It showed prevalence in Maharashtra, Madhya Pradesh, and Andhra Pradesh. It causes about 90% yield loss in the infected field, whereas in Jodhpur, Rajasthan, chilli leaf curl diseases cause 14–100% loss in the field (Senanayake et al. 2012). *Chilli leaf curl virus, Chilli leaf curl India virus* (Saeed et al. 2017), *Chilli leaf curl Ahmedabad virus (*Bhatt et al. 2016), *Chilli leaf curl Vellanad virus* (Kumar et al. 2012a, b, c), *Chilli leaf curl Kanpur virus, Tomato leaf curl Joydebpur virus* (Shih et al. 2006), *Tomato leaf curl New Delhi virus* (Hussain et al. 2000), *Pepper leaf curl Bangladesh virus, Rhynchosia leaf curl virus*, and *Tomato leaf curl virus* are major begomoviruses infecting chilli in India (Kumar et al. 2012a, b, c, 2015).

An association of satellite molecules enhanced the severity of disease incidence in the field (Kumar et al. 2011). Approximately six satellite molecules are found from India recording the association with DNA-A component. Generally, chilli leaf curl virus is an old world monopartite begomovirus. Chilli leaf curl betasatellite, tomato leaf curl Joydebpur betasatellite, tomato leaf curl Bangladesh betasatellite, radish leaf curl betasatellite, and tomato leaf curl Ranchi betasatellite are isolated with DNA-A component of chilli leaf curl virus. Among all these satellite molecules, tomato leaf curl Bangladesh is more prevalent and frequently involved with ChiLCV (Kumar et al. 2015). Many chilli-infecting isolates of begomoviruses are a combination of two or more begomoviruses. Apart from infection on chilli plants and on weed, they can also infect important vegetable crops, for example, tomato, bitter gourd, eggplant, petunia, Mentha, and papaya (George et al. 2014; Saeed et al. 2014; Senanayake et al. 2012; Raj et al. 2008, 2010; Nehra and Gaur 2014).

6.2.5 Cassava

The first published record of the disease happened only in 1966 by Alagianagalingam and Ramakrishnan (1966). Later on, Malathi and Shrinivasaan reported severe cassava mosaic diseases in 1983 (Malathi and Sreenivasan 1983). In the Indian subcontinent, Indian cassava mosaic virus (ICMV) and its recombinant species, Sri Lanka cassava mosaic virus (SLCMV), are the most threatening species. Recently, cassava plants were infected with cassava mosaic virus in Ratnagiri, Reunion, Cambodia, etc. (Wang et al. 2015). Interestingly, India cassava mosaic virus and Sri Lanka cassava mosaic virus are restricted to the Indian subcontinent only.

Affected plants showed discoloration of pale green tissue to the mosaic pattern, stunted growth, and distorted curl leaves (Legg et al. 2015). For fulfillment of Koch's postulate, agrobacterium-based infection study was done on *N. clevlandi* and *N. glutinosa*. Infected plants showed symptoms of stem swelling and leaf rolling. ICMV DNA-A alone can give expression of leaf curling by biolistic transfection. Studies showed that the SLCMV is more virulent compared to ICMV (Saunders et al. 2002). Cassava mosaic virus has wide host range. It can easily transmit to *Nicotiana*, *Nicandra physalodes*, and *Petunia hybrid.* Jose et al. (2008) did transreplication studies on *Solanaceae* family members among which 39 species developed symptoms upon infection with SLCMV. Infectivity studies of SLCMV were also done on *N. amplexicaulis*, *N. nudicaulis*, and *N. benavidesii.* These plants were easily infected by SLCMV (Jose A. et al. 2008). In natural condition, ICMV was also reported from bitter gourd, jatropha, and mulberry (Rajinimala and Rabindran 2007; Sherry 2016; Aswathanaryana et al. 2007; Gao et al. 2010). There are reports of recombination events occurring between SLCMV and ICMV. When pusedo-recombination was done between ICMV DNA-A and SLCMV DNA-A, recombinant molecule induced significant disease symptoms in *N. benthamiana* (Rothenstein et al. 2006), whereas pseudo-recombinants of ACMV and ICMV were not too infectious to induce disease symptoms in *N. benthamiana*.

6.2.6 Cucurbits

Yellow vein mosaic disease on cucurbits is a serious threat to its cultivation (Maruthi et al. 2007). Begomovirus infects many cucurbits from different parts of the world. Watermelon, squash, pumpkin, chyote, cucumber, etc., are the host of begomovirus (Sohrab et al. 2006). They mostly belong to new world begomovirus since they share less relationship with satellite molecules. Mosaic patterns on leaves, vein yellowing, leaf curling, vein clearing, stunting of stem, etc., are general symptoms appearing on cucurbits (Tiwari et al. 2011, 2012). *Pumpkin yellow vein mosaic virus* (Muniyappa et al. 2003), *Squash leaf curl China virus* (Saritha et al. 2011; Singh et al. 2009), *Tomato leaf curl New Delhi virus* (Zaidi et al. 2017), *tomato leaf curl Palampur virus* (Namrata et al. 2010; Ali et al. 2010), *Chayote yellow mosaic virus* (Mandal et al. 2000), and *Coccinia mosaic virus* (Nagendran et al. 2016) are the major viruses infecting cucurbits in India . Mixed infections of more than one begomoviruses, *Tomato leaf curl Palampur virus, Squash leaf curl China virus,* and *Tomato leaf curl New Delhi virus*, were reported from Varanasi, India, on pumpkin (Jaiswal et al. 2012). Chayote yellow mosaic virus, infecting Chayote (*Sechium edule*), shares maximum nucleotide identity with previously characterized Tomato leaf curl New Delhi virus (ToLCNDV). ChaYMV also infects other members of cucurbits, namely bitter gourd, cucumber, and squash (Mandal et al. 2004).

6.2.7 Cotton

Cotton leaf curl virus is categorized as the most devastating virus belonging to the begomovirus genus. It is responsible for causing the most threatening effect in the world, especially to the Indian subcontinent. India, Pakistan, and Sri Lanka face a huge economic loss due to the infection of cotton leaf curl virus (Narula et al. 1999). Pakistan faces around 30% loss due to cotton leaf curl disease infection (Ahmad et al. 2018; Hassan et al. 2016). First report of cotton leaf curl virus infecting cotton was from Nigeria in 1912. A cotton leaf curl disease epidemic in the Indian subcontinent was first reported in 1967 from Multan, Pakistan (Husain 1932; Hussain and Ali 1975; Hassan et al. 2016). The begomoviral strain that acts as a causal agent of cotton leaf curl diseases was classified based on genomic identity into three strains, namely cotton leaf curl Multan virus, cotton leaf curl Burewala virus, and cotton leaf curl Kokhran virus (Chowdareddy et al. 2005; Radhakrishnan et al. 2004; Kirthi et al. 2004; Kumar et al. 2010; Rajagopalan et al. 2012; Zaffalon et al. 2012). In India, cotton leaf curl diseases were prevalent during 1993–1996 in northwestern India from Rajasthan, Punjab, and Haryana. Recombination between cotton leaf curl Multan virus and cotton leaf curl Kokhran virus is known as cotton leaf curl Rajasthan virus (Kumar et al. 2010; Rajagopalan et al. 2012; Zaffalon et al. 2012). It is designated as a different strain from India. Apart from these strains, Cotton leaf curl Bangalore virus, Cotton leaf curl Allahabad virus, Cotton leaf curl Barasat virus, and Cotton leaf curl Shahdadpur virus were reported from India (Datta et al. 2017; Sattar et al. 2013; Briddon 2003; Zhou et al. 1998; Mansoor et al. 1999). Apart from these, cotton plants are also sensitive to infection by tomato leaf curl Bangalore virus and tomato leaf curl Patna virus (Kirthi et al. 2004). Systemic reports of cotton leaf curl virus in India were available from Punjab and neighboring states in 1990s (Briddon and Markham 2000; Briddon 2003). Disease epidemic was spread and became prevalent toward south India and reported from the home garden in Bengaluru (Nateshan et al. 1996). It also infected okra, tomato, cotton, and hibiscus plants. It is responsible for causing around 10–80% loss in seed production in different varieties of cotton.

There is no evidence of association of cotton leaf curl virus with DNA-B component from India but DNA-B component from ICMV and SLCMV showed association with DNA-A component of cotton leaf curl virus (Sattar et al. 2013). Cotton leaf curl Multan virus and cotton leaf curl Kokhran virus share 78–79% identity between them. It shows a close association with satellite molecules. The recombination between CLCuKoV and CLCuMuV strains resulted in Rajasthan, Shahdadpur, and Burewala strain. They are more virulent compared to the parent strand. Rajasthan strand is more prevalent in India and might originate from India. Burewala strain is more virulent in India and Pakistan and suspected to originate in Pakistan and travel via vector *B. tabaci* and enter India (Kumar et al. 2010).

6.2.8 Papaya

Papaya leaf curl disease was first reported from Tamil Nadu in India in 1939 (Thomas and Krishnaswami 1939), whereas it was noticed in Pakistan in 1997 (Nadeem et al. 1997). Initially, on the basis of symptom appearance on leaves, it is known as papaya leaf crumple disease. It shows symptoms like vein enation, stunted growth, and deformed and leathery leaf. Infected plants did not produce fruits (Summanwar and Ram 1993; Singh-Pant et al. 2012). Papaya leaf curl virus has wide host range of plant families, for example, *Apocynanceae, Malvaceae, Ephorbeaceae, Caricaceae, and Asteraceae* (Kumar et al. 2009; Srivastava et al. 2013; Varun et al. 2017). Furthermore, virus outbreak was observed in northern India and east India initially, which was then spread over Haryana, Maharashtra, Uttar Pradesh, Karnataka, Tamil Nadu, and Andhra Pradesh (Surekha et al. 1977; Pandey and Marathe 1986; Raj et al. 2008; Krishnareddy et al. 2010).

6.2.9 Mesta

Mesta, affected by begomovirus, was first reported from India in 2005 (Chatterjee et al. 2005). The symptoms were yellowing of vein and entire lamina turned into yellow. The similar kind of symptoms was found from Uttar Pradesh and West Bengal in 2007 and 2009, respectively (Ghosh et al. 2007; Das et al. 2008a, b; Roy et al. 2009). Mesta yellow vein mosaic virus and mesta yellow vein mosaic Bahraich virus are affecting in India. It can efficiently transmit up to 85% in *H. sabdariffa*. The association of betasatellite has been found in India in field condition (Chatterjee and Ghosh 2007). There was an association of cotton leaf curl betasatellite with MeYVMV.

6.2.10 Radish

Leaf curl disease in radish was first observed in India in 2003. It was observed in both kitchen garden and field from eastern Uttar Pradesh. The disease appeared as upward and downward curling in leaf, leaf distortion, and vein enation in infected plants (Singh et al. 2007, 2010). The disease incidence was noted to be 10–40% in field. It is a whitefly-transmitted disease. Scanning electron microscopy and PCR confirmed the presence of begomovirus. They also reported the presence of DNA-B genome component and associated betasatellite molecule from the infected plants. Nonhost infection of radish leaf curl virus (RLCuV) was reported in okra plant in Bihar (Kumar et al. 2012a, b, c). Associated satellite molecules (alphasatellite and betasatellite) increased symptom severity. Nonhost infection was also seen in tobacco plants in field condition (Singh et al. 2012). The begomovirus genome experiences frequent recombination event which is a major factor for enormous viral genome diversity. As of other begomovirus, RLCuV also undergoes recombination, pseudo-recombination, and mutation. These strategies of virus made them feasible for surviving in multiple and nonhost plants (Singh et al. 2012).

6.2.11 Jatropha

Production of jatropha is increasing day by day due to its economic importance. The most important value of the plant is its role in fuel production. It is native to India, America, and Caribbean countries. The natural infection of begomovirus was by jatropha mosaic virus in Jamaica and Puerto Rico (Roye et al. 2006). It has also been reported from Kenya and Nigeria (Kashina et al. 2013). The disease incidence was more than 40% in *Jatropha curcas*. The infected plants were showing symptoms of leaf curling, leaf blustering, leaf mosaic pattern, leaf distortion, etc. In India, it spread over almost all jatropha-growing regions. Up to 25% of disease incidence in field condition has been reported from Uttar Pradesh. Jatropha mosaic India virus, jatropha leaf yellow mosaic Katarniaghat virus, and jatropha leaf crumple India virus have been reported to date (Srivastava et al. 2015; Snehi et al. 2016). Apart from these, jatropha plants also experience nonhost infection by Indian cassava mosaic virus, and croton yellow mosaic virus associated with betasatellite has also been reported in *J. gossypifolia* (Gao et al. 2010; Narayana et al. 2007).

7 Conclusion

Yield constraint, tremendous losses, and economical outbreaks lead researchers and policy makers to gain interest in geminiviruses. Sugar beet infection by beet curly top virus, cassava infection by African cassava mosaic virus, cotton infection by cotton leaf curl virus, bean golden mosaic virus of common bean, maize infection by maize streak virus, and finally tomato infection by tomato leaf curl virus are past pandemics that cause huge loss in the production of respective crops in one or the other parts of the world. Geminivirus infection is one of the major limiting factors for the production of cash crops. This chapter provides a review on geminivirus infections in the Indian subcontinent. Virus database and availability of full-length sequences are major prerequisites to develop long-term stable resistant variety against virus attack. Understanding host–pathogen interactions and mechanism of defense stratagems could be an important future aspect.

References

Ahmad A, Yasin NA, Ibrahim A, Shahzadi I, Gohar M, Bashir Z, Khan J, Khan WU, Waheed A (2018) Modelling of cotton leaf curl viral infection in Pakistan and its correlation with meteorological factors up to 2015. Clim Develop 10(6):520–525

Ahmad M, Harwood RF (1973) Studies on whitefly transmitted yellow mosaic urd bean (Phaseolus mungo). Plant Dis Rep 57(9):800–802

Akhtar S, Khan A, Briddon R (2014) A distinct strain of chickpea chlorotic dwarf virus infecting pepper in Oman. Plant Dis 98:286–286

Alagianagalingam MN, Ramakrishnan K (1966) Cassava mosaic in India. S Indian Hortic 14:71–72

Ali I, Malik AH, Mansoor S (2010) First report of tomato leaf curl Palampur virus on Bitter Gourd in Pakistan. Plant Dis 94(2):276

Aswathanarayana DS, Rangaswamy KT, Shankarappa KS, Maruthi MN, Reddy LCN, Rekha AR, Murthy KKV (2007) Distinct begomoviruses closely related to cassava mosaic viruses cause Indian jatropha mosaic disease. Int J Virol 3:1–11

Bernardo P, Muhire B, François S, Deshoux M, Hartnady P, Farkas K, Kraberger S, Filloux D, Fernandez E, Galzi S, Ferdinand R, Granier M, Marais A, Monge Blasco P, Candresse T, Escriu F, Varsani A, Harkins GW, Martin DP, Roumagnac P (2016) Molecular characterization and prevalence of two capulaviruses: alfalfa leaf curl virus from France and Euphorbia caput-medusae latent virus from South Africa. Virology 493:142–153

Bhatt BS, Chahwala FD, Rathod S, Singh AK (2016) Identification and molecular characterization of a new recombinant begomovirus and associated betasatellite DNA infecting Capsicum annuumin India. Arch Virol 161:1389–1394

Biswas KK, Varma A (2000) Identification of variants of mungbean yellow mosaic geminivirus by host reaction and nucleic acid spot hybridization. Indian Phytopathology 53(2):134–141

Borah BK, Dasgupta I (2012) Begomovirus research in India: A critical appraisal and the way ahead. J Biosci 37(4):791–806

Briddon RW (2003) Cotton leaf curl disease, a multicomponent begomovirus complex. Mol Plant Pathol 4(6):427–434

Briddon RW, Markham PG (2000) Cotton leaf curl virus disease. Virus Res 71(1–2):151–159

Brown JK, Fauquet CM, Briddon RW, Zerbini M, Moriones E, Navas-Castillo J (2012) Geminiviridae. In: King AMQ, Adams MJ, Carstens EB, Lefkowitz EJ (eds) Virus taxonomy – ninth report of the international committee on taxonomy of viruses. Associated Press, Elsevier Inc, London, pp 351–373

Chakraborty S, Pandey PK, Banerjee MK, Kalloo G, Fauquet CM (2003) Tomato leaf curl Gujarat virus, a new Begomovirus species causing a severe leaf curl disease of tomato in Varanasi, India. Phytopathology 93(12):1485–1495

Chatchawankanphanich O, Maxwell DP (2002) Tomato leaf curl Karnataka virus from Bangalore, India, appears to be a recombinant Begomovirus. Phytopathology 92(6):637–645

Chatterjee A, Roy A, Padmalatha KV, Malathi VG, Ghosh SK (2005) Occurrence of a Begomovirus with yellow vein mosaic disease of mesta (Hibiscus cannabinus L. and H. sabdariffa L.). Aust Plant Pathol 34:609–610

Chatterjee A, Ghosh SK (2007) Association of a satellite DNA β molecule with mesta yellow vein mosaic disease. Virus Genes 35(3):835–844

Chigurupati P, Sambasiva RK, Jain RK, Mandal B (2012) Tomato leaf curl New Delhi virus is associated with pumpkin leaf curl: a new disease in northern India. Indian J Virol 23(1):42–45

Chowdareddy RV, Muniyappa V, Colvin J, Seal S (2005) A new begomovirus isolated from*Gossypiumbarbadense* in southern India. Plant Pathol 54:570

Das S, Ghosh R, Paul S, Roy A, Ghosh SK (2008a) Complete nucleotide sequence of a monopartite begomovirus associated with yellow vein mosaic disease of mesta from North India. Arch Virol 153:1791–1796

Das S, Roy A, Ghosh R, Paul S, Acharyya S, Ghosh SK (2008b) Sequence variability and phylogenetic relationship of betasatellite isolates associated with yellow vein mosaic disease of mesta in India. Virus Genes 37:414–424

Datta S, Budhauliya R, Das B, Gopalakrishnan R, Sharma S, Chatterjee S, Vanlalhmuaka, Raju PS, Veer V (2017) Rebound of cotton leaf curl Multan virus and its exclusive detection in cotton leaf curl disease outbreak, Punjab (India), 2015. Sci Rep 7:17361

Dhanraj KS, Seth ML (1968) Enation in *Capsicum annuum*L (chilli) caused by a new strain of leaf curl virus. Indian J Hortic 25:70–71

Duffy S, Holmes EC (2009) Validation of high rates of nucleotide substitution in geminiviruses: phylogenetic evidence from east African cassava mosaic viruses. J Gen Virol 90:1539–1547

Duffy S, Shackleton LA, Holmes EC (2008) Rates of evolutionary change in viruses: patterns and determinants. Nat Rev Genet 9:267–276

Evans D, Jeske H (1993) DNA – B facilitates but is not essential for the spread of Abutilon mosaic virus in agroinoculated *Nicotiana benthamiana*. Virology 194(2):752–757

Gao S, Qu J, Chua N, Ye J (2010) A new strain of Indian cassava mosaic virus causes a mosaic disease in the biodiesel crop *Jatrophacurcas*. Arch Virol 155:607–612

George B, Kumar RV, Chakraborty S (2014) Molecular characterization of chilli leaf curl virus and satellite molecule associated with leaf curl disease of Amaranthus spp. Virus Genes 48:397–401

Ghosh R, Paul S, Roy A, Mir JI, Ghosh SK, Srivastava RK, Yadav US (2007) Occurrence of begomovirus associated with yellow vein mosaic disease of kenaf (*Hibiscus cannabinus*) in northern India. Plant Health Prog (Online)

Haider M, Afghan S, Riaz H, Tahir M, Javed MA, Rashid N, Iqbal J (2011) Identification of two sugarcane mosaic virus (SCMV) variants from naturally infected sugarcane crop in Pakistan. Pak J Bot 43:1157–1162

Hameed U, Zia-Ur-Rehman M, Herrmann H-W, Haider MS, Brown JK (2014) First report of okra enation leaf curl virus and associated cotton leaf curl Multan betasatellite and cotton leaf curl Multan alphasatellite infecting cotton in Pakistan: a new member of the cotton leaf curl disease complex. Plant Dis 98(10):1447

Hameed U, Zia-Ur-Rehman M, Ali S, Haider M, Brown J (2017) First report of chickpea chlorotic dwarf virus infecting cucumber in Pakistan. Plant Dis 101:848

Hanley-Bowdoin L, Bejarano ER, Robertson D, Mansoor S (2013) Geminiviruses: masters at redirecting and reprogramming plant processes. Nat Rev Microb 11:777–788

Hassan F, Qayyum A, Malik W, Maqbool A, Hassan M, Abdur RM, Shoaib M, Muhammad S, Ahmad S, Ahmad L, Arshad M (2016) Cotton leaf curl virus (CLCuV) disease in Pakistan: A critical review. J Appl Sci Bus Econ 3(1):8–14

Honda Y (1986) Mungbean yellow mosaic virus. Trop Agric Res Ser 19:121–127

Honda Y, Iwaki M, Saito Y (1983) Mechanical transmission, purification and some properties of whitefly-borne mungbean yellow mosaic virus in Thailand. Plant Dis 67:801–804

Horn NM, Reddy SV, Roberts IM, Reddy DVR (1993) Chickpea chlorotic dwarf virus, a new leafhopper-transmitted geminivirus of chickpea in India. Ann Appl Biol 122:467–479

Horn NM, Reddy SV, Reddy DVR (1994) Virus-vector relationships of chickpea chlorotic dwarf geminivirus and the leafhopper *Orosiusorientalis* (*Hemiptera*: *Cicadellidae*). Ann Appl Biol 124:441–450

Husain M (1932) Leaf curl in cotton and other plants. Nature 129:792

Hussain T, Ali M (1975) A review of cotton diseases in Pakistan. Pak Cottons 19:71–86

Hussain M, Mansoor S, Iram S, Zafar Y, Briddon RW (2000) First report of *Tomato leaf curl New Delhi virus* affecting chilli pepper in Pakistan. New Dis Rep 9:20

Islam M, Khan S, Borna R (2012) Molecular characterization of mungbean yellow mosaic disease and coat protein gene in mungbean varieties of Bangladesh. Plant Tissue Cult Biotech 22(1): 73–81

Jaiswal N, Saritha RK, Datta D, Singh M, Dubey RS, Rai AB, Rai M (2012) Mixed infections of begomoviruses in pumpkins with yellow vein mosaic disease in North India. Arch Phytopathol Plant Protect 45(8):938–941

Jose J, Usha R (2000) Extraction of geminiviral DNA from a highly mucilaginous plant (*Abelmoschusesculentus*). Plant Mol Biol Rep 18:349–355

Jose J, Usha R (2003) Bhendi yellow vein mosaic disease in India is caused by association of aDNA β satellite with begomovirus. Virology 305:310–317

Jose A, Makeshkumar T, Edison S (2008) Host range of Sri Lankan cassava mosaic virus. J Root Crops 34:21–25

Kanakala S, Sakhare A, Verma HN, Malathi VG (2013a) Infectivity and the phylogenetic relationship of a mastrevirus causing chickpea stunt disease in India. Eur J Plant Pathol 135:429–438

Kanakala S, Verma HN, Vijay P, Saxena DR, Malathi VG (2013b) Response of chickpea genotypes to agrobacterium-mediated delivery of chickpea chlorotic dwarf virus (CpCDV) genome and identification of resistance source. Appl Microbiol Biotechnol 97:9491–9501

Kashina BD, Alegbejo MD, Banwo OO, Nielsen SL, Nicolaisen M (2013) Molecular identification of a new begomovirus associated with mosaic disease of *Jatropha curcas* L. in Nigeria. Arch Virol 158:511–514

Kirthi N, Priyadarshini CGP, Sharma P, Maiya SP, Hemalatha V, Sivaraman P, Dhawan P, Rishi N, andSavithri H.S. (2004) Genetic variability of begomoviruses associated with cotton leaf curl disease originating from India. Arch Virol 149:2047–2057

Kothandaraman S, Devadason A, Ganeshan MV (2016) Seed-borne nature of a begomovirus, Mung bean yellow mosaic virus in black gram. Appl Microbiol Biotechnol 100(4):1925–1933

Kraberger S, Harkins GW, Kumari SG, Thomas JE, Schwinghamer MW, Sharman M, Collings DA, Briddon RW, Martin DP, Varsani A (2013) Evidence that dicot-infecting mastreviruses are particularly prone to inter-species recombination and have likely been circulating in Australia for longer than in Africa and the Middle East. Virology 444(1–2):282–291

Krishnareddy M, Venkataravanappa V, Madhuvanthi B, Jalali S (2010) Molecular characterization of begomoviruses associated with papaya leaf curl disease in India. Acta Hortic 851:465–472

Kulkarni GS (1924) Mosaic and other related diseases of crops in the Bombay presidency. Poona Agric Coll Mag 6:12

Kumar Y, Hallan V, Zaidi AA (2008) Molecular characterization of a distinct bipartite begomovirus species infecting tomato in India. Virus Genes 37(3):425–431

Kumar J, Kumar A, Khan JA, andAminuddin A. (2009) First report of papaya leaf curl virus naturally infecting tobacco in India. J Plant Pathol 91:97–112

Kumar A, Kumar J, Khan JA (2010) Sequence characterization of cotton leaf curl virus from Rajasthan: phylogenetic relationship with other members of geminiviruses and detection of recombination. Virus Genes 40:282–289

Kumar Y, Hallan V, Zaidi A (2011) Chilli leaf curl Palampur virus is a distinct begomovirus species associated with a betasatellite. Plant Pathol 60:1040–1047

Kumar J, Kumar A, Singh SP, Roy JK, Lalit A, Parmar D, Sharma NC, Tuli R (2012a) First report of *Radish leaf curl virus* infecting okra in India. New Dis Rep 25:9

Kumar J, Singh SP, Kumar J, Tuli R (2012b) A novel mastrevirus infecting wheat in India. Arch Virol 157:2031–2034

Kumar RV, Singh AK, Chakraborty S (2012c) A new monopartite begomovirus species, *Chilli leaf curl Vellanad virus*, and associated betasatellites infecting chilli in the Vellanad region of Kerala, India. New Dis Rep 25:20

Kumar J, Kumar J, Singh SP, Tuli R (2014a) Association of satellites with a mastrevirus in natural infection: complexity of wheat dwarf India virus disease. J Virol 88(12):7093–7104

Kumar J, Kumar J, Singh SP, Tuli R (2014b) βC1 is a pathogenicity determinant: not only for begomoviruses but also for a mastrevirus. Arch Virol 159:3071–3076

Kumar R, Singh AK, Singh AK, Yadav T, Basu S, Kuswaha N, Chattopadhyay B, Chakraborty S (2015) Complexity of begomovirus and betasatellite populations associated with chilli leaf curl disease in India. J Gen Virol 96(10):3143–3158

Kumar S, Srivastava A, Jaidi M, Raj SK (2016) Molecular characterization of a begomovirus, alphasatellite and recombinant betasatellite associated with leaf curl disease of Parthenium-hysterophorus. Plant Dis 100:2299–2305

Kumar RV, Prasanna HC, Singh AK, Ragunathan D, Garg GK, Chakraborty S (2017) Molecular genetic analysis and evolution of begomoviruses and betasatellites causing yellow mosaic disease of bhendi. Virus Genes 53:275–285
Kumari P, Singh AK, Chattopadhyay B, Chakraborty S (2010) Molecular characterization of a new species of Begomovirus and betasatellite causing leaf curl disease of tomato in India. Virus Res 152(1–2):19–29
Legg JP, Lava KP, Makeshkumar T, Tripathi L, Ferguson M, Kanju E, Ntawuruhunga P, Cuellar W (2015) Cassava virus diseases: biology, epidemiology, and management. Adv Virus Res 91: 85–142
Malathi VG, Sreenivasan MA (1983) Association of gemini particles with cassava mosaic disease in India. J Root Crops 9:69–73
Mandal B, Mandal S, Sohrab SS, Pun KB, Varma A (2000) A new yellow mosaic disease of chayote in India. New Dis Rep 9:22
Mandal B, Mandal S, Sohrab SS, Pun KB, Varma A (2004) A new yellow mosaic disease of chayote in India. Plant Pathol 53:797
Mansoor S, Khan SH, Bashir A, Saeed M, Zafar Y, Malik KA, Briddon RW, Stanley J, Markham PG (1999) Identification of a novel circular single stranded DNA associated with cotton leaf curl disease in Pakistan. Virology 259(1):190–199
Mansoor S, Amin I, Hussain M, Zafar Y, Bull S, Briddon RW, Markham PG (2001) Association of a disease complex involving a begomovirus, DNA 1 and a distinct DNA beta with leaf curl disease of okra in Pakistan. Plant Dis 85(8):922
Mansoor S, Briddon RW, Zafar Y, Stanley J (2003) Geminivirus disease complexes: an emerging threat. Trends Plant Sci 8(3):128–134
Mansoor S, Zafar Y, Briddon RW (2006) Geminivirus disease complexes: the threat is spreading. Trends Plant Sci 11(5):209–212
Manzoor MT, Ilyas M, Shafiq M, Haider MS, Shahid AA, Briddon RW (2014) A distinct strain of chickpea chlorotic dwarf virus (genus Mastrevirus, family Geminiviridae) identified in cotton plants affected by leaf curl disease. Arch Virol 159(5):1217–1221
Maruthi MN, Rekh AR, Muniyappa V (2007) Pumpkin yellow vein mosaic disease is caused by two distinct begomoviruses: complete viral sequences and comparative transmission by an indigenous *Bemisiatabaci* and the introduced B-biotype. EPPO Bull 37:412–419
Mishra MD, Raychaudhri SP, Jha A (1963) Virus causing leaf curl of chilli (*Capsicum annum* L.). Indian J Microbiol 3:73–76
Muniyappa V, Reddy HR (1976) Studies on the yellow mosaic disease of horsegram (DolichosbiflorusLinn.). I. Virus vector relationships. Mysore J Agric Sci 10:605–610
Muniyappa V, Veeresh GK (1984) Plant virus diseases transmitted by whiteflies in Karnataka. Proc Indian Acad Sci (Anim Sci) 93:397–406
Muniyappa V, Rajeshwari R, Bharathan N, Reddy DVR, Nolt BL (1987) Isolation and characterization of a geminivirus causing yellow mosaic disease of horsegram (*Macrotylomauniflorum* (Lam.) Verdc.) in India. J Phytopathol 119(1):81–87
Muniyappa V, Maruth MN, Babith CR, Colvi J, Briddo RW, Rangaswamy KT (2003) Characterization of pumpkin yellow vein mosaic virus from India. Ann Appl Biol 142:323–331
Nadeem A, Mehmood T, Tahir M, Khalid S, Xiong Z (1997) First report of papaya leaf curl disease in Pakistan. Plant Dis 81(11):1333
Nagendran K, Satya VK, Mohankumar S, Karthikeyan G (2016) Molecular characterization of a distinct bipartite Begomovirus species infecting ivy gourd (Coccinia grandis L.) in Tamil Nadu, India. Virus Genes 52(1):146–151
Nahid N, Amin I, Mansoor S, Rybicki EP, van der Walt E, Briddon RW (2008) Two dicot-infecting mastreviruses (family Geminiviridae) occur in Pakistan. Arch Virol 153:1441–1451
Namrata J, Saritha RK, Datta D, Singh M, Dubey RS, Rai AB, Rai M (2010) Molecular characterization of Tomato leaf curl Palampur virus and Pepper leaf curl betasatellite naturally infecting Pumpkin (Cucurbita moschata) in India. Indian J Virol 21(2):128–132

Narayana DSA, Rangaswamy KT, Shankarappa KS, Maruthi MN, Reddy CNL, Rekha AR, Murthy KVK (2007) Distinct begomoviruses closely related to cassava mosaic viruses cause Indian jatropha mosaic disease. Int J Virol 3:1–11

Nariani TK (1960) Yellow mosaic of mung (*Phaseolus aureus*). Indian Phytopathol 13:24–29

Narula AM, Monga D, Chauhan MS, Raj S (1999) Cotton leaf curl virus disease in India-the challenge ahead. J Cotton Res Dev 13:129–138

Nateshan MM, Muniyappa V, Swanson MM, Harrison BD (1996) Host range, vector relations and serological relationships of cotton leaf curl virus from southern India. Ann Appl Biol 128: 233–244

Nehra C, Gaur RK (2014) Molecular characterization of chilli leaf curl viruses infecting new host plant Petunia hybridain India. Virus Genes 50:58–62

Nene YL (1973) Viral diseases of some warm weather pulse crops in India. Plant Dis Rep 57: 463–467

Nicaise V (2014) Crop immunity against viruses: outcomes and future challenges. Front Plant Sci 5:660

Pandey PK, Marathe TS (1986) A leaf crinkle disease of papaya in Maharashtra. J Maharashtra Agric Univ 11:105–106

Pasumarthy KK, Choudhury NR, Mukherjee SK (2010) *Tomato leaf curl Kerala virus* (ToLCKeV) AC3 protein forms a higher order oligomer and enhances ATPase activity of replication initiator protein (rep/AC1). Virol J 7:128

Pun KB, Doraiswamy S (1999) Effect of age of okra plantson susceptibility to okra yellow vein mosaic virus. Indian J Virol 15:57–58

Radhakrishnan G, Malathi VG, Varma A (2004) Biological characterization of an isolate of cotton leaf curl Rajasthan virus from northern India and identification of source of resistance. Indian Phytopathol. 57(2):174–180

Raj SK, Snehi SK, Khan MS, Singh R, Khan AA (2008) Molecular evidence for association of with leaf curl disease of papaya (*Carica papaya*L.) in India. Aust Plant Dis Notes 3(1):152–155

Raj SK, Snehi SK, Khan MS, Tiwari AK, Rao GP (2010) First report of pepper leaf curl Bangladesh virus strain associated with bitter gourd (*Momordicacharantia*L.) yellow mosaic disease in India. Aust Plant Dis Notes 5:14–16

Rajagopalan PA, Naik A, Katturi P, Kurulekar M, Kankanallu RS, Anandalakshmi R (2012) Dominance of resistance-breaking cotton leaf curl Burewala virus (CLCuBuV) in northwestern India. Arch Virol 57(5):855–868

Rajinimala N, Rabindran R (2007) First report of Indian cassava mosaic virus on bittergourd (*Momordicacharantia*) in Tamil Nadu. India Austral Plant Dis Notes 2:81–82

Rothenstein D, Haible D, Dasgupta I, Dutt N, Patil BL, Jeske H (2006) Biodiversity and recombinationof cassava-infecting begomoviruses from southern India. Arch Virol 151:55–69

Roy A, Sanchalika A, Subha D, Raju G, Sujay P, Ram Kumar S, Subrata Kumar G (2009) Distribution, epidemiology and molecular variability of the begomovirus complexes associated with yellow vein mosaic disease of mesta in India. Virus Res 141:237–246

Roye M, Collins A, Maxwell D (2006) First report of a begomovirus associated with the common weed *Jatropha gossypifolia* in Jamaica. Plant Pathol 55:286

Saeed ST, Khan A, Kumar B, Ajayakumar PV, andSamad A. (2014) First report of chilli leaf curl India virus infecting *Menthaspicata*(Neera) in India. Plant Dis 98(1):164

Saeed ST, Kumar B, Shasany AK, Samad A (2017) Molecular identification of *Chilli leaf curl India virus* along with betasatellite molecule causing leaf curl disease of menthol mint (*Mentha arvensis* var. Kosi) in India. J Gen Plant Pathol 83:333

Saikia AK, Muniyappa V (1986) Epidemiology and control of tomato leaf curl virus. In: National seminar on whitefly transmitted plant virus diseases, June, 25–27, IARI, New Delhi, pp 30–31

Saritha RK, Bag TK, Loganathan M, Rai AB, Rai M (2011) First report of *Squash leaf curl china virus*causing mosaic symptoms on summer squash (*Cucurbita pepo*) grown in Varanasi district of India. Plant Dis 44:179–185

Sattar MN, Kvarnheden A, Saeed M, Briddon RW (2013) Cotton leaf curl disease-an emerging threat to cotton production worldwide. J Gen Virol 94:695–710

Saunders K, Salim N, Mal VR, Malathi VG, Briddon SR, Markham PG, Stanley J (2002) Characterisation of Sri Lankan cassava mosaic virus and Indian cassava mosaic virus: evidence for acquisition of a DNA B component by a monopartite begomovirus. Virology 293:63–74

Senanayake DMJB, Mandal B, Lodha S, Varma A (2007) First report of chilli leaf curl virus affecting chilli in India. Plant Pathol 56:343

Senanayake DMJB, Varma A, Mandal B (2012) Virus-vector relationships, host range, detection and sequence comparison of chilli leaf curl virus associated with an epidemic of leaf curl disease of chilli in Jodhpur. India J Phytopathol 160:146–155

Serfraz S, Amin I, Akhtar KP, Mansoor S (2015) Recombination among begomoviruses on malvaceous plants leads to the evolution of okra enation leaf curl virus in Pakistan. J Phytopathol 163:764–776

Seth ML, Singh DV, Raychaudhuri SP (1972) Bajra (pearl millet) streak: a leafhopper-borne cereal virus in India. Plant Dis Rep 56:424–428

Shahid MS, Ikegami M, Natsuaki KT (2012) First report of Mungbean yellow mosaic India virus on Lima bean affected by yellow mosaic disease in Nepal. Aust Plant Dis Notes 7(1):85–89

Sherry AVM (2016) A new variant of cassava mosaic virus causes mulberry mosaic disease in India. Int J Plant Anim Environ Sci 6:83–83

Shih SL, Tsai WS, Green SK, Singh D (2006) First report of *Tomato leaf curl Joydebpur virus* infecting chilli in India. New Dis Rep 14:17

Singh RN (1979) Natural infection of bean (Phaseolus vulgaris) by mung bean yellow mosaic virus. Indian J Mycol Plant Pathol 9(2):124–126

Singh SJ (1996) Assessment of losses in okra due to enation leaf curl virus. Indian J Virol 12 (1):51–53

Singh AK, Chattopadhyay B, Pandey PK, Chakraborty S (2007) A new begomovirus species causing leaf curl disease of radish in India. Plant Dis 91(8):1053

Singh AK, Mishra KK, Chattopadhyay B, Chakraborty S (2009) Biological and molecular characterization of a begomovirus associated with yellow vein mosaic disease of pumpkin from northern India. Virus Genes 39(3):359–370

Singh AK, Chattopadhyay B, Chakraborty S (2010) Biology and interactions of two distinct monopartite begomoviruses and betasatellites associated with radish leaf curl disease in India. Virol J 9:43

Singh MK, Haq QMR, Mandal B, Varma A (2012) Evidence of the association of *Radish leaf curl virus* with tobacco yellow leaf curl disease in Bihar, India. Indian J Virol 23(1):64–69

Singh-Pant P, Pant P, Mukherjee SK, andMazumdar-Leighton S. (2012) Spatial and temporal diversity of begomoviral complexes in papayas with leaf curl disease. Arch Virol 157(7): 1217–1232

Snehi SK, Prihar SS, Gupta G, Singh V, Raj SK (2016) The current status of new emerging begomovirus diseases on *Jatropha* species from India. J Plant Pathol Microbiol 7:357

Sohrab SS, Mandal B, Ali A, Varma A (2006) Molecular diagnosis of emerging begomovirus diseases in cucurbits occurring in northern India. Indian J Virol 17:88–95

Srivastava A, Raj SK, Kumar S, Snehi SK (2013) New record of papaya leaf curl virus and ageratum leaf curl beta-satellite associated with yellow vein disease of aster in India. New Dis Rep 28:6

Srivastava A, Kumar S, Jaidi M, Raj SK (2015) Molecular characterization of a new begomovirus associated with leaf yellow mosaic disease of Jatropha curcas in India. Arch Virol 160(5): 1359–1362

Summanwar AS, Ram RD (1993) Virus diseases of papaya. In: Chaddha KL, Pareek OPM (eds) Advances in horticulture. Fruit crops: part 3, vol 3. Publishing House, New Delhi, pp 1439–1446

Surekha SK, Mathur K, Shukla DD (1977) Virus disease on papaya (*Caricapapaya*) in Udaipur. Indian J Mycol Plant Pathol 7:115–121

Teng PS (1985) A comparison of simulation approaches to epidemic modeling. Annu Rev Phytopathol 23:351–379

Thomas KM, Krishnaswami CS (1939) Leaf crinkle – a transmissible disease of papaya. Curr Sci 8:316

Tiwari AK, Snehi SK, Singh R, Raj SK, Rao GP, Sharma PK (2011) Molecular identification and genetic diversity among six Begomovirus isolates affecting cultivation of cucurbitaceous crops in Uttar Pradesh, India. Arch Phytopathol Plant Protect 45(1):1–7

Tiwari AK, Rao GP, Khan MS, Pandey N, Raj SK (2012) Detection and elimination of Begomovirus infecting *Trichosanthesdioica* (pointed gourd) plants in Uttar Pradesh, India. Arch Phytopathol Plant Protect 45(9):1–6

Tsai WS, Shih SL, Rauf A, Safitri R, Hidayati N, Huyen BTT, Kenyon L (2013) Genetic diversity of legume yellow mosaic begomoviruses in Indonesia and Vietnam. Ann Appl Biol 163(3): 367–377

Uppal BN, Varma PM, Capoor SP (1940) Plant viruses online-Bhendi yellow vein mosaic bigeminivirus. Curr Sci 9:227

Vanitharani R, Karthikeyan AS, Anuradha S, Veluthambi K (1996) Genome homologies among geminiviruses infecting vigna, cassava, acalypha, croton and vernonia. Curr Sci 70(1):63–69

Varma A, Malathi VG (2003) Emerging geminivirus problems: A serious threat to crop production. Ann Appl Biol 142(2):145–164

Varma A, Dhar AK, Mandal B (1992) MYMV transmission and control in India. In: Green SK, Kim D (eds) Mungbean yellow mosaic disease. Proceedings of an international workshop, 2–3 Jul 1991, Bangkok, Thailand. AVRDC, Shanhua, Tainan, Taiwan. Publication No. 92–373. pp 8–27

Varsani A, Navas-Castillo J, Moriones E, Cecilia Hernández-Zepeda C, Idris A, Brown JK, Zerbini FM, Martin DP (2014) Establishment of three new genera in the family Geminiviridae: Becurtovirus, Eragrovirus and Turncurtovirus. Arch Virol 159(8):2193–2203

Varsani A, Roumagnac P, Fuchs M, Navas-Castillo J, Moriones E, Idris A, Briddon RW, Rivera-Bustamante R, MuriloZerbini F, Martin DP (2017) Capulavirus and Grablovirus: two new genera in the family Geminiviridae. Arch Virol 162:1819–1831

Varun P, Ranade SA, Saxena S (2017) A molecular insight into papaya leaf curl—a severe viral disease. Protoplasma 254(6):2055–2070

Vasudeva RS, Samraj J (1948) A leaf curl disease of tomato. Phytopathology 38:364–369

Venkataravanappa V, CNL R, Devaraju A, Jalali S, Reddy MK (2013) Association of a recombinant cotton leaf curl Bangalore virus with yellow vein and leaf curl disease of okra in India. Indian J Virol 24:188–198

Verma PM (1955) Persistence of yellow-vein mosaic virus of *Abelmoschusesculentus*. Indian J Agric Sci 25:293–302

Wang HL, Cui XY, Wang XW, Liu SS, Zhang ZH, Zhou XP (2015) First report of Sri Lankan cassava mosaic virus infecting cassava in Cambodia. Plant Dis 100:1029

Williams FJ, Grewal JS, Amin KS (1968) Serious and new diseases of pulse crops in India in 1966. Plant Dis Rep 52(4):300–304

Zaffalon V, Mukherjee SK, Reddy VS, Thompson JR, Tepfer M (2012) A survey of geminiviruses and associated satellite DNAs in the cotton-growing areas of northwestern India. Arch Virol 157(3): 483–495

Zaidi SS, Martin DP, Amin I, Farooq M, Mansoor S (2017) Tomato leaf curl New Delhi virus: a widespread bipartite begomovirus in the territory of monopartite begomoviruses. Mol Plant Pathol 18(7):901–911

Zhou X, Liu Y, Robinson DJ, Harrison BD (1998) Four variants among Pakistani isolates of cotton leaf curl virus and their affinities to DNA-A of geminivirus isolates from okra. J Gen Virol 79: 915–923

Zia-Ur-Rehman M, Hameed U, Herrmann H-W, Iqbal M, Haider M, Brown J (2015) First report of chickpea chlorotic dwarf virus infecting tomato crops in Pakistan. Plant Dis 99:1287–1287

Zia-Ur-Rehman M, Hameed U, Ali C, Haider M, Brown J (2017) First report of chickpea chlorotic dwarf virus infecting okra in Pakistan. Plant Dis 101:1336–1336

Geminivirus Occurrence in Australia, China, Europe, and the Middle Eastern Countries

Adel Ali Mohammed Al Shihi

Abstract

Geminiviruses (family: Geminiviridae) are plant pathogenic viruses with single-stranded DNA (ssDNA) genome. Geminiviruses are classified into nine genera: Begomovirus, Mastrevirus, Curtovirus, Becurtovirus, Topocuvirus, Turncurtovirus, Capulavirus, Grablovirus, and Eragrovirus. Begomoviruses constitute the largest number of viruses in Geminiviridae family infecting most economically important crops in Australia, China, Europe, and the Middle East countries. Crops that have been infected with begomoviruses belong to the families, Malvaceae (cotton and okra), Cucurbitaceae (melon, watermelon, squash, and gourds), Euphorbiaceae (cassava), Solanaceae (tobacco, potato, tomato, and pepper), and Fabaceae (soybean, cowpea, common bean, and mungbean). Mastreviruses infect chickpea and pepper crops in Australia, Oman, Yemen, Jordan, Syria, and Iraq. Becurtoviruses infect some crops like sugar beet and tomato in Iran. Capulaviruses have been recorded in France and Finland infecting Alfalfa and *Plantago* plants, respectively. The geminiviruses pose a great challenge to the countries by their fast spread and infecting economic crops. Cooperation among these countries in exchanging information and adopting the most up-to-date system in quarantine can prevent further introduction of new viruses into new geographic regions.

1 Introduction

The viruses can be defined as "entities whose genomes are elements of nucleic acid that replicate in living cells using cellular synthetic machinery and causing the synthesis of specialized elements that can transfer the viral genome to other cells"

A. A. M. Al Shihi (✉)
Department of Plant Protection, Ministry of Agriculture and Fisheries Wealth, Muscat, Oman

R. V. Kumar (ed.), *Geminiviruses*, https://doi.org/10.1007/978-3-030-18248-9_4

(Luria et al. 1978). The International Committee on Taxonomy of Viruses (ICTV) has approved 3 orders, 73 families, 9 subfamilies, 287 genera, and ~1950 species of viruses (Briddon et al. 2008). Out of these, the plant viruses constitute 20 families, 88 genera, and around 750 species. More than 90% of plant viruses have ssRNA genomes and the remaining have DNA genomes, including both ssDNA and dsDNA. Caulimoviruses (family *Caulimoviridae*) are dsDNA viruses, whereas nanoviruses (*Nanoviridae*) and geminiviruses (*Geminiviridae*) are ssDNA viruses. The viruses show a wide range of genome sizes: the largest genomes of known virus are the mimiviruses and the smallest are the circoviruses. The family Geminiviridae has a worldwide impact on agricultural production that is ongoing. The diseases caused by geminiviruses represent serious constraints to agriculture. The name of geminivirus was derived when virus particles, which have a unique twinned quasi-isometric morphology, were isolated from maize which had streak symptoms and beet which showed curly top symptoms (Bock et al. 1974; Mumford 1974). This attribute provided the name geminivirus, symbolizing twins (Harrison 1977).

Because of the great losses caused by geminiviruses, they have become the subject of concern worldwide (Briddon et al. 2001). These viruses encode a few genes for their replication and depend mostly on their host proteins for their replication (Hanley-Bowdoin et al. 1999). The geminivirus was established as a group in 1979 (Matthews 1979). It was upgraded to the family *Geminiviridae* in 1995 (Murphy et al. 1995). They infect both monocots, such as wheat and maize, and dicots, such as tomato and cassava (Hanley-Bowdoin et al. 1999). It was reported that geminivirus has emerged in the Middle East and subsequently extended to the Mediterranean basin, Asia, Africa, and America (Czosnek and Laterrot 1997; Freitas-Astua et al. 2002; Varma and Malathi 2003). Several theories have been proposed for the recent distribution of geminiviruses around the world. One theory implicates the import of ornamental plants (Polston et al. 1999); another proposes that geminivirus was spread due to the introduction of infected nonsymptomatic tomato plants from the Eastern Mediterranean region into the Dominican Republic of the Caribbean islands (Brown and Bird 1992). In this chapter, I discuss about the geographic distribution of different geminiviruses in Australia, China, and some European and Middle East countries.

1.1 Geminiviruses

There are more than 199 recognized species of geminivirus in which 181 belong to the genus *Begomovirus* and more than 670 complete sequences are deposited in databases (Fauquet et al. 2008). Based on their host range, genome organization, and insect vector, geminiviruses are classified into nine genera: Begomovirus, Mastrevirus, Curtovirus, Becurtovirus, Topocuvirus, Turncurtovirus, Capulavirus, Grablovirus, and Eragrovirus (Stanley et al. 2005; Fauquet et al. 2008; Brown et al. 2012; Adams et al. 2013). Begomoviruses constitute the largest group of geminiviruses (Briddon et al. 2001; Mansoor et al. 2008). Based on their genomes, they are divided into two groups: monopartite (single component of size 2.8 kb) and bipartite (two components of about the same size known as DNA-A and DNA-B; (Stanley et al. 2005; Fig. 1). Bipartite begomoviruses include *Tomato golden mosaic virus* (TGMV), *Tobacco yellow crinkle virus* (TbYCV), *Tomato leaf curl New Delhi virus* (ToLCNDV), *African cassava mosaic virus* (ACMV), and (Padidam et al. 1995). Monopartite begomoviruses include *Tomato yellow leaf curl virus* (TYLCV), *Tomato leaf curl virus* (ToLCV), and *Tomato yellow leaf curl Sardinia virus* (TYLCSV); even they lack DNA-B but they can induce disease in plants due to the differences in their gene functions (Briddon and Stanley 2006).

Two subgenomic molecules are associated with monopartite begomoviruses: betasatellite and alphasatellite. Betasatellite is defined as a satellite that has no sequence homology to monopartite begomovirus (helper virus) and is entirely dependent on it for replication (Mayo et al. 2005). The first DNA satellite isolated from tomato crops was infected with the monopartite begomovirus, *Tomato leaf curl virus* (ToLCV), which has no open reading frame (ORF) (Dry et al. 1997; Behjatnia et al. 1998). DNA-βs are widely distributed in the Old World (OW) and absent in the New World (NW) (Briddon et al. 2008; Fig. 1). Alphasatellites are the second group of DNA molecules that have a conserved structure and genome size of ~1380 nt. These molecules are associated with monopartite begomoviruses along with betasatellites in the same host (Mansoor et al. 1999; Saunders and Stanley 1999; Briddon et al. 2004; Fig. 1).

The *Mastrevirus* genus includes leafhopper-transmitted viruses, which have monopartite genomes infecting both monocotyledonous and dicotyledonous plants (Boulton 2002; Nahid et al. 2008). *Maize streak virus* (MSV) and W*heat dwarf virus* (WDV) are two well-studied members of the genus. The *Curtovirus* genus includes leafhopper-transmitted viruses, which have monopartite genomes and infect dicots. The curtovirus genome consists of circular ssDNA molecule of about 3.0 kb (Hur et al. 2007). *Beat curly top virus* (BCTV) is a well-studied member in this genus. The *Topocovirus* genus contains treehopper-transmitted viruses which have monopartite genomes. The only known topocuvirus is *Tomato pseudo-curly top virus* (TPCTV), which was isolated from Florida (Briddon et al. 1996). Becurtoviruses have close similarities to the *Curtovirus* genus in terms of their biological properties. An example is *Beet curly top Iran virus* (BCTIV). *Eragrovirus* genus has a single member *Eragrostis curvularia streak virus* (ECSV). The CP of this virus is very close to the CP of viruses in *Mastrevirus* genus. *Turncurtovirus* genus has only one virus, *Turnip curly top virus* (TCTV). The genome organization of geminivirus genera and the genes they encode are clarified in Fig. 1.

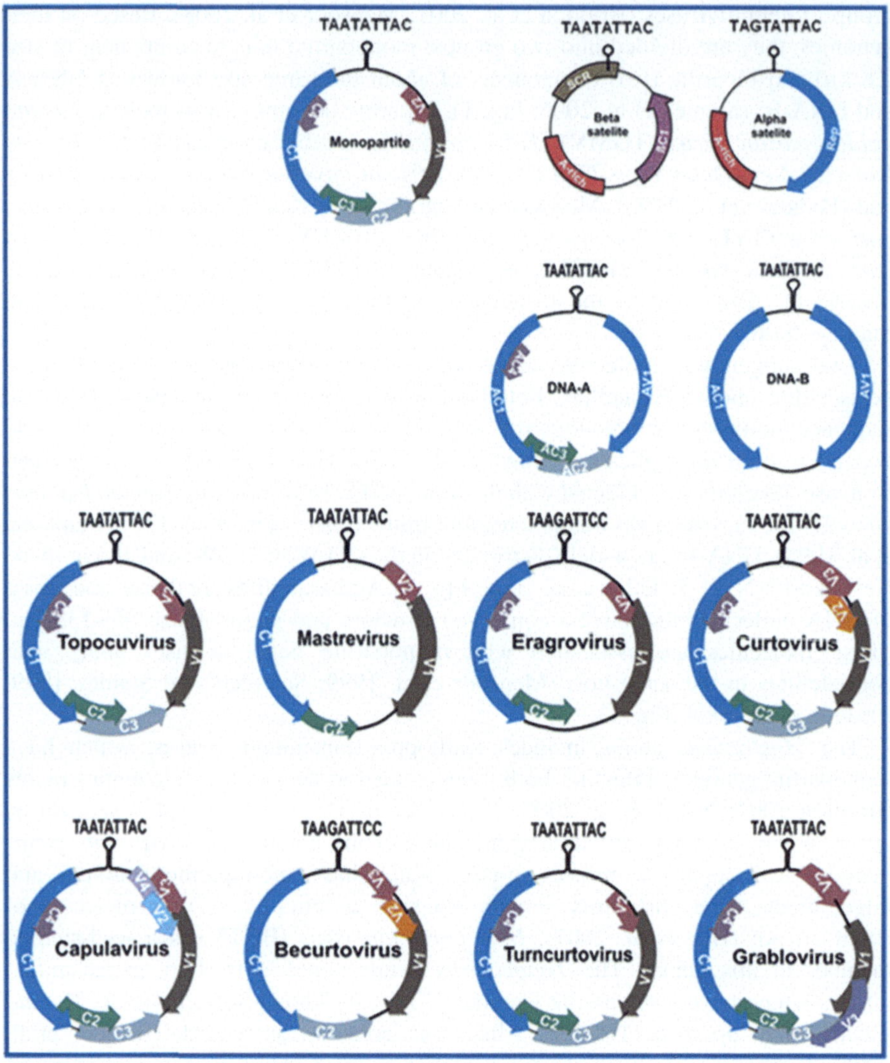

Fig. 1 *Genome organization of different geminivirus genera*. The ORFs (V1, V2, V3, C1, C2, C3, C4.) are coded according to the function of their genes (C1, replication associated protein; C2, transcriptional activator protein; C3, replication enhancer protein; C4, symptom determinant; V1, capsid protein; V2, movement protein;V3, a protein involved in regulating the ss/ds DNA ratio; AV1, nuclear shuttle protein; AC1, movement protein). The position of the stem-loop containing the conserved sequence located in the intergenic region is shown (TAAGATTCC sequence for Becurtoviruses and Eragoviruses; TAATATTAC for other genera). Genome map of Beta satellite consist of Adenine rich sequence (A-rich), sequence common region (SCR) and hairpin structure having the nonanucleotide sequence TAATATTAC. Genome map of Alphasatellites consists of one large gene in the virion-sense (Rep), adenine rich sequence (A-rich) and a hairpin structure containing, the nonanucleotide sequence TAGTATTAC

1.2 Geminivirus Evolution

Genetic variation can arise in the genome of geminiviruses through mutation, recombination, and pseudorecombination (Seal et al. 2006). The rate of mutation is very low in DNA as compared to RNA viruses. Isnard et al. (1998) reported that mutation in *Maize streak virus* (MSV) has occurred at frequencies of about 10^{-4}–10^{-5} throughout their genome. Pseudorecombination has been reported to occur among begomoviruses in different countries. It describes the exchange between the genome components of DNA-A and DNA-B (Pita et al. 2001; Ramos et al. 2003; Idris and Brown 2005). Experimentally, it was reported that the exchange between components of *Tomato mottle Taino virus* (ToMoTV) pseudorecombines with *Potato yellow mosaic virus* (PYMV) but not with *Tomato mottle virus* (ToMoV) (Ramos et al. 2003). The DNA-A component of some geminiviruses can form association with some DNA-B of other viruses and can cause infection when co-inoculated with each other (Karthikeyan et al. 2004). Recombination is the process by which the segments of one nucleotide strand are incorporated into segments of other nucleotide strands during replication process. Recombination is common under natural field conditions among geminiviruses (Zhou et al. 1997; Padidam et al. 1995; Al Shihi et al. 2014). Recombination has been reported to occur between DNA-A molecules of different begomoviruses. For example, Zhou et al. (1997) reported that *Cassava mosaic virus* (CMV) is a recombinant between *African cassava mosaic virus* (ACMV) and *East African cassava mosaic virus* (EACMV). In addition, Monci et al. (2002) stated a recombinant between *Tomato yellow leaf curl Sardinia virus* (TYLCSV) and *Tomato yellow leaf curl virus* (TYLCV). Similarly, Al Shihi et al. (2014) reported a recombinant between *Tomato leaf curl Oman virus* (ToLCOMV) and *Croton yellow vein mosaic virus* (CroYVMV) and hence the name *Tomato leaf curl Barka virus* (ToLCBrV).

1.3 Disease Symptoms

Yield losses due to infection by TYLCD have become a major threat to crop production in the Middle East, Southeast Asia, and Europe (Czosnek and Laterrot 1997; Moriones and Navas-Castillo 2000). Fiallo-Olive et al. (2013) reported that about 40 different countries (about seven million hectares) are subjected to the attack by geminiviruses. There are different types of symptoms resulting from geminivirus infection. Up-curling of leaves and reduction of upper leaf size is common when hot pepper, okra, and papaya plants are infected with geminivirus. Leaf curling, mosaic-like pattern, and general stunting of plant can be seen on plants like squash, radish, and tomato when infected with geminiviruses. The disease symptoms can vary from slight to severe depending on several factors such as plant stage, time of infection, type of virus strain, and type of vector (Fig. 2).

Fig. 2 *Geminivirus disease symptoms*. (**a**) Papaya infected with TYLCV and ChLCV; (**b**) Okra infected with OLCOMV; (**c**) Hot pepper infected with ChLCV; (**d**) Tomato infected with ToLCBrV and; (**e**) Radish infected with TYLCV and ChLCV, and (**f**) Squash infected with TYLCV (Source: Adel Al Shihi)

2 Geminiviruses Occurrence in Australia, China, Europe, and the Middle East Countries

2.1 Geminiviruses in Australia

Geminiviruses (*Geminiviridae*) are economically important pathogens which lead to serious losses in food crops worldwide. Agriculture in tropical and subtropical regions is mostly under danger, especially those growing crops such as tomatoes, beans, peppers, cucurbits, and cassavas (Brown 1994). *They* limit crop production in several regions in the world, including Australia (Behjatnia et al. 1998).

In Australia, a monopartite *Begomovirus*, *Tomato leaf curl virus* (TLCV), was reported in 1970 in the Northern Province of the country causing severe losses to tomato crops (Behjatnia et al. 1998). It is having a ssDNA genome of 2766 nt encoding six open reading frames (Dry et al. 1997). The whitefly (*B. tabaci* biotype B) was recorded in Australia for the first time since 1994 (Gunning et al. 1995). In addition to their high efficiency in begomovirus transmission, they cause significant damage through direct feeding on some crops such as soybeans, sunflowers, tomatoes sweet potatoes, cotton, cucurbits, and eggplants.

Areas infected with TLCV have till now been away from intensive horticultural areas being located on the east coast of Queensland and currently crop losses are limited to a relatively small area around Darwin (Stonor et al. 2003). Whitefly inoculation was done to a group of plants and weed species common in northern Australia to test their susceptibility to TLCV. Out of 58 species tested, only 11 species showed symptoms typical to begomovirus infection, but 47 species failed to show symptoms, and when tested molecularly using probe hybridization, no TLCV DNA was detected (Stonor et al. 2003). This screening by whitefly inoculation might suggest that TLCV may have a narrow host range. The area where TLCV (Northern area) occurs is separated from southern areas by a distinct climatic region. This region is represented by a long dry period extending from April to November (Anonymous 1988).

Another virus in *Mastrevirus* genus is named as *Tobacco yellow dwarf virus* (TYDV) which is recorded only in Australia and causes significant diseases in bean (*Phaseolus vulgaris*) (Hill 1937) and tobacco (*Nicotiana tabacum*) (Ballantyne 1968). It occurs in all states of Australia and transmitted by the leafhopper vector (*Orosius argentatus*) or by grafting 30 species in seven dicotyledonous families (Helson 1951; Hill and Mandryk 1954; Thomas and Bowyer 1979). The common symptoms that result from TYDV infection on bean (*Phaseolus vulgaris*) are reduction in the growth rate of the first trifoliolate leaf, down-curling of the trifoliolate leaf margin, and vascular necrosis of the stem. In tobacco, the symptoms which are seen when plants are infected with TYDV are down-curling of the tips and margins of the youngest leaves, chlorosis, and stunting of the whole plant. TYDV can be distinguished from other viruses by its geminate particles and leafhopper vector. Other geminiviruses, like *Tobacco leaf curl virus* (TbLCV) and *Beet curly top virus* (BCTV), have many hosts in common with TYDV. However, TbLCV is transmitted by the whitefly (*Bemisia tabaci*). BCTV causes similar symptoms to TYDV in several hosts and is also transmitted by leafhoppers, so the two viruses can be distinguished easily by using serological tests. In addition to the recognized species (TYDV), two more distinct species of mastrevirus are known to infect dicotyledonous crops in Australia including *chickpea Chlorosis virus* (CpCV-A, CpCV-B) and *Chickpea red leaf virus* (CpRLV; Thomas et al. 2010). These mastreviruses infect chickpea, bean, and tobacco crops. The dicot-infecting mastreviruses, biologically, serologically, and phylogenetically, constitute a distinct group in comparison to monocot-infecting mastreviruses (Brown et al. 2012). The important strains of dicot-infecting mastreviruses from eastern Australia are TYDV, CpCV-A, CpCV-B, and *Chickpea red leaf virus* (CpRLV) (Thomas et al. 2010). It was believed that Australia could be the hotspot of dicot-infecting mastrevirus diversity and more mastrevirus diversity could exist in chickpea and maybe other cultivated host species in Australia. Schwinghamer et al. (2010) suggested that the geographical distribution of distinct dicot-infecting mastreviruses is overlapping broadly in eastern Australia.

2.2 Geminiviruses in China

In China, geminiviruses have been emerging as serious plant pathogens in many areas in recent years, and several begomovirus species and strains have been reported infecting tomato, squash, tobacco, and *Malvastrum coromandelianum*. Some of these viruses were found to be associated with betasatellite molecules. Begomovirus–DNAβ complex was found to be associated with tomato leaf curl disease in Guangxi province, China. Monopartite begomoviruses that have been associated with betasatellite are *Tobacco leaf curl virus* (TbLCV), *Tomato leaf curl*

China virus (ToLCCNV), *Tobacco leaf curl Yunnan virus* (TLCYNV), and *Tobacco curly shoot virus* (TbCSV) (Meng et al. 2012). In China, there are three types of whitefly biotypes, *Bemisia tabaci* biotype B, biotype Q, and Biotype Cv. Biotypes B and Q are well known for their high efficiency in transmitting viruses (Cui et al. 2004). Infectivity assay showed that *Tomato leaf curl China betasatellite* (ToLCCNB) is required for inducing disease symptoms in the tested plants. This role coincides with DNA-β species associated with *Cotton leaf curl Multan virus*, *Ageratum yellow vein virus* (AYVV), and TYLCCNV (Briddon et al. 2001; Cui et al. 2004).

Zhou et al. (2003) reported that begomoviruses isolated from some crops including tobacco, tomato, and weed species in China (Yunnan) were found to be associated with DNA betasatellites, and the complete nucleotide sequences were found to be 1333–1355 nt. DNA betasatellites associated with begomoviruses from the same region are clustered closely, but begomovirus isolates from different regions were more distantly related. *Ageratum yellow vein China virus* (AYVCNV) retained more concentrations in infected leaves in the presence of DNA betasatellites. It was suggested that DNA betasatellites may have a direct effect on viral DNA replication, probably by providing proper cellular functions (Liu et al. 1999; Nagar et al. 1995). Another hypothesis is that they may facilitate the systemic movement of viral DNA within the plant, hence enhancing the level of viral DNA in infected tissues (Xiong et al. 2007).

Alphasatellites were identified in begomovirus-infected plants in Yunnan and all crops (tobacco, tomato, and squash) infected with alphasatellites also have been found to be infected with betasatellites (Xie et al. 2010). They were divided into three types based on phylogenetic tree of the complete nucleotide sequences. The first type was associated with *Tomato yellow leaf curl China virus* (TYLCCNV)/*Tomato yellow leaf curl China betasatellite* (TYLCCNB) complex. The second type was associated with *Tobacco curly shoot virus* (TbCSV)/*Tobacco curly shoot betasatellite* (TbCSB) complex. The third type was associated with TbCSV/*Ageratum yellow vein betasatellite* (AYVB) complex (Xie et al. 2010). It was confirmed that unlike betasatellites, alphasatellites are self-replicating in host plants because they encode a rolling-circle replication initiator protein; however, they require helper begomoviruses for movement in plants and insect transmission as well. They may play an important role in the epidemiology of begomovirus and betasatellite complexes, but more studies need to be conducted to clarify this role.

In China, geminivirus species has obvious geographical characteristics, that is different regions have different virus strains. The geminiviruses occur in high incidence in Yunnan, but due to the isolated mountain valleys, its distribution is discontinuous rather than continuous (Jing et al. 2016). After amplifying the whole genome of DNA-A, cloning, and sequencing analysis, the results revealed the presence of a number of begomoviruses such as *Malvastrum yellow vein Yunnan virus* (MYVYnV), *Chinese squash leaf curl virus*, *Squash leaf curl China virus*, (SLCCNV), *Sweet potato leaf curl virus* (SPLCV), *Tomato yellow leaf curl China virus* (TYLCCNV), Yunnan chilli leaf curl virus (Pepper leaf curl Yunnan virus (PeLCYnV), *Crassocephalum yellow vein virus* (CYVV), *Yunnan Tobacco leaf curl virus, and Tobacco leaf curl Yunnan virus* (TbLCYnV) (Meng et al. 2012).

The distribution of geminiviruses has showed the dominance of some strains in some regions more than others. For instance, in northern district, there are four geminivirus species, of which *Tomato yellow leaf China curl virus* (TYLCCNV) is the most dominant type; in South Central of Yuanjiang, there are four geminiviruses, of which the dominant species is TYLCCNV, followed by *Pepper leaf curl Yunnan virus* (PeLCYnV); in the western district where climate is humid, four geminiviruses species have been identified of which the dominant species is *Sweet potato leaf curl virus* (SPLCV); and in the southern of Lancang River Basin four geminiviruses have been identified as well, among them *Tobacco curly shoot virus* (TbCSV) and *Tobacco leaf curl Yunnan virus* (TLCYnV). The occurrence of begomoviruses in some regions like Sichuan is increasing more. The mixed infection of several begomoviruses like TYLCCNV/TYLCCNB and *Papaya leaf curl China virus* (PaLCuCNV) was identified in tomato (Jing et al. 2016).

2.3 Geminiviruses in Europe

Over the past years, surveys in the main tomato production area of Sicily Ragusa province in Italy confirmed the presence of TYLCV (Accotto et al. 2000). Tomato (*Solanum Lycopersicon*) crops in Sardinia and Sicily have been severely affected by yellow leaf curl disease. Accotto et al. (2000) reported that TYLCV has spread quickly in an area where the other viral species like *Tomato yellow leaf curl Sardinia virus* (TYLCSV) causes yellow leaf curl disease. In Portugal (Algarve), disease symptoms on some vegetable crops include plant stunting, leaf curling, and yellowing. Louro et al. (1996) reported that up to 100% of tomato crops were affected and yield was severely reduced due to TYLCV infection. In Spain, severe leaf yellowing has occurred in tomato (*Solanum Lycopersicon*) crops in southern Spain, and this outbreak was associated with high populations of the whitefly *Bemisia tabaci*. Symptoms including leaf interveinal yellowing that developed initially on lower leaves and then progressed to the upper leaves of tomato have been seen (Moriones et al. 1993). In Spain, sweet potato (*Ipomoea batatas*) and *Ipomoea indica* plants were found to be infected with sweepoviruses. They comprise a monophyletic group of begomoviruses which have been known to infect sweet potato (*Ipomoea batatas*) and other species of the family Convolvulaceae (Lozano et al. 2016). Lozano et al. (2016) reported that sweepoviruses infecting *Ipomoea* sp. in Spain were associated with small molecules named as deltasatellites (ToLCV-sat). In September 2013, in the province of Almeria, Spain symptoms including leaf chlorotic mottling and vein distortion on middle leaves were seen in tomato (*Solanum lycopersicum* L.) growing in a greenhouse. Nearby greenhouse having zucchini squash plants (*Cucurbita pepo* L.) showed leaf curling symptoms and chlorotic mottling on intermediate leaves. The results revealed the presence of ToLCNDV, which has been known to infect tomato crops in India for the last two decades, in both samples of tomato and zucchini squash crops (Padidam et al. 1995). Recently, ToLCNDV was reported to infect zucchini squash crops in Italy (Panno et al. 2016). Pepper plants infected with begomoviruses like symptoms exhibiting light mosaic leaf distortion, interveinal and leaf chlorosis, and upward curling of leaf margins combined with large population of the whitefly *Bemisia tabaci* were observed in Basilicata region in Italy. The molecular analysis confirmed the presence of *Tomato yellow leaf curl Sardinia virus* (TYLCSV) in the infected pepper plants (Fanigliulo et al. 2008). In France, TYLCV was reported to infect tomato in a single field in the Camargue district in 1999 (Dalmon et al. 2005; Lefeuvre 2010). *Alfalfa leaf curl virus* (Genus: *Capulavirus*), which is transmitting through *Aphis craccivora* (Roumagnac et al. 2015), was isolated from alfalfa crop showing leaf curl disease in France (Varsani et al. 2017). In Greece, tomato crops grown in greenhouses in several places in Crete, Attiki, and southern Peloponnese showed severe symptoms of TYLCV. Infected plants were infested with high populations of *Bemisia tabaci*. Partial sequencing indicated the identity of TYLCV strain (Avgelis et al. 2001). In Finland, the sequence analysis confirmed the presence of *Plantago lanceolata latent virus* which was infecting flowering plant called *Plantago lanceolata*; this virus belongs to the genus *Capulavirus* (Varsani et al. 2017).

2.4 Geminiviruses in the Middle East Countries

Geminiviruses have emerged as a problem for agriculture in some Middle East countries such as Oman, Iran, Saudi Arabia, Yemen, Jordan, Syria, Kuwait, and Iraq. Generally, begomoviruses constitute the largest number of geminiviruses spreading in the Middle East countries infecting many economically important crops then mastreviruses coming in the second rank in their distribution.

In Oman, the presence of geminivirus was detected first in 1993, when symptoms of tomato leaf curl disease were seen on some tomato and papaya crops (Ministry of Agriculture and Fisheries, Government of Oman). Begomoviruses constitute the largest number of isolated geminiviruses in Oman and affect the most economically important crops in the country. Both types of begomoviruses either with a monopartite or bipartite genome are present in Oman. Most monopartite begomoviruses are associated with DNA satellites. Most begomoviruses that have been identified in Oman are not native to the region. Till now, two betasatellites have been identified in Oman: *Tomato leaf curl betasatellite* (ToLCB) and *Okra leaf curl Oman betasatellite* (OLCOMB) (Al Shihi 2017). The distribution of geminiviruses in Oman is concentrated mostly in Al Batinah Governorate, which constitutes about 85% of agricultural area in Oman. The geminiviruses that have been isolated and characterized in this region belong to begomovirus genus including *Tomato leaf curl Al Batinah virus* (ToLCABV; Khan et al. 2014), *Tomato leaf curl Oman virus* (ToLCOMN; Khan et al. 2008), *Chilli leaf curl virus Oman* (ChLCV-OM; Khan et al. 2013), *Tomato leaf curl Barka Virus* (ToLCBrV; Al Shihi et al. 2014), and *Cotton leaf curl Gezira virus-Al Batinah* (CuLCGV-Al Batinah; Al Shihi et al. 2017). All these begomoviruses were isolated from tomato crops plus ChLCV-OM, which infect both tomato and pepper crops. *Tomato yellow leaf curl virus* (TYLCV-OM) was identified in most regions in Oman including northern part of Oman, Musandam Governorate. This virus constitutes high identity to the Iranian strain (TYLCV-IL) (Khan et al. 2008). Most of these viruses are monopartite begomoviruses and are associated with betasatellites (ToLCB). In the Southern region of Oman, Dhofar Governorate, two monopartite begomoviruses have been identified, *Chilli leaf curl Multan virus* (ToLCMuV) and *Tomato leaf curl Sudan virus* (ToLCSDV) infecting tomato and pepper (Al Shihi 2017). Some bipartite begomoviruses have been seen in some crops such as watermelon, cassava, and bean. *Watermelon chlorotic stunt virus* has been isolated from watermelon (Khan et al. 2012), *East African cassava mosaic virus* isolated from Cassava (Khan et al. 2013), and *Mungbean yellow mosaic Indian virus* from bean (Shahid et al. 2017). One mastrevirus named as *Chickpea chlorotic dwarf virus* (CpCDV) had been isolated from pepper which was collected from Al Sharqia Governorate (Akhtar et al. 2014). This wide distribution of begomoviruses in Oman refers to several factors such as presence of whitefly (*Bemisia tabaci* Genn; Al Shihi and Khan 2013) biotype B which is known as highly aggressive in transmitting begomoviruses worldwide (Brown et al. 2012). In addition, internal transport of plants and plant products among different governorates without proper inspection help to spread pests and diseases. The global movement of agricultural products plays a major role in introducing geminiviruses, and this was indicated to the virus origin. Most begomoviruses that have been identified in Oman have their origin outside the country (Al Shihi 2017). Farmers mostly use F1 tomato hybrid seeds which can offer moderate protection against begomoviruses, and they can get a good protection if floating row cover (AGRYL) is used from the seedling till flowering stage. Al Shihi et al. (2016) stated that covering tomato crops with floating row cover for 6–7 weeks can minimize tomato leaf curl disease and maximize the yield.

The geminivirus infection was detected on several crops in Saudi Arabia. The first report of infection was published in 1957 in tomato crops which showed mosaic-like symptoms grown under field condition (Talhouk 1957). The most important crops that have been infected with begomoviruses in Saudi Arabia are tomato, beans, okra, squash, and cucumber (Idris et al. 2012; Alhudiab et al. 2014). In Jeddah, Sohrab et al. (2016b) reported that *Tomato leaf curl Sudan virus* and *Tomato yellow leaf curl virus* cause leaf curling and yellowing. Okra (*Abelmoschus esculentus* L.) crops showing disease symptoms like leaf curling and whole plant stunting have been reported in Hofuf and Al-Hassa Governorates. Molecular analysis confirmed the presence of *Cotton leaf curl Gezira virus* (CLCuGV) which shares 89% identity with CLCuGV-Egypt isolate (Idris et al. 2014). Bean (*Phaseolus vulgaris* L.) crops showed disease symptoms such as dwarfing, leaf malformation, and vein yellowing, grown under field conditions in Al-Hassa, Hofuf, Eastern Province of Saudi Arabia (Ghanem et al. 2003). The serological analysis confirmed the presence of begomoviruses which has been named as *Bean dwarf mosaic virus* Saudi Arabian isolate (BDMV-SA) (Ghanem et al. 2003). Recently, TYLCV was isolated from cucumber (*Cucumis sativus*) crops which showed mosaic-like symptoms on the leaves (Sohrab et al. 2016b). Weeds can act as alternative hosts for begomoviruses; just recently the natural occurrence of begomovirus on a weed called Corchorus has been reported from Saudi Arabia (Sohrab 2016a).

Iran is one of the main countries growing all kinds of vegetables in the world. The total area harvested with vegetable accounts for about 811,616 hectare (Ha) and yield 264,367 hectogram Hg/Ha. Recently, geminiviruses cause significant losses to many vegetable crops and most of these viruses belong to the genus begomovirus (Farzadfar et al. 2002; Ayazpour 2014). Some begomoviruses like *Tomato yellow leaf curl virus* (TYLCV) infect important crops such as potato, tomato, and cucurbits and can cause significant yield losses. TYLCV was first reported in 1996 from tomato crops grown in the southern provinces of Iran (Bushehr, Khuzestan, Hormozgan, Sistan-va- Baluchestan, and Kerman) (Hajimorad et al. 1996). TYLCV virus was also detected in other plant species such as cucumber (*Cucumis sativus*), pepper (*Capsicum annuum*), alfalfa (*Medicago sativa*), cowpea (*Vigna unguiculata*), cantaloupe (*Cucumis melo* var. cantalupensis), and red pepper (*Capsicum sp.*) (Hosseinzadeh and Garivani 2014; Azadvar et al. 2016; Bananej 2016). Five strains of TYLCV have been identified in Iran including TYLCV-IL, TYLCV-IR, TYLCV-Bou, TYLCV-Ker, and TYLCV-OM. Among these strains, TYLCV-IL is considered the most damaging strain worldwide, and it is present in different provinces of Iran (Lefeuvre et al. 2010; Pakniat et al. 2010). *Tomato Leaf Curl Palampur Virus* (ToLCPMV) is a bipartite begomovirus which was isolated from tomato fields located in the southern region of the country in 2006 (Hormozgan Province). *Tomato Leaf Curl New Delhi Virus* (ToLCNDV) is another destructive bipartite begomovirus species infecting melons in Iran (Yazdani-Khameneh et al. 2013). ToLCNDV is also infecting other crops including tomato, potato, pepper, and cucurbit plants (Hussain et al. 2005). *Tomato Yellow Leaf Curl Iran Virus* (TYLCIRV) is another strain of TYLCV infecting tomato crops which showed typical yellow leaf curl symptoms in the following provinces: Iranshahr, Sistan, and Baluchestan. *Okra Enation Leaf Curl Virus* (OELCuV) was isolated from papaya crops showing leaf curl disease in Bahu Kalat, Zarabad in Sistan-va-Baluchestan. Therefore, papaya was listed as a new species in the natural host range of OELCuV (Bananej et al. 2016). *Watermelon Chlorotic Stunt Virus* (WmCSV) was first identified in Yemen and then in Sudan (Bedford et al. 1994). In 1998, watermelon was found to be severely infected with begomovirus-like symptoms in the south of Iran; the virus was isolated and characterized from plants through molecular analysis. The sequence analysis confirmed the presence of *Watermelon chlorotic stunt virus* (Bananej et al. 1998). *Beet curly top Iran virus* (BCTIV) is a major geminivirus (Genus: *Becurtovirus*) of sugar beet in Iran. Nine genomes of new BCTIV isolates were characterized and sequenced. These genomes were isolated from crops such as cowpea, bean, tomato, and sugar beet showing leaf curling, yellowing, and swelling of veins. The BCTIV is distributed in some fields in north-eastern Iran (Khorasan Razavi, Northern Khorasan), north-western Iran (East and West Azerbaijan), and southern Iran (Fars) provinces (Kardani et al. 2013). The presence of *Bemisia tabaci* in different parts of Iran combined with different climatic conditions seems to encourage the potential spread of these viruses in many new areas of the country (Shahbazi et al. 2010).

Anfoka et al. (2016) reported that tomato plants in Jordan were infected with new begomovirus strain named as *Tomato yellow leaf curl Axarquia virus*. Another virus,

Chickpea chlorotic dwarf virus (*Mastrevirus* genus), was reported to infect some crops like chickpea and pepper in Yemen, Jordan, Iraq, and Syria (Akhtar et al. 2011; Kumari et al. 2006). In Kuwait, TYLCV is widespread in tomato fields causing a devastating disease since 1993 (Montasser et al. 1999). In Yemen, TYLCV is increasing in tomato-growing regions since the 1970s. It is present in the Abayan and Hadramaut Governorates. Based on partial sequencing, the results indicated that TYLCV from Yemen (TYLCYV) is distinct from other TYLCV isolates (Bedford et al. 1994) (Fig. 3).

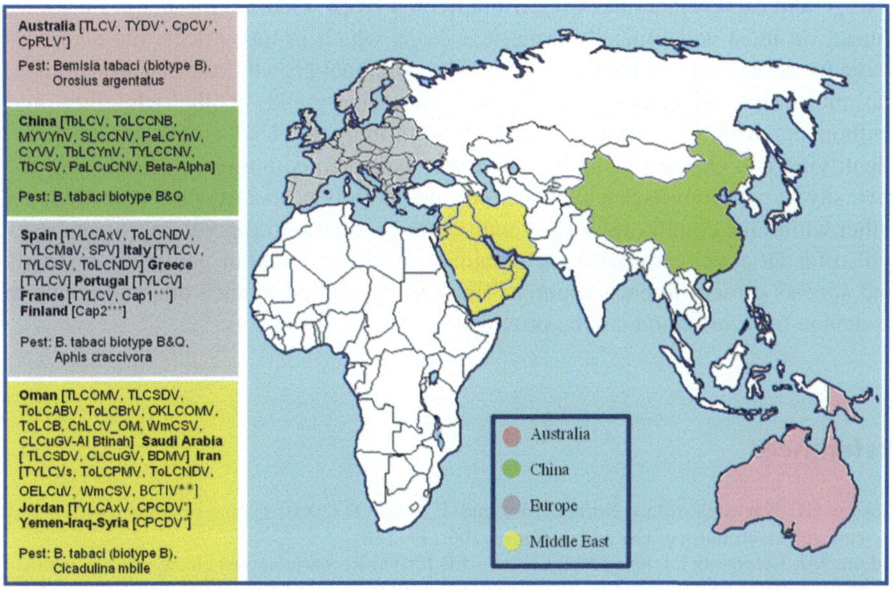

Fig. 3 *World map along with colored boxes showing the geminivurs species and strains in Australia, China, Europe and Middle East countries. TLCV Tomato leaf curl virus*; *TYDV** Tobacco yellow dwarf virus, *CPCV** Checkpea chlorosis virus, *CPRLV** Checkpea red leaf virus, Bemisia tabaci biotype B. *TbLCV* Tobacco leaf curl virus, *ToLCCNB* Tomato leaf curl China beta satellite, *MYVYnV* Malvastrum yellow vein Yunnan virus, *SLCCNV* Squash leaf curl China virus, *TYLCCNV* Tomato yellow leaf curl China virus, *PeLCYnV* Pepper leaf curl Yunnan virus, *TbLCYnV* Tobacco leaf curl Yunnan virus, *PaLCuCNV* Papaya leaf curl China virus, *CYVV* China yellow vein virus, *SPV* Sweet potato leaf curl virus, *TYLCV* Tomato yellow leaf curl virus, *TYLCAxV* Tomato yellow leaf curl Axarquia virus, *TYLCSV* Tomato yellow leaf curl Sardinia virus, *TYLCMaV* Tomato yellow leaf curl Malaga virus, *ToLCNDV* Tomato leaf curl New Delhi virus, Cap1***Alfalfa leaf curl virus; Cap2*** Plantago lanceolata latent virus; *ToLCABV* Tomato leaf curl Al Batinah virus, *ToLCBrV* Tomato leaf curl Barka virus, *ToLCSDV* Tomato leaf curl Sudan virus, *CpCDV** Chickpea chlorotic dwarf virus, *ChLCMuV* Chilli leaf curl Multan virus, *ChLCV* Chilli leaf curl virus, *CLCuGV* Cotton leaf curl Gezira virus, *OLCOMV* Okra leaf curl Oman virus, *ToLCPMV* Tomato Leaf Curl Palampur Virus, *BDMV* Bean dwarf mosaic virus. *OELCuV* Okra enation leaf curl virus, *WmCSV* Watermelon chlorotic stunt virus, *BCTIV*** Beet curly top Iran virus, *Mastrevirus, **Becurtovirus, ***Capulavirus, No star means Begomoviruses (Source: Adel Al Shihi)

3 Future Aspects

Available information reveals that these disease complexes are expanding rapidly in terms of their geographical distribution and host range. For instance, ToLCNDV was originally a major problem in India but now it is spreading and causing extensive damage in Spain, Italy, and Iran. In some countries in Europe and the Middle East, new virus strains are emerging and their host range is expanding to other new crops. The presence of such a diverse population of geminiviruses in some regions, combined with the ability of these viruses to exchange their genetic material by recombination, will increase the probability of evolution of new viruses which may emerge and cause epidemics in new unaffected crops. Geminiviruses have a strong impact on most economically important crops which in turn affects the economy value for some crops in most countries. The continual growth in international trade, the movement of infected plants, and the widespread of the whiteflies and leafhoppers will facilitate the spread of geminiviruses. Under any circumstance, identifying and characterizing geminiviruses will help countries in determining the diversity of geminiviruses which later can aid to apply proper quarantine procedures either within the country regions or with other countries. Computer-based databases will offer an excellent choice for obtaining information about geminivirus strains and species present in each country. This can be applied in each quarantine where exchange of information can be provided easily.

References

Accotto GP, Navas-Castillo J, Noris E, Moriones E, Louro D (2000) Typing of *Tomato yellow* leaf curl viruses in Europe. Eur J Plant Pathol 106:179–186

Adams MJ, Lefkowitz EJ, King AM, Carstens EB (2013) Recently agreed changes to the international code of virus classification and nomenclature. Arch Virol 158:2633–2639

Akhtar KP, Ahmad M, Shah TM, Atta BM (2011) Transmission of *Chickpea chlorotic dwarf virus* in chickpea by the leafhopper *Orosius albicinctus* (distant) in Pakistan—short communication. Plant Prot Sci 47:1–4

Akhtar S, Khan AJ, Singh AS, Briddon RW (2014) Identification of a disease complex involving a novel monopartite begomovirus with beta- and alphasatellites associated with okra leaf curl disease in Oman. Arch Virol 159:1199–1205

Al Shihi AAM (2017) Status of Begomovirus in Oman. In: Saxena S, Tiwari A (eds) Begomoviruses: occurrence and management in Asia and Africa. Springer, Singapore

Alhudiab K, Alaraby W, Rezk A (2014) Molecular characterization of tomato yellow leaf curl disease associated viruses in Saudi Arabia. Int J Virol 10:192–203

Al Shihi AA, Khan AJ (2013) Identification of whitefly (*Bemicia tabaci* Genn.) biotypes and associated bacterial symbionts in Oman. J Plant Sci 8:39–44

Al Shihi AAM, Khan AJ, Akhtar S, Lima ATM, Zerbini FM, Briddon RW (2014) Occurrence of a new recombinant begomovirus species infecting tomato in the Al-Batinah region of Oman. Plant Pathol 63:1177–1184

Al Shihi AA, Al Sadi AM, Al-Said FA, Ammara U, Deadman ML (2016) Optimizing the floating row cover period to minimize the incidence of tomato yellow leaf curl disease and maximize the yield of tomato. Ann Appl Biol 168:328–336

Al Shihi AA, Al Sadi AM, Deadman M., Briddon RW, Shahid MS (2017) Identification of a distinct strain of *Cotton leaf curl Gezira virus* infecting tomato crop in Oman. J Phytopathol (First Published Online: 8 Dec 2017)

Anfoka G, Al-Talb M, Ahmad H, Fatima (2016) A new isolate of tomato yellow leaf curl axarquia virus associated with tomato yellow leaf curl disease in Jordan. J Plant Pathol 98:145–149

Anonymous (1988) Climate atlas of Australia. Bureau of Meteorology, Department of Administrative Services, Canberra

Avgelis AD, Roditakis N, Dovas CI, Katis NI, Varveri C, Vassilakos N, Bem F (2001) First report of *Tomato yellow leaf curl virus* on tomato crops in Greece. Plant Dis 85:678

Ayazpour K (2014) Alphabetic list of plant viruses and viroids reported from Iran. Islamic Azad University, Jahrom, Jahrom Branch

Azadvar M, Namvar P, Darini A (2016) Study on control methods of tomato yellow leaf curl disease in Southern Kerman. Final project of Agricultural Extension, Education and Research Organization, Project No. 14–70-16-9152. Iranian Research Institute of Plant Protection (IRIPP) Tehran, Tehran

Ballantyne B (1968) Summer death of beans. Agric Gaz NSW 79:486–489

Bananej K (2016) An analysis on the status of tomato yellow leaf curl disease. Appl Entomol Phytopathol 84:157–174

Bananej K, Kheyr-Pour A, Ahoonmanesh A (1998) Identification of *watermelon chlorotic stunt virus,* WmCSV in Iran. In: Proceedings of the 13th Iranian Plant Protection Congress, Karaj, p 194

Bananej K, Kraberger S, Varsani A (2016) Okra enation leaf curl virus in papaya from Iran displaying severe leaf curl symptoms. J Plant Pathol 98:637–639

Bedford ID, Briddon RW, Brown JK, Rosell RC, Markham PG (1994) Geminivirus transmission and biological characterization of *Bemisia tabaci* (Gennadius) biotypes from different geographical regions. Ann Appl Biol 125:311–325

Behjatnia SAA, Dry IB, Rezaian MA (1998) Identification of the replication-associated protein binding domain within the intergenic region of tomato leaf curl geminivirus. Nucleic Acids Res 26:925–931

Bock KR, Guthrie EJ, Woods RD (1974) Purification of maize streak virus and its relationship to viruses associated with streak diseases of sugar cane and *Panicum maximum.* Ann Appl Biol 77:289–296

Boulton MI (2002) Functions and interactions of mastrevirus gene products. Physiol Mol Plant Pathol 60:243–255

Briddon RW, Stanley J (2006) Sub-viral agents associated with plant-infecting single-stranded DNA viruses. Virology 344:198–210

Briddon RW, Lunness P, Bedford ID, Chamberlin LCL, Mesfin T, Markham PG (1996) A streak disease of pearl millet caused by a leafhopper-transmitted geminivirus. Eur J Plant Pathol 102:397–400

Briddon RW, Mansoor S, Bedford ID, Pinner MS, Saunders K, Stanley J, Zafar Y, Malik KA, Markham PG (2001) Identification of DNA components required for induction of cotton leaf curl disease. Virology 285:234–243

Briddon RW, Bull SE, Amin I, Mansoor S, Bedford ID, Rishi N, Siwatch SS, Zafar Y, Abdel-salam AM, Markham PG (2004) Diversity of DNA 1: a satellite-like molecule associated with monopartite begomovirus-DNA β complexes. Virology 324:462–474

Briddon RW, Brown JK, Moriones E, Stanley J, Zerbini M, Zhou X, Fauquet CM (2008) Recommendations for the classification and nomenclature of the DNA-β satellites of begomoviruses. Arch Virol 153:763–781

Brown JK (1994) Current status of *Bemisia tabaci* as a plant pest and virus vector in agroecosystems worlwide. FAO Plant Protect Bull 42:3–32

Brown JK, Bird J (1992) Whitefly-transmitted geminiviruses and associated disorders in the Americas and the Caribbean basin. Plant Dis 76:220–225

Brown JK, Fauquet CM, Briddon RW, Zerbini FM, Moriones E, Navas-Castillo J (2012) Geminivirdae. In: King AMQ, Adams MJ, Carstens EB, Lefkowitz EJ (eds) Virus taxonomy-ninth report of the internation committee on taxonomy of viruses. Associated Press, Elsevier Inc., London, pp 351–373

Cui X, Tao X, Xie Y, Fauquet CM, Zhou X (2004) A DNAβ associated with *Tomato yellow leaf curl China virus* is required for symptom induction. J Virol 78:13966–13974

Czosnek H, Laterrot H (1997) A worldwide survey of tomato yellow leaf curl viruses. Arch Virol 142:1391–1406

Dalmon A, Bouyer S, Cailly M (2005) First report of tomato chlorosis virus and tomato infectious chlorosis virus in France. Plant Dis 89:1243

Dry I, Krake LR, Rigden JE, Rezaian MA (1997) A novel subviral agent associated with a geminivirus: the first report of a DNA satellite. Proc Natl Acad Sci USA 94:7088–7093

Fanigliulo A, Pacella R, Comes S, Crescenzi A (2008) First record of *Tomato yellow leaf curl Sardinia virus* (TYLCSV) on pepper in Italy. Commun Agric Appl Biol Sci 73:297–302

Farzadfar S, Golnaraghi AR, Pourrahim R (2002) Plant viruses of Iran (in English). Saman Co, Tehran

Fauquet, C. M. Briddon, R. W. Brown, J. K. Moriones, E. Stanley, J. Zerbini, M. Zhou, X. (2008) Geminivirus strain demarcation and nomenclature. Archives of Virology 153 (4):783–821

Fiallo-Olive E, Hamed A, Navas-Castillo J, Moriones E (2013) *Cotton leaf curl Gezira alphasatellite* associated with *Tomato leaf curl Sudan virus* approaches the expected upper size limit in the evolution of alphasatellites. Virus Res 178:506–510

Freitas-Astua J, Purcidfull DE, Polston JE, Hiebert E (2002) Traditional and transgenic strategies for controlling tomato-infecting begomovirus. Fitopatol Bras 27:437–449

Ghanem GAM, Al-Ajlan AM, Abdulsalam KS (2003) A whitefly-transmitted geminivirus infecting bean (*Phaseolus vulgaris* L.) plants in Saudi Arabia. Egypt J Phytopathol 31:1–15

Gunning RV, Bryne FJ, Conde BD, Connelly MI, Hergstrom K, Devonshire AL (1995) First report of *Bemisia tabaci* (Hemiptera: Aleyrodidae) in Australia. J Aust Entomol Soc 34:116

Hajimorad MR, Kheyr-Pour A, Alavi V, Ahoonmanesh A, Bahar M, Rezaian MA, Gronenborn B (1996) Identification of whitefly transmitted tomato yellow leaf curl geminivirus from Iran and a survey of its distribution with molecular probes. Plant Pathol 45:418–425

Hanley-Bowdoin L, Settlage SB, Orozco BM, Nagar S, Robertson D (1999) Geminviruses: models for plant DNA replication, transcription, and cell cycle regulation. Crit Rev Plant Sci 18:71–106

Harrison BD (1977) Ecology and control of viruses with soil-inhabiting vectors. Annu Rev Phytopathol 15:331–360

Helson GAH (1951) The transmission of whiches broom disease of lucerne by the common brown leafhoppers, *Orosius argentatus* (Evans). Aust J Sci Res B Biol Sci 4:115–124

Hill (1937) Yellow dwarf of tobacco in Australia, I. symptoms. J Council Sci Indus Res 10:228–230

Hill AV, Mandryk M (1954) A study of virus diseases "big bud" of tomato and "yellow dwarf" of tobacco. Aust J Agric Res 5:617–625

Hosseinzadeh M, Garivani M (2014) Emerging two distinct groups of the *Tomato yellow leaf curl virus*-severe strain (TYLCV-IL) variants in Iran. Trakia J Sci 12:149–161

Hur J, Kenneth JB, Lee S, Keith RD (2007) Transcriptional activator elements for *Curtovirus* C1 expression reside in the 3′ coding region of ORF C1. Mol Cells 23:80–87

Hussain M, Mansoor S, Iram S, Fatima AN, Zafar Y (2005) The nuclear shuttle protein of tomato leaf curl New Delhi virus is a pathogenicity determinant. J Virol 79:4434–4439

Idris AM, Brown JK (2005) Evidence for interspecific-recombination for three monopartite begomoviral genomes associated with the tomato leaf curl disease from Central Sudan. Arch Virol 150:1003–1012

Idris AM, Abdullah NM, Brown JK (2012) Leaf curl diseases of two Solanaceous species in Southwest Arabia are caused by a monopartite begomovirus evolutionarily most closely related to a species from the Nile Basin and unique suite of betasatellites. Virus Res 169:296–300

Idris AM, Al-Saleh Piatek MJ, Al-Shahwan I, Ali S, Brown JK (2014) Viral metagenomics: analysis of begomoviruses by illumina high throughput sequencing. Virus 6:1219–1236

Isnard M, Granier M, Frutos R, Reynaud B, Peterschmitt M (1998) Quasispecies nature of three maize streak virus isolates obtained through different modes of selection from a population used to assess response to infection of maize cultivars. J Gen Virol 79:3091–3099

Jing C, Wang C, Li K, Wu G, Sun X, Qing L (2016) Molecular identification of tobacco leaf curl disease in Sichuan province of China. Virol J 13:4

Kardani GS, Heydarnejad J, Zakiaghl M, Mehrvar M, Kraberger S, Varsani A (2013) Diversity of *Beet curly top Iran virus* isolated from different hosts in Iran. Virus Genes 46:571–575

Karthikeyan AS, Vanitharani R, Balaji V, Anuradha S, Thillaichidambaram P, Shivaprasad PV, Parameswari C, Balamani V, Saminathan M, Veluthambi K (2004) Analysis of an isolate of *Mungbean yellow mosaic virus* (MYMV) with a highly variable DNA B component. Arch Virol 149:1643–1652

Khan AJ, Idris AM, Al-Saady NA, Al-Mahruki MS, Al-Subhi AM, Brown JK (2008) A divergent isolate of *Tomato yellow leaf curl virus* from Oman with an associated DNA beta satellite: an evolutionary link between Asian and the middle eastern virus-satellite complexes. Virus Genes 36:169–176

Khan AJ, Akhtar S, Briddon RW, Ammara U, Al-Matrooshi AM, Mansoor S (2012) Complete nucleotide sequence of watermelon chlorotic stunt virus originating from Oman. Viruses 4:1169–1181

Khan AJ, Akhtar S, Al-Zaidi AM, Singh AK, Briddon RW (2013) Genetic diversity and distribution of a distinct strain of chili leaf curl virus and associated betasatellite infecting tomato and pepper in Oman. Virus Res 177:87–97

Khan AJ, Akhtar S, Singh AK, Al-Shehi AA, Al-Matrushi AM, Ammara U, Briddon RW (2014) Recent evolution of a novel begomovirus causing tomato leaf curl disease in the Al-Batinah region of Oman. Arch Virol 159:445–455

Kumari SG, Makkouk KM, Attar N (2006) An improved antiserum for sensitive serologic detection of *chickpea chlorotic dwarf virus*. J Phytopathol 154:129–133

Lefeuvre P, Martin DP, Harkins G, Lemey P, Gray AJA, Meredith S (2010) The spread of *Tomato Yellow Leaf Curl Virus* from the Middle East to the world. PLoS Pathog 6:1–12

Liu L, Saunders K, Thomas CL, Davies JW, Stanley J (1999) *Bean yellow dwarf virus* RepA, but not rep, binds to maize retinoblastoma protein, and the virus tolerates mutations in the consensus binding motif. Virology 256:270–279

Louro D, Noris E, Veratti F, Accotto GP (1996) First report of *Tomato yellow leaf curl virus* in Portugal. Plant Dis 80:1079

Lozano G, Trenado HP, Fiallo-Olivé E, Chirinos D, Geraud-Pouey F, Briddon RW, Navas-Castillo J (2016) Characterization of non-coding DNA satellites associated with Sweepoviruses (genus Begomovirus, Geminiviridae) – definition of a distinct class of Begomovirus-associated satellites. Front Microbiol 7:162

Luria SE, Darnell JEJ, Baltimore D, Cambell A (1978) General virology. Wiley, New York, pp 1–7

Mansoor S, Khan SH, Bashir A, Saeed M, Zafar Y, Malik KA, Briddon RW, Stanley J, Markham PG (1999) Identification of a novel circular single-stranded DNA associated with cotton leaf curl disease in Pakistan. Virology 259:190–199

Mansoor S, Amin I, Briddon RW (2008) Cotton leaf curl disease. In: Mahy BWJ, Van Regenmortel MHV (eds) Encyclopedia of virology, vol 5. Elsevier, Oxford, pp 563–569

Matthews REF (1979) Classification and nomenclature of viruses. Third report of the international committee on taxonomy of viruses. Intervirology 12:132–296

Mayo MA, Leibowitz MJ, Palukaitis P, Scholthof KBG, Simon AE, Stanley J, Taliansky M (2005) Satellites. In: Fauquet CM, Mayo MA, Maniloff J, Desselberger U, Ball LA (eds) VIIIth report of the international committee on taxonomy of viruses. Virus taxonomy. Elsevier/Academic Press, London, pp 1163–1169

Meng J, Li Z, Wei M (2012) Molecular identification of the causal agents causing tobacco leaf curl disease in some regions of Guangxi. Plant Prot 2:37–41

Monci F, Sanchez-Campos S, Navas-Castillo J, Moriones E (2002) A natural recombinant between the geminiviruses *Tomato yellow leaf curl Sardinia virus* and *Tomato yellow leaf curl virus* exhibits a novel pathogenic phenotype and is becoming prevalent in Spanish populations. Virology 303:317–326

Montasser MS, Al-Sharidah A, Ali NY, Nakhla MK, Farag BL, Maxwell DP (1999) A single DNA component of tomato yellow leaf curl geminivirus causing epidemics in the State of Kuwait. Kuwait J Sci Eng 25:127–142

Moriones E, Navas-Castillo J (2000) *Tomato yellow leaf curl virus*, an emerging virus complex causing epidemics worldwide. Virus Res 71:123–134

Moriones E, Arnó J, Accotto GP, Noris E, Cavallarin L (1993) First report of *Tomato yellow leaf curl virus* in Spain. Plant Dis 77:953

Mumford DL (1974) Purification of curly top virus. Phytopathology 64:136–142

Murphy FA, Fauquet CM, Bishop DHL, Ghabrial SA, Jarvis AW, Martelli GP, Mayo MA, Summers MD (1995) Virus taxonomy: sixth report of the international committee on taxonomy of viruses. Springer, New York, p 585

Nagar S, Pedersen TJ, Carrick KM, Hanley-Bowdoin L, Robertson D (1995) A geminivirus induces expression of a host DNA-synthesis protein in terminally differentiated plant-cells. Plant Cell 7:705–719

Nahid N, Amin I, Mansoor S, Rybicki EP, Van Der Walt E, Briddon RW (2008) Two dicot-infecting mastreviruses (family *Geminiviridae*) occur in Pakistan. Arch Virol 153:1441–1451

Padidam M, Beachy RN, Fauquet CM (1995) Classification and identification of geminiviruses using sequence comparisons. J Gen Virol 76:249–263

Pakniat A, Behjatnia SAA, Kharazmi S, Shahbazi M, Izadpanah K (2010) Molecular characterization and construction of an infectious clone of a new strain of *Tomato yellow leaf curl virus* in southern Iran. Iran J Plant Pathol 46:101–115

Panno S, Lacono G, Davino M, Marchione S, Zappardo V, Bella P, Tomassoli L, Accotto GP, Davino S (2016) First report of *Tomato leaf curl New Delhi virus* affecting zucchini squash in an important horticultural area of southern Italy. New Dis Rep 33:6

Pita JS, Fondong VN, Sangare A, Otim-Nape GW, Ogwal S, Fauquet CM (2001) Recombination, pseudorecombination and synergism of geminiviruses are determinant keys to the epidemic of severe cassava mosaic disease in Uganda. J Gen Virol 82:655–665

Polston JE, Mcgovern RJ, Brown LG (1999) Introduction of *Tomato yellow leaf curl virus* in Florida and implications for the spread of this and other geminiviruses of tomato. Plant Dis 83:984–988

Ramos PL, Guevara Gonzalez RG, Peral R, Ascencio-Ibañez JT, Polston J, Argüello-Astorga GR, Vega-Arreguín JC, Rivera-Bustamante RF (2003) *Tomato mottle Taino virus* pseudorecombines with PYMV but not with ToMoV: implications for the delimitation of cis- and trans-acting replication specificity determinants. Arch Virol 148:1697–1712

Roumagnac P, Granier M, Bernardo P, Deshoux M, Ferdinand R, Galzi S, Fernandez E, Julian C, Abt I, Filloux D, Mesleard F, Varsani A, Blanc S, Martin DP, Peterschmitt M (2015) *Alfalfa leaf curl virus*: an aphid-transmitted geminivirus. J Virol 89:9683–9688

Saunders K, Stanley J (1999) A nanovirus-like component associated with yellow vein disease of *Ageratum conyzoides*: evidence for interfamilial recombination between plant DNA viruses. Virology 264:142–152

Schwinghamer M, Thomas J, Schilg M, Parry J, Dann E, Moore K, Kumari S (2010) Mastreviruses in chickpea (*Cicer arietinum*) and other dicotyledonous crops and weeds in Queensland and northern New South Wales, Australia. Australas Plant Pathol 39:551–561

Seal SE, Jeger MJ, Van Den Bosch F, Maramorosch K, Shatkin AJ, Thresh JM (2006) Begomovirus evolution and disease management. Adv Virus Res 67:297–316

Shahbazi M, Behjatnia SAA, Alichi M, Bananej K, Izadpanah K (2010) Identification of *Bemisia tabaci* biotypes in Iran based on ITS1 region of ribosomal DNA and DNA polymorphism. In: Proceedings of the 19th Iranian Plant Protection Congress, Tehran, Iran, p 551

Shahid MS, Briddon RW, Al-Sadi AM (2017) Identification of *Mungbean yellow mosaic Indian virus* associated with tomato leaf curl Betasatellite infecting *Phaseolus vulgaris* in Oman. J Phytopathol 165:204–211

Sohrab SS (2016a) The role of Corchorus in spreading of *Tomato yellow leaf curl virus* on tomato in Jeddah, Saudi Arabia. Virusdisease 27:19–26

Sohrab SS, Yasir M, El-Kafrawy SA, Abbas AT, Mousa MAA, Bakhashwain AA (2016b) Association of *Tomato leaf curl Sudan virus* with leaf curl disease of tomato in Jeddah, Saudi Arabia. Virusdisease 19:1–9

Stanley J, Bisaro DM, Briddon RW, Brown JK, Fauquet CM, Harrison BD, Rybicki EP, Stenger DC (2005) Geminiviridae. In: Fauquet CM, Mayo MA, Maniloff J, Desselberger U, Ball LA (eds) Virus taxonomy, VIIIth report of the ICTV. Elsevier/Academic Press, London

Stonor J, Hart P, Gunther M, DeBarro P, Rezaian M (2003) *Tomato leaf curl geminivirus* in Australia: occurrence, detection, sequence diversity and host range. Plant Pathol 52:379–388

Talhouk AMS (1957) Diseases and insects pests of crops in the eastern province of Saudi Arabia. Arabian American Oil Company, Dammam, p 87

Thomas JE, Bowyer JW (1979) Properties of tobacco yellow dwarf and bean summer death viruses. Phytopathology 70:214–217

Thomas J, Parry J, Schwinghamer M, Dann E (2010) Two novel mastreviruses from chickpea (Cicer arietinum) in Australia. Arch Virol 155:1777–1788

Varma A, Malathi VG (2003) Emerging geminivirus problems: a serious threat to crop production. Ann Appl Biol 142:145–164

Varsani A, Roumagnac P, Fuchs M, Navas-Castillo J, Moriones E, Idris A, Briddon RW, Rivera-Bustamante R, Murilo ZF, Martin DP (2017) Capulavirus and Grablovirus: two new genera in the family Geminiviridae. Arch Virol 162:1819–1831

Xie Y, Wu P, Liu P, Gong H, Zhou X (2010) Characterization of alphasatellites associated with monopartite begomovirus/betasatellite complexes in Yunnan, China. Virol J 7:178

Xiong Q, Fan S, Wu J, Zhou X (2007) Ageratum yellow vein China virus is a distinct begomovirus species associated with a DNAβ molecule. Phytopathology 97:405–411

Yazdani-Khameneh S, Golnaraghi AR, Rakhshandehroo F (2013) Report of a new Begomovirus on melon in Iran. New Dis Rep 28:17

Zhou X, Liu Y, Calvert L, Munoz C, Otim-Nape GW, Robinson DJ, Harrison BD (1997) Evidence that DNA-A of a geminivirus associated with sever cassava mosaic disease in Uganda has arisen by interspecific recombination. J Gen Virol 78:2101–2111

Zhou XP, Xie Y, Tao XR, Zhang ZK, Li ZH, Fauquet CM (2003) Characterization of DNAβ associated with begomoviruses in China and evidence for co-evolution with their cognate viral DNA-A. J Gen Virol 84:237–247

Mastreviruses in the African World: Harbouring Both Monocot and Dicot Species

Avinash Marwal, Rakesh Kumar Verma, Megha Mishra, Rajesh Kumar, and R. K. Gaur

Abstract

Mastreviruses are the main causal viral agent of a variety of plant diseases in African continent posing a serious threat to economically important plant species, residing/harbouring chiefly in the uncultivated ones. *Mastreviruses* are known to have numerous diverse carrier vectors infecting either monocotyledons or dicotyledons plants. More than 1950 *Mastreviruses* sequences are publically available in nucleotide database of NCBI, majority (~850) of them are reported from Africa alone. All the known *Mastreviruses* encompass a monopartite genomic nature (ranges from 2.5 to 2.7 kb) encoding four genes. Reports and evidences suggest a strong intra- or inter-specific recombination among the identified *Mastreviruses*, but the extent to which it creates diversity is still a challenging task to understand and get a clear insight. Such diversity is also supported by gene acquisition and mutations (especially point mutations and small insertions or deletions). Current study focuses on the molecular diversity analysis and genomic characterization of the reported *Mastreviruses* from the African continent.

1 Introduction

Mastreviruses are a well-known genus of the family *Geminiviridae* known to infect a wide range of plant species in tropical and sub-tropical African continent, causing great economic crop losses from 30 to 100%. The *Geminiviridae* family has been

A. Marwal
Department of Biotechnology, Mohanlal Sukhadia University, Udaipur, Rajasthan, India

R. K. Verma · M. Mishra · R. Kumar · R. K. Gaur (✉)
Department of Biosciences, Faculty of Arts, Science and Commerce, Mody University, Lakshmangarh, Sikar, Rajasthan, India

R. V. Kumar (ed.), *Geminiviruses*, https://doi.org/10.1007/978-3-030-18248-9_5

clustered into nine diverse genera such as *Mastrevirus*, *Begomovirus*, *Topocuvirus*, *Curtovirus*, *Becurtovirus*, *Turncurtovirus*, *Eragrovirus*, *Capulavirus* and the recently joined member is the *Grablovirus*. The representing member of this genus (*Mastrevirus*) is the *Maize streak virus*, from where the name is derived is enclosed in a twin icosahedral symmetry, as supported by electron microscopy. All the known *Mastreviruses* sequences deposited in the GenBank from across the globe include a monopartite (Old World *Geminivirus*) circular single-stranded DNA molecule whose size ranges from ~2.5 to ~2.8 kb. Studies and facts recommend a strong intra or inter-specific recombination among the identified *Mastreviruses*, but the extent to which it creates diversity is still a challenging task to understand and to get a clear insight (Muhire et al. 2013; Prajapat et al. 2012; Varsani et al. 2008a; Shepherd et al. 2010).

In accordance with *Mastrevirus*, another genus *Begomovirus* leads to the vast devastating diseases of the dicot plants in the African continent; the major one is the cassava mosaic disease (CMD) reported in Cassava (*Manihot esculenta*). Cassava forms the leading and staple food for over 200 million sub-Saharan Africans known to be infected by *Cassava mosaic virus*. Cassava mosaic diseases are transmitted by *Bemisia tabaci* (Whiteflies), and these *Begomoviruses* are having bipartite genomic constitution, that is, DNA-A and DNA-B molecules accompanied by high rates of nucleotide substitution (Duffy and Holmes 2009; Bock and Woods 1983).

To date, a total of eight different and distinct species of this *Begomovirus* have been endemic to African continent causing epidemic outbreaks. These eight species are *East African cassava mosaic virus* (EACMV), *East African cassava mosaic Malawi virus* (EACMMV), *East African cassava mosaic Cameroon virus* (EACMCV), *South African cassava mosaic virus* (SACMV), *East African cassava mosaic Kenya virus* (EACMKV), *Cassava mosaic Madagascar virus* (CMMGV), *East African cassava mosaic Zanzibar virus* (EACMZV) and *African cassava mosaic Burkina Faso virus* (ACMBFV). Numerous strains of the *Cassava mosaic virus* result from inter- or intra-species recombination (Bruyn et al. 2012; Bull et al. 2006; Legg and Fauquet 2004; Tiendrébéogo et al. 2012).

In Africa, the best reliable and effectual technique employed to tackle against *Mastreviruses* is host plant resistance. Its magnitude resides in its enormous genetic diversity and flexibility to a broad assortment of African ecological zones. With periodic outbreaks, diseases caused by *Mastreviruses* are one of the most severe biotic constraints in the African continent (Bigarré et al. 1999; Schnippenkoetter et al. 2001; Halley-Stott et al. 2007; Harkins et al. 2009; Kraberger et al. 2012).

Ample information is available regarding geographical distribution, transmission, epidemiology and symptomatology of all known *Mastrevirus* diseases and their recombinant strains in Africa seeks requirement for better understanding in effective disease control and management. This form the basis of our study i.e. highlighting molecular diversity analysis and genomic characterization of the reported *Mastreviruses* from the African continent and nearby islands.

2 *Mastreviruses*: Distribution and Diversity

At present, there are a total 1950 *Mastreviruses* sequences publically available in GenBank, NCBI, as reported across the globe. Out of which nearly 845 sequences are reported from the African continent alone which are divided among 16 different *Mastreviruses* (https://www.ncbi.nlm.nih.gov/nuccore/?term=mastrevirus+africa). These 15 *Mastreviruses* are as follows: *Maize streak virus* (683/845), *Panicum streak virus* (40/845), *Chickpea chlorotic dwarf virus* (31/845), *Urochloa streak virus* (9/845), *Sugarcane streak virus* (4/845), *Sugarcane streak Reunion virus* (3/845), *Bean yellow dwarf virus* (2/845), *Eragrostis streak virus* (2/845), *Axonopus compressus streak virus* (2/845), *Eragrostis minor streak virus* (3/845), *Saccharum streak virus* (2/845) and *Sugarcane chlorotic streak virus* (4/845). All these *Mastreviruses* are called as African streak viruses, abbreviated as AfSV making it the widest known geographical range of viruses (Table 1).

Table 1 All the known and reported *Mastreviruses* from different region of African continent

S. no.	Infecting *Mastrevirus*	Reported African country
1	*Maize streak virus*	South Africa, Kenya, Nigeria, Cameroon, Burkina Faso, Zimbabwe, Madagascar, Lesotho, Mozambique, Uganda, Reunion, Burundi and Mauritius
2	*Panicum streak virus*	Kenya, Zimbabwe, Namibia, Mozambique, South Africa, Nigeria and Central African Republic
3	*Chickpea chlorotic dwarf virus*	Sudan, Eritrea, Burkina Faso, South Africa and Tunisia
4	*Urochloa streak virus*	Nigeria
5	*Sugarcane streak virus*	South Africa
6	*Bean yellow dwarf virus*	South Africa
7	*Eragrostis streak virus*	Zimbabwe
8	*Sugarcane streak Reunion virus*	Nigeria, Reunion and Zimbabwe
9	*Axonopus compressus streak virus*	Nigeria
10	*Maize streak Reunion virus*	Reunion and Nigeria
11	*Eragrostis minor streak virus*	Namibia
12	*Saccharum streak virus*	South Africa and Reunion
13	*Sugarcane chlorotic streak virus*	Nigeria
14	*Sugarcane white streak virus*	Sudan
15	*Sugarcane streak Egypt virus*	Egypt

2.1 *Maize Streak Virus*

From time immemorial, the first ever disease caused by *Maize streak virus* was reported by Fuller in 1901; it was later in 1925 that Storey proposed the name maize streak (Fuller 1901; Storey 1925). Leafhoppers are the major carrier vector of *Maize streak virus* requiring 6–12 h latent period for transmission in a persistent manner. Since the advent of genera *Mastrevirus* of the *Geminiviridae* family, numerous *Mastreviruses* has been identified, whereas the prime one is *Maize streak virus*, known very well and most characterized for its ability and pathogenicity to infect large assortment of the member of *Poaceae* family and the very first host of this virus was maize plants (*Zea mays*). *Maize streak virus* alone poses 11 isolates ranging from isolate A to isolate K naturally infecting crops and weed plants. All of them belong to the Old World, none to be residing in the New World, and it has been recommended a best way for assigning the classification of existing *Mastreviruses* and the newly discovered viruses by means of species demarcation tool (SDT) simply by reducing the value one from the Hamming distance. A threshold value of 78 has been set as per the tool; surprisingly, the strain A of *Mastrevirus* itself got divided into six sub-categories/sub-types (Power 2000; Thottappily et al. 1993; Prajapat et al. 2014; Sahu et al. 2015; Bigarré et al. 1999; Palmer and Rybicki 1998).

Maize streak virus is the representing member of *Mastreviruses* performing replication within the host cell nucleus through rolling circle and recombination dependent mechanisms; thus it encodes for four genes (which expressed/transcribed bidirectionally), responsible for the vitality of this virus by capturing the host plant cell machinery. Out of the four genes, two are residing on the virion-sense DNA strand [coat protein (CP) plus movement protein (MP)] and the rest two on the complementary DNA strand [replication-associated protein (REP) and a replication-associated protein-A (REP-A)]; further the genome is coupled with a short and a long noncoding intergenic region assisting in viral gene expression and replication (Monjane et al. 2011; Alegbejo et al. 2002; Das et al. 2011; Marwal et al. 2018c; Nehra et al. 2018; Willment et al. 2001, 2002; Marwal et al. 2016a; Thottappilly et al. 1993).

Mastreviruses have a diverse host choice (both monocotyledons and dicotyledons plants) from cultivated (crops and ornamentals) to uncultivated ones (weeds and grasses), which are transmitted among different host plants all the way through an insect vector leafhopper belonging to *Cicadulina* genus (Homoptera: family *Cicadellidae*). From five different genera, a total of around 14 different leafhopper species are known to be the principal carrier of *Mastreviruses*. As the name suggests, the symptoms caused by *Mastrevirus* are streak-like patterns on the infected plants and rendering plants to develop seeds or cobs. Based on their degree of pathogenicity, *Mastreviruses* are classified into highly pathogenic and milder pathogenic strains designated as MSV-A and MSV-B, respectively. To combat the crop losses, it is vital to have swift and accurate method for recognition of the principal pathogen, harshness of disease and mechanism of virulence. Such harmful *Mastreviruses* have been thoroughly characterized at the molecular, serological and in silico level

(Magenya et al. 2008; Liu et al. 1998; Varsani et al. 2009a, b; Wu et al. 2008; Marwal et al. 2017; Martin and Shepherd 2009).

Begomoviruses have two genomic components, namely DNA-A and DNA-B. The DNA-B is responsible for virus movement across the plant cells, mainly transmitted into phloem sieve tubes and concentrated into the mesophyll cell (Lucy et al. 1996). In connection to it, MSV has no B component (because Old World category), the requirements for virus movement are likely to be similar, and MSV V1, or V1 and V2, may share the functions of BR1 and BL1 of the B component gene. The coat protein gene (cp) of *Begomoviruses* is accountable for just coat protein; the counterpart MSV supports its envelope development as well as helps in systemic infection and also helps in the ability to bind to the single-stranded and double-stranded DNA (Liu et al. 1997b). Collectively, the *Begomoviruses* and *Maize streak virus* share the same mechanism for cell-to-cell movement of its infectious particles mediated via their encoded proteins and associated proteins. MSV has more influence of the Southern, Western and Eastern African fields (Martin et al. 1999; Thottappilly et al. 1993; Bosque-Perez et al. 1998; Kotlizky et al. 2000).

To understand the functions of various ORFs present in the *Mastrevirus* genome, the lone member of the dicot-infecting *Mastrevirus*, that is, *Bean yellow dwarf virus*, was considered for giving a better overview of ORFs functional similarities/ differences in monocot- and dicot-infecting *Mastreviruses*. To take up this challenge, individual ORFs were abridged by introducing mutations or via insertion of a stop codon. Mutant form of ORF V1 was allowed to grow in both protoplasts and the complete plants, revealing the confinement nature in the protoplast and no systemic contamination in tested plant, suggesting the ORF V1 role in virus movement. Similar results were also supporting the ORF V2 function for coat protein formation, ORF C1 for replication and ORF C2 for assisting in replication; finally, the scientific group concluded the conserved nature of the ORFs in both monoecious and dioecious plants (Liu et al. 1998; Zhan et al. 1993; Wright et al. 1997; Woolston et al. 1989; Boulton et al. 1989, 1993; Dickinson et al. 1996).

In a study, 12 MSVs were identified in maize plant and uncultivated grasses from the African continent, which were subjected to diversity analysis using the basic technique of PCR followed by restriction fragment length polymorphism (RFLP). A great diversity was found among the isolated viruses and strong recombination was too recovered (Martin et al. 2001a, b). Padidam and co-workers worked on a large number of *Geminivirus* sequences for detection of possible recombinants using GENECONV method among different genera of the family *Geminiviridae*, including *Begomoviruses*, *Curtoviruses* and *Mastreviruses* from across the globe, and remarkably identified 420 recombinants. The highest rate of such recombinations occurred among the *Mastreviruses* strains and isolates, robustly supporting interspecies recombination, which also confirms recombination among New Word (Americas) and Old World (Africa and Asia) *Geminiviruses* (Padidam et al. 1999; Shepherd et al. 2005; van der Walt et al. 2009).

Sub-Saharan African countries are more prone to *Geminiviruses* infections, especially *Mastreviruses*. For this, a novel virus species has been identified, that is, *Euphorbia caput-medusae latent virus* (EcmLV) from a grass species *Euphorbia caput-medusae* showed maximum (~72%) genomic similarity with *Mastreviruses*. The identified EcmLV consists of seven ORFs in comparison to four ORFs of *Mastreviruses*, out of which four reside in the sense strand and three in the complementary strand (Bernardo et al. 2013). The increasing occurrence of *Maize streak virus* demands efforts to study their diversity in order to anticipate and monitor outbreaks as well as to understand the evolutionary forces driving the emergence of novel strains (Bosque-Pérez 2000; Bernardo et al. 2013; Martin et al. 2001a, b).

2.2 *Panicum Streak Virus*

One of the first ever report of *Panicum streak virus* dated back to the year 1992. The virus was identified in Kenya infecting *Panicum maximum* plants. *Panicum maximum* is native to the African continent. The virus was also characterized to fulfil the Koch postulates by constructing the infectious clone, but the symptoms revealed were milder than the observed ones in the natural conditions. The *Panicum streak virus* was successfully transmitted among maize plants using the carrier vector *Cicadulina mbila*. *Mastreviruses* are known to code for four ORFs whereas *Panicum streak virus* encodes an extra fifth ORF (Briddon et al. 1992).

The host range of *Panicum streak virus* is not yet well recognized and later the same virus was rediscovered in 2001 in South Africa, making it indigenous to the African continent. Such viruses are a potent threat to food security, especially the chief ones like maize, barley and wheat crops. The genomic features of this virus are well matched with the rest of the *Mastreviruses*; the only difference lies in the nature of their stem loop arrangement and the present motifs. In contrast, the study employed by Schnippenkoetter and group publicized the negative results for Koch postulates, wherein the carrier vector failed to cause visible disease symptoms in the test plants (Schnippenkoetter et al. 2001).

An in-depth diversity study of *Panicum streak virus* with *Maize streak virus* described analogous outcomes. A year later in 2002, 23 more *Panicum streak virus* were identified across the entire African subcontinent by thorough survey especially from Southern, Central Africa and part (Varsani et al. 2008b; Oluwafemi et al. 2007).

2.3 *Chickpea Chlorotic Dwarf Virus*

Chickpea chlorotic dwarf virus (CpCDV) is a dicot-infecting *Mastrevirus* that was first identified in chickpea (*Cicer arietinum*) in India but has since been shown to occur across the Indian subcontinent, the Middle East, North Africa and the Arabian Peninsula (Muhire et al. 2013) with wide host range including chickpea, faba bean, lentil, sugar beet, French bean, cotton, *Sesbania bispinosa*, pepper and watermelon (Kraberger et al. 2015). According to ICTV report, there are six known species of dicot-infecting *Mastreviruses* (Muhire et al. 2013) in which one species CpCDV has been found only in the Middle East (including Turkey), Africa and India and the remaining five species (*Chickpea red leaf virus*, *Chickpea yellows virus*, *Chickpea chlorosis virus*, *Chickpea chlorosis Australia virus* and *Tobacco yellow dwarf virus*) are found in Australia (Kraberger et al. 2013).

Recent studies on diversity, phylogeny and distribution of CpCDV suggest that there are 12 strains (A–L) (Zaagueri et al. 2017). In yet another instance, the same *Chickpea chlorotic dwarf virus* was reported from Burkina Faso causing curling, reduced size and yellowing of leaves disease in papaya (*Carica papaya*) and tomato (*Solanum lycopersicum*) plants (Ouattara et al. 2017).

2.4 *Urochloa Streak Virus*

Nigeria again supported a novel *Mastrevirus* from its place known to be *Urochloa streak virus* isolated from wild grass *Urochloa deflexa*. The virus showed disease symptoms typical to that of *Panicum streak virus*, that is, white line streaks in the leaves. In regard to earlier finding that Mastreviruses bear single iteron responsible for ori recognition, whereas the *Urochloa streak virus* has an extra part for recognition. Moreover the recombinational software failed to recover any evolved recombinant for this virus both for the case of intra and inter species studies (Oluwafemi et al. 2008).

2.5 *Sugarcane Streak Virus*

For the first time, *Sugarcane streak virus* was identified from sugarcane crops at Regional Sugarcane research station of South Africa. While performing multiple sequences of cloned viruses, it was found that there were two transversions in the genome: each in the coat protein and replication-associated protein coding genes. Further nucleotide sequence identity ranges from 63 to 73% and the sequences were analysed through alignment software and phylogenetic tree tool, suggesting that the Sugarcane *streak virus* clades with the rest of the *Mastreviruses* of the African continent (Hughes et al. 1993).

2.6 *Bean Yellow Dwarf Virus*

Liu et al. collected symptomatic plants from Malelane in the Mpumalanga region of South Africa and reported a new *Mastrevirus*, that is, *Bean yellow dwarf virus* (BeYDV) isolated from French bean (*Phaseolus vulgaris* cv. Bonus) showing stunting, chlorosis and leaf curl symptoms. The scientific group observed that it has highest nucleotide sequence identity (65%) with *Tobacco yellow dwarf virus* from Australia, whereas the currently performed BLAST analysis of BeYDV nucleotide sequence (Y11023) showed maximum genome-wide identity with *Chickpea chlorotic dwarf virus* (99%) reported from Pakistan; now it might be due to the availability of more number of *Mastreviruses* sequences in GenBank (Liu et al. 1997a).

A variant of *Bean yellow dwarf virus* has been documented by Plant Protection Research Institute (PPRI), Pretoria, South Africa, infecting *Phaseolus vulgaris*, but the kind of symptoms it exhibited was of gentle type as compared to the severity of BeYDV; therefore, the new strain was termed as *Bean yellow dwarf virus*-mild (BeYDV-m) and has lesser extent of systemic spread of its genomic DNA, even confirmed through agro-inoculation studies where lower concentration of viral DNA was documented (Halley-Stott et al. 2007; Boulton 2000).

2.7 *Eragrostis Streak Virus*

In yet another remarkable study in Sothern African region and the La Reunion province, a new *Mastrevirus* strain was found in Zimbabwe and was characterized from the *Eragrostis curvula* plant. *Eragrostis curvula* is a monocotyledon plant belonging to the *Poaceae* family of wild grasses. Such grasses are mostly considered as weed throughout the continent, acting as an alternative host/reservoir of disease causing *Mastreviruses*. Based on the host plant it harbours, the virus was named as *Eragrostis streak virus*. Considering the sequence analysis, the identified virus shared less than 77% nucleotide identity with the known and deposited *Mastreviruses*.

As per the recombination analysis, the results supported the virus to be an emerged recombinant from the *Sugarcane streak virus* (SSV) and *Sugarcane streak Egypt virus* (SSEV). Both the parents are of different geographical origin, former reported from African continent and the latter from an Asian country. The scientific group (Shepherd et al. 2008) that identified *Eragrostis streak virus* strongly suggested that the Rep gene was under high recombination events. At last, the group with their results suggested that *Eragrostis streak virus* might be having a wider host range falling in both perennial and annual wild African grasses (Shepherd et al. 2008).

2.8 *Sugarcane Streak Reunion Virus, Axonopus compressus Streak Virus* and *Maize Streak Reunion Virus*

Axonopus compressus streak virus (ACSV) and *Maize streak Reunion virus* have been isolated from grass *Axonopus compressus* exhibiting streak symptoms in a number of Nigerian maize fields in 2007. It is scientifically dissimilar (<63% genome-wide sequence identity) to be considered a distinct virus. The nucleotide sequence of ACSV (KJ437671) was typical to *Mastreviruses* having all ORFs (MP, CP, RepA and Rep) with one exceptional feature, one more intron (total of two) in the rep gene. The same study was further extended in millet species *Eleusine coracana* from Reunion islands and a *Sugarcane streak Reunion virus* was identified and similar analysis were performed as in the case of *Axonopus compressus streak virus* (Oluwafemi et al. 2014; Peterschmitt et al. 1996).

2.9 *Eragrostis minor Streak Virus*

Eragrostis minor streak virus (EMSV) was discovered in Namibia, Southern Africa, in 2009 from grass *Eragrostis minor* showing typical symptoms of *Mastreviruses*. *Eragrostis minor* plant was sampled in the Caprivi region of Namibia exhibiting chlorotic discontinuous streaks running along the major leaf veins and identified a new *Mastrevirus* EMSV. Sequence alignment analysis of the virus genome showed <75% sequence identity (*Miscanthus streak virus*) with other known *Mastreviruses*. Further when Rep protein was checked for similarity, it revealed less than 65% identity with *Miscanthus streak virus* reported from Japan and 47 to 51% amino acid identity with the rest of the reported *Mastreviruses*.

2.10 *Saccharum Streak Virus*

In 2008, Lawry et al. (2009) screened the sugarcane fields for evidence of *Mastrevirus* infections in the KwaZulu-Natal province of South Africa and recognized a new *Mastrevirus*, that is, *Saccharum streak virus* (SacSV) which shares less than 66% identity with any other *Mastrevirus*, but is most closely related to *Urochloa streak virus* (USV). The genome sizes of SacSV were 2744 bp and have all typical ORFs of *Mastreviruses* and its associated conserved inverted repeat sequences. The group has also identified four binding motifs along the sequenced genome of *Saccharum streak virus*. The research group was unable to detect any evidence of inter-species recombination in SacSV genome (Lawry et al. 2009).

2.11 *Sugarcane Chlorotic Streak Virus*

Recently, a novel virus has joined the *Mastrevirus* group from Nigeria, recovered from sugarcane plants sampled from seven fields in the country. The symptomatic nature was severe chlorotic streaks and exhibited short statured host plants. The identified virus was named as *Sugarcane chlorotic streak virus* and derived from the recombination between two different *Mastreviruses*, that is, *Eragrostis streak virus* and *Urochloa streak virus* sharing 61–67% similarity with known and reported *Mastreviruses* (Yahaya et al. 2016).

2.12 *Sugarcane Streak Egypt Virus* and *Sugarcane White Streak Virus*

Mastreviruses are not only characterized from natural infection for diversity analysis but even they are subjected to analysis by performing quarantine trials. People have used polymerase chain reactions (PCR), next generation sequencing (NGS) and virion-associated nucleic acids (VANA) methods to identify potential harmful pathogens crossing international boundaries. Such studies were conducted on sugarcane plants sent for international trade (export/import) from Egypt and were positive for two new *Mastrevirus* strains: given the name *Sugarcane streak Egypt Virus* and *Sugarcane white streak Virus*. These *Mastreviruses* were negative for normal experiments employed in quarantine strategies and even detected in Sudan cultivated sugarcane plants. The authors strongly suggest that such practices should be included in general routine quarantine stations (Candresse et al. 2014; Bigarré et al. 1999).

Such diversity and demography of *Mastreviruses* in different regions of Africa, infecting cultivated and uncultivated host plants, pose a greater pressure on the epidemiology; most of them are concentrated in the southern part of the continent. Out of them *Maize streak virus* has maximum reported cases of viral diseases but shows less inter-strain recombination (Fig. 1). These are further supported by making laboratory chimaeric virus of two different *Maize streak virus* (A and B)

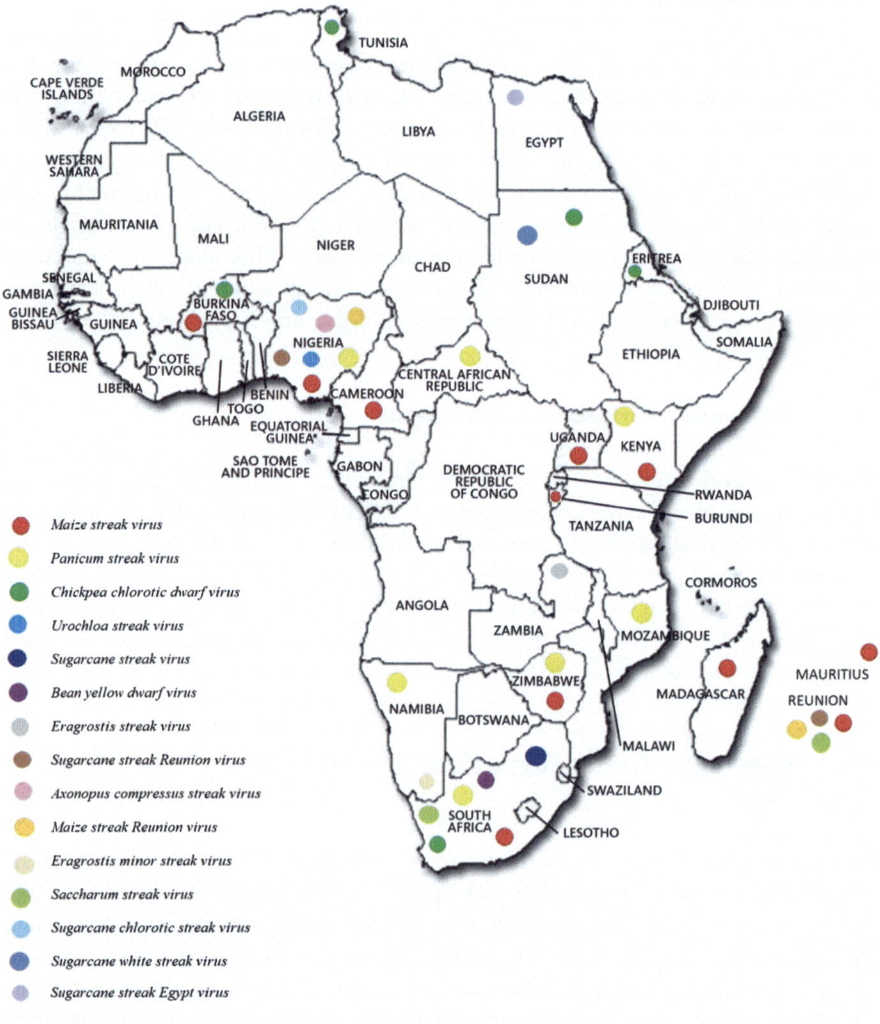

Fig. 1 Distribution diversity map of *Mastreviruses* in African continent highlighted with different colours

by reverting the movement and coat protein genes, but when left for replication in a suitable host the virus recombines to form the originally similar sequence (van der Walt et al. 2009; van Antwerpen et al. 2008; Dabrowski 1987; Damsteegt 1983; Padidam et al. 1999).

3 *Mastreviruses* Management: Ray of Hope

Many conventional tactics of virus control emphasize vector management by pesticides, activating natural predators or the use of physical barriers (Legg et al. 2014). Other methods like weed management, virus free planting material and removal of infected plants have also been implemented for disease control (Loebenstein and Katis 2014). The most effective approach of improving the plant cellular immunity against viruses is utilizing genetic resistance (Whitham and Hajimorad 2016). Numerous mechanisms have been developed and introduced artificially in plants to successfully determine engineered virus resistance (Sahu and Prasad 2015; Konate and Traore 1992).

3.1 RNAi-Mediated Resistance

RNA silencing is a technique which regulates the expression of many genes in a sequence-specific manner and acts as natural antiviral system to provide immunity by using small interfering RNA (siRNA) of 21–23 nt (Zvereva and Pooggin 2012). Main tools involved in the RNA silencing mechanism include the RNA-dependent RNA polymerase (RDR), argonaute (AGO) and ribonuclease Dicer (Bisaro 2006). In this process, double-stranded short interfering RNA (siRNA) induces the post-transcriptional degradation of homologous transcripts. Ribonuclease Dicer cleaves the long ds RNA and produces 21–23 nt long siRNA (Saxena et al. 2013). The siRNA then combines with various proteins having endoribonuclease activity and forms RNA induced silencing complex (RISC). Activated RISC then reaches to mRNA and binds with its complementary sequence and causes cleavage of targeted mRNA, and induces gene silencing (Vanitharani et al. 2004).

First report of siRNA generated by geminivirus infection came from investigation of siRNA extracts from tomato plant infected by *Tomato yellow leaf curl virus* (TYLCV) (Lucioli et al. 2003). Up to 99% reduction of Rep transcripts and 66% decline in viral DNA were observed in *African cassava mosaic virus* (ACMV) by using RNAi technology. RNAi technique was successfully used to silence the AC2 protein of *Mungbean yellow mosaic India virus* (Marwal et al. 2016b; Marwal and Gaur 2017).

3.2 Crispr/Cas9-Based Resistance Against DNA Viruses

Genome engineering has emerged as an important tool to improve many organisms, including crop plants, by introducing numerous traits of interest via site-specific modification of the genome by using site-specific nucleases. There are four major classes of SSNs: meganucleases, zinc finger nucleases (ZFNs), transcription activator-like effector nucleases (TALENs) and clustered regularly interspaced palindromic repeats/CRISPR-associated 9. CRISPR/Cas9 is a two-component system which entails Cas9 nuclease and a virus-based guide RNA (gRNA) that targets the specific site of DNA (Hsu et al. 2014).

CRISPR/Cas9 system allows manipulating any genomic sequence by using a short stretch of guide RNA. For sequence-specific silencing and cleavage of pathogenic foreign DNA by Cas proteins, CRISPR systems depend on CRISPR RNA (crRNA) and transactivating crRNAs (tracrRNA) (Gaur et al. 2018). This system is based on the type II CRISPR/Cas immune system in bacteria that protects the bacteria from invading DNA viruses. The crRNAs, in turn, anneal to transactivating crRNAs (tracrRNA) and create guide RNAs (gRNA) that direct sequence-specific cleavage and silencing of foreign invading DNA through Cas proteins (Jinek et al. 2012).

The CRISPR/Cas system uses short fragments of foreign DNA spacers that incorporate into the CRIPSR loci and are successively processed into CRISPR RNAs (crRNAs) by the transcription. The crRNAs then combine with transactivating crRNAs (tracrRNA) and generate guide RNAs (gRNA) that causes sequence-specific cleavage and silencing of pathogenic gene through Cas proteins. CRISPR/Cas9 system has been effectively used in controlling various geminiviruses 87% reduction in targeted viral load was reported in *N. benthamiana* by using various sgRNA from *Bean yellow dwarf virus*. In *N. benthamiana* plant sgRNA construct aiming stem loop structure in IR showed better resistance against tomato leaf curl virus (Baltes et al. 2015; Marwal et al. 2018a; Ali et al. 2015).

4 Conclusion

Management for virus infecting crop plants is attaining magnitude with the increased spread of viruses and threats of their epidemics (Gilbertson et al. 2011; Khurana and Marwal 2016). The extent of evolution of *Mastreviruses* is predominantly quite rapid due to highest rate of recombination in this genus, forming newer strains by overpowering the host genes viable for confrontation/resistance. Therefore, there is a demand for an improved management of viruses employed by a series of strategies; in fact such practices relied on the ecology of the virus. Many approaches have been used to decrease crop losses due to geminiviruses; only a few are effective in their management (Ausher 1997; Marwal et al. 2018b; Hilje and Stansly 2008).

This manuscript describes the geographical distribution, transmission, epidemiology, greater diversity and recombination among the known African *Mastreviruses*, suggesting that the plants are under serious threat to such pathogens during cropping season. From all the above-mentioned *Mastreviruses*, the most and the prominently studied is the *Maize streak virus* (MSV) of Africa due to its overwhelming impact on *Zea mays* crop cultivation (Varsani et al. 2008a, b; Manzoor et al. 2013), whereas the weeds and grasses species serve as a reservoir alternate host of these viruses during the off season, as supported by recent studies on *Sugarcane streak mastrevirus* in which sugarcane crops act as a reservoir host of cereal-infecting *Mastreviruses* residing in the northern Guinea savannah region of Nigeria (Yahaya et al. 2016; Prajapat et al. 2013; Isnard et al. 1998; Nehra et al. 2016; Kyetere et al. 1999).

Previous studies suggest that *Mastreviruses* show high mutation rates than expected and are thus skilled in quick host adaptation through recombination (Lawry et al. 2009; Prajapat et al. 2011; Pinner and Markham 1990). It is strictly noteworthy that the *Mastreviruses* established even in small region show recombination and diversity throughout the rest of the African continent, despite differences in their distribution pattern which might imitate unfairness during sample collection or due to geographical barricade in the course of their movement around the African continent (Nehra et al. 2014; Sahu et al. 2014; Nehra et al. 2019; Oluwafemi et al. 2008).

Expansion of agriculture in the African continent has also resulted in the emergence and spread of numerous diseases and insect pests. It is possible that suppression of leafhopper populations, either via biological control or with other natural or traditional methods, may help reduce the spread of the *Mastreviruses* suppressing their diversity and epidemics. Identifying *Mastreviruses* is becoming more difficult with globalization of trade and it will be to our benefit to investigate further the devastating nature of *Mastreviruses*, as this would constitute a novel epidemiological adaptation for *Mastreviruses* having the capability to produce new virus strains definitely influencing agricultural practices of poor farmers. Hence timely identification or detection of *Mastreviruses* for implementation of quarantine policies for crop protection is a prerequisite.

Acknowledgements The authors are thankful to Science and Engineering Research Board—Department of Science and Technology, New Delhi, India for the financial assistance (File No. YSS/2015/000265 and EMR/2016/000579).

References

Alegbejo MD, Olojede SO, Kashina BD, Abo ME (2002) Maize streak mastrevirus in Africa: distribution, transmission, epidemiology, economic significance and management strategies. J Sustain Agric 19(4):35–45

Ali Z, Abulfaraj A, Idris A, Ali S et al (2015) CRISPR/Cas9-mediated viral interference in plants. Genome Biol 16:238

Ausher R (1997) Implementation of integrated pest management in Israel. Phytoparasitica 25:119–141

Baltes NJ, Hummel AW, Konecna E, Cegan R et al (2015) Conferring resistance to *Geminiviruses* with the CRISPR–Cas prokaryotic immune system. Nat Plants 1:15145

Bernardo P, Golden M, Akram M, Naimuddin et al (2013) Identification and characterisation of a highly divergent *Geminivirus*: evolutionary and taxonomic implications. Virus Res 177:35–45

Bigarré L, Salah M, Granier M, Frutos R et al (1999) Nucleotide sequence evidence for three distinct sugarcane streak *Mastreviruses*. Arch Virol 144:2331–2344
Bisaro DM (2006) Silencing suppression by *Geminivirus* proteins. Virology 344:158–168
Bock K, Woods R (1983) Etiology of African cassava mosaic disease. Plant Dis 67:994–995
Bosque-Pérez NA (2000) Eight decades of *Maize streak virus* research. Virus Res 71:107–121
Bosque-Perez NA, Olojede SO, Buddenhagen IW (1998) Effect of *Maize streak virus* disease on the growth and yield of maize as influenced by varietal resistance levels and plant stage at time of challenge. Euphytica 101:307–317
Boulton M (2000) Agrobacterium-mediated transfer of *Geminiviruses* to plant tissues. In: Jones H (ed) Methods in molecular biology: plant gene transfer and expression protocols. Humana Press, Totowa, NJ, p 77
Boulton MI, Steinkellner H, Donson J, Markham PG et al (1989) Mutational analysis of the virion-sense genes of *Maize streak virus*. J Gen Virol 70:2309–2323
Boulton MI, Pallaghy CK, Chatani M, MacFarlane S et al (1993) Replication of *Maize streak virus* mutants in maize protoplasts: evidence for a movement protein. Virology 192:85–93
Briddon RW, Lunness P, Chamberlin LCL, Pinner MS et al (1992) The nucleotide sequence of an infectious insect-transmissible clone of the *Geminivirus Panicum streak virus*. J Gen Virol 73:1041–1047
Bruyn A, Villemot J, Lefeuvre P, Villar E, Hoareau M, Harimalala M, Abdoul-Karime AL, Abdou-Chakour C, Reynaud B, Harkins GW, Varsani A, Martin DP, Lett JM (2012) East African cassava mosaic-like viruses from Africa to Indian Ocean islands: molecular diversity, evolutionary history and geographical dissemination of a bipartite *Begomovirus*. BMC Evol Biol 12:228
Bull SE, Briddon RW, Sserubombwe WS, Ngugi K, Markham PG, Stanley J (2006) Genetic diversity and phylogeography of cassava mosaic viruses in Kenya. J Gen Virol 87:3053–3065
Candresse T, Filloux D, Muhire B, Julian C, Galzi S, Fort G, Bernardo P, Daugrois JH, Fernandez E, Martin DP, Varsani A, Roumagnac P (2014) Appearances can be deceptive: revealing a hidden viral infection with deep sequencing in a plant quarantine context. PLoS One 9(7):e102945
Dabrowski ZT (1987) *Cicadulina ghaurii* (Hem., Euscelidae): distribution, biology and *Maize streak virus* (MSV) transmission. J Appl Entomol 103:489–496
Damsteegt VD (1983) *Maize streak virus*: I. Host range and vulnerability of maize germplasm. Plant Dis 67:734–737
Das S, Marwal A, Choudhary DK, Gupta VK, Gaur RK (2011) Mechanism of RNA interference (RNAi): current concept. In: International Proceedings of Chemical, Biological and Environmental Engineering. International Conference on Food Engineering and Biotechnology, vol 9, pp 240–245
Dickinson VJ, Halder J, Woolston CJ (1996) The product of *Maize streak virus* ORF V1 is associated with secondary plasmodesmata and is detected with the onset of viral lesions. Virology 220:51–59
Duffy S, Holmes EC (2009) Validation of high rates of nucleotide substitution in geminiviruses: phylogenetic evidence from East African cassava mosaic viruses. J Gen Virol 90:1539–1547
Fuller C (1901) Mealie variegation. In: 1st report of the government entomologist, Natal, 1899–1900. P. Davis & Sons, Government Printers, Pietermaritzburg
Gaur RK, Verma RK, Khurana SMP (2018) Genetic engineering of horticultural crops: present and future. In: Rout GR, Peter KV (eds) Genetic engineering of horticultural crops. Academic, San Diego, CA, pp 23–46
Gilbertson RL, Rojas M, Natwick E (2011) Development of integrated pest management (IPM) strategies for whitefly (*Bemisia tabaci*) transmissible geminiviruses. Chapter 12. In: Thompson WMO (ed) The whitefly, *Bemisia tabaci* (Homoptera: Aleyrodidae). Interaction with geminivirus-infected host plants. Springer, Dordrecht, pp 323–356. https://doi.org/10.1007/978-94-007-1524-0_12

Halley-Stott RP, Tanzer F, Martin DP, Rybicki EP (2007) The complete nucleotide sequence of a mild strain of *Bean yellow dwarf virus*. Arch Virol 152:1237–1240

Harkins GW, Martin DP, Duffy S, Monjane AL et al (2009) Dating the origins of the maize-adapted strain of *Maize streak virus*, MSV-A. J Gen Virol 90:3066–3074

Hilje L, Stansly PA (2008) Living ground covers for management of *Bemisia tabaci* (Gennadius) (Homoptera: *Aleyrodidae*) and *Tomato yellow mottle virus* (ToYMoV) in Costa Rica. Crop Prot 27:10–16

Hsu PD, Lander ES, Zhang F (2014) Development and applications of CRISPR-Cas9 for genome engineering. Cell 157:1262–1278

Hughes FL, Rybicki EP, Kirby R et al (1993) Complete nucleotide sequence of sugarcane streak monogeminivirus. Arch Virol 132:171–182

Isnard M, Granter M, Frutos R, Reynaud B, Peterschmitt M (1998) Quasi-species nature of the *Maize streak virus* isolates obtained from a population used to assess maize cultivar response to infection. J Gen Virol 79:3091–3099

Jinek M et al (2012) A programmable dual-RNA-guided DNA endonuclease in adaptive bacterial immunity. Science 337:816–821

Khurana SMP, Marwal A (2016) Recent developments towards detection & diagnosis for management of plant viruses. Indian Phytopathol 69(4s):30–34

Konate G, Traore O (1992) Reservoir hosts of *Maize streak virus* (MSV) in the Sudan–Sahel zone: identification and spatio-temporal distribution. Phytoprotection 73:111–117

Kotlizky G, Boulton MI, Pitaksutheepong C, Davies JW et al (2000) Intracellular and intercellular movement of maize streak *Geminivirus* V1 and V2 proteins transiently expressed as green fluorescent protein fusions. Virology 274:32–38

Kraberger S, Thomas JE, Geering AD et al (2012) Australian monocot-infecting *Mastrevirus* diversity rivals that in Africa. Virus Res 169:127–136

Kraberger S, Harkins GW, Kumari SG, Thomas JE et al (2013) Evidence that dicot-infecting *Mastreviruses* are particularly prone to inter-species recombination and have likely been circulating in Australia for longer than in Africa and the Middle East. Virology 444:282–291

Kraberger S, Kumari SG, Hamed AA, Gronenborn B et al (2015) Molecular diversity of chickpea chlorotic dwarf virus in Sudan: high rates of intraspecies recombination a driving force in the emergence of new strains. Infect Genet Evol 29:203–215

Kyetere DT, Ming R, McMullen MD, Pratt RC, Brewbaker J, Musket T (1999) Genetic analysis of tolerance to *Maize streak virus* in maize. Genome 42:20–26

Lawry R, Martin DP, Shepherd DN, Antwerpen TV et al (2009) A novel sugarcane-infecting *Mastrevirus* from South Africa. Arch Virol 154:1699–1703

Legg JP, Fauquet CM (2004) Cassava mosaic *Geminiviruses* in Africa. Plant Mol Biol 56:585–599

Legg JP, Shirima R, Tajebe LS, Guastella D et al (2014) Biology and management of Bemisia whitefly vectors of cassava virus pandemics in Africa. Pest Manag Sci 70:1446–1453

Liu L, van Tonder T, Pietersen G, Davies JW et al (1997a) Molecular characterization of a subgroup I *Geminivirus* from a legume in South Africa. J Gen Virol 78:2113–2117

Liu H, Boulton MI, Davies JW (1997b) *Maize streak virus* coat protein binds single- and double-stranded DNA in vitro. J Gen Virol 78:1265–1270

Liu L, Davies JW, Stanley J (1998) Mutational analysis of *Bean yellow dwarf virus*, a *Geminivirus* of the genus *Mastrevirus* that is adapted to dicotyledonous plants. J Gen Virol 79:2265–2274

Loebenstein G, Katis N (2014) Control of plant virus diseases seed-propagated crops. In: Gad L, Nikolaos K (eds) Advance virus research. Academic, Cambridge, MA, p 11

Lucioli A, Noris E, Brunetti A, Tavazza R et al (2003) *Tomato yellow leaf curl sardinia virus* Rep-derived resistance to homologous and heterologous *Geminiviruses* occurs by different mechanisms and is overcome if virus-mediated transgene silencing is activated. J Virol 77 (12):6785–6798

Lucy AP, Boulton MI, Davies JW, Maule AJ (1996) Tissue specificity of *Zea mays* infection by *Maize streak virus*. Mol Plant-Microbe Interact 9:22–31

Magenya OEV, Mueke J, Omwega C (2008) Significance and transmission of *Maize streak virus* disease in Africa and options for management: a review. Afr J Biotechnol 7(25):4897–4910

Manzoor MT, Ilyas M, Shafiq M, Haider MS et al (2013) A distinct strain of *Chickpea chlorotic dwarf virus* (genus *Mastrevirus*, family *Geminiviridae*) identified in cotton plants affected by leaf curl disease. Arch Virol 159(5):1217–1221

Martin DP, Shepherd DN (2009) The epidemiology, economic impact and control of maize streak disease. Food Secur 1:305–315

Martin DP, Willment JA, Rybicki EP (1999) Evaluation of *Maize streak virus* pathogenicity in differentially resistant *Zea mays* genotypes. Phytopathology 89(8):695–700

Martin DP, Willment JA, Billharz R, Velders R et al (2001a) Sequence diversity and virulence in *Zea mays* of *Maize streak virus* isolates. Virology 288:247–255

Martin DP, Linderme D, Lefeuvre P, Varsani A (2001b) *Eragrostis minor streak virus*: an Asian streak virus in Africa. Arch Virol 156:1299

Marwal A, Gaur RK (2017) Understanding functional genomics of PTGS silencing mechanisms for *Tobacco streak virus* and other ilarviruses mediated by RNAi and VIGS. Chapter 24. In: Singh DP (ed) Plant-microbe interactions in agro-ecological perspectives. Volume 1: fundamental mechanisms, methods and functions. Springer, Singapore. ISBN: 978-981-10-5812-7

Marwal A, Prajapat R, Gaur RK (2016a) First report of recombination analysis of betasatellite and aplhasatellite sequence isolated from an ornamental plant marigold in India: an in silico approach. Int J Virol 12:10–17

Marwal A, Gaur RK, Khurana SMP (2016b) RNAi mediated gene silencing against plant viruses. Chapter 11. In: Chowdappa P, Sharma P, Singh D, Misra AK (eds) Perspectives of plant pathology in genomic era. Today and Tomorrow's Printers and Publishers, New Delhi, pp 235–254. ISBN 10: 8170195268 ISBN 13: 9788170195269

Marwal A, Mishra M, Sekhsaria C, Gaur RK (2017) Computational analysis and predicting ligand binding site in the *Rose leaf curl virus* and its betasatellite proteins: a step forward for antiviral agent designing. In: Saxena S, Tiwari A (eds) *Begomoviruses*: occurrence and management in Asia and Africa. Springer, Singapore, pp 157–168

Marwal A, Gaur RK, Khurana SMP (2018a) Possible approaches for developing different strategies to prevent transmission of *Geminiviruses* to important crops. Chapter 18. In: Gaur RK, Khurana SMP, Dorokhov Y (eds) Plant viruses: diversity, interaction and management. CRC Press, Boca Raton, FL. ISBN: 9781138061514

Marwal A, Verma RK, Khurana SMP, Gaur RK (2018b) Molecular interactions between plant viruses and their biological vectors. Chapter 12. In: Gaur RK, Khurana SMP, Dorokhov Y (eds) Plant viruses: diversity, interaction and management. CRC Press, Boca Raton, FL. ISBN: 9781138061514

Marwal A, Kumar R, Khurana SMP, Gaur RK (2018c) Complete nucleotide sequence of a new geminivirus isolated from *Vitis vinifera* in India: a symptomless host of Grapevine red blotch virus. Virus Dis. https://doi.org/10.1007/s13337-018-0477-x

Monjane AL, Harkins GW, Martin DP, Lemey P et al (2011) Reconstructing the history of *Maize streak virus* strain dispersal to reveal diversification hot spots and its origin in Southern Africa. J Virol 85:9623–9636

Muhire B, Martin DP, Brown JK, Navas-Castillo J et al (2013) A genome-wide pairwise-identity-based proposal for the classification of viruses in the genus *Mastrevirus* (family *Geminiviridae*). Arch Virol 158:1411–1424

Nehra C, Marwal A, Gaur RK (2016) Diversity and phylogeny of begomovirus populations and their managements. Acta Microbiol Bulg 32(2):108–113

Nehra C, Marwal A, Verma RK, Gaur RK (2018) Molecular characterization of begomoviruses DNA-A and associated beta satellites with new host *Ocimum sanctum* in India. Proc Natl Acad Sci India Sect B Biol Sci. https://doi.org/10.1007/s40011-018-1006-9

Nehra C, Sahu AK, Marwal A, Gaur RK (2014) Natural occurrence of Clerodendron yellow mosaic virus on Bougainvillea in India. New Dis Rep BSPP 30:19

Nehra C, Verma RK, Mishra M, Marwal A, Sharma P, Gaur RK (2019) Papaya yellow leaf curl virus: a newly identified begomovirus infecting *Carica papaya* from the Indian Subcontinent. J Hortic Sci Biotechnol. https://doi.org/10.1080/14620316.2019.1570827

Oluwafemi S, Jackai LE, Alegbejo MD (2007) Comparison of transmission abilities of four Cicadulina species vectors of maize streak virus from Nigeria. Ent Exp Et App 124:235–239

Oluwafemi S, Varsani A, Monjane AL, Shepherd DN et al (2008) A new *African streak virus* species from Nigeria. Arch Virol 153:1407–1410

Oluwafemi S, Kraberger S, Shepherd DN, Martin DP et al (2014) A high degree of *African streak virus* diversity within Nigerian maize fields includes a new *Mastrevirus* from *Axonopus compressus*. Arch Virol 159(10):2765–2770

Ouattara A, Tiendrébéogo F, Lefeuvre P, Hoareau M et al (2017) New strains of *Chickpea chlorotic dwarf virus* discovered on diseased papaya and tomato plants in Burkina Faso. Arch Virol 162:1791–1794

Padidam M, Sawyer S, Fauquet CM (1999) Possible emergence of new *Geminiviruses* by frequent recombination. Virology 265:218–225

Palmer KE, Rybicki EP (1998) The molecular biology of *Mastreviruses*. Adv Virus Res 50.183–234

Peterschmitt M, Granier M, Frutos R, Reynaud B (1996) Infectivity and complete nucleotide sequence of the genome of a genetically distinct strain of *Maize streak virus* from Réunion Island. Arch Virol 141:1637–1650

Pinner MS, Markham PG (1990) Serotyping and strain identification of *Maize streak virus* isolates. J Gen Virol 71:1635–1640

Power AG (2000) Insect transmission of plant viruses: a constraint on virus variability. Curr Opin Plant Biol 3:336–340

Prajapat R, Marwal A, Bajpai V, Gaur RK (2011) Genomics and proteomics characterization of alphasatellite in weed associated with begomovirus. Int J Plant Pathol 2(1):1–14

Prajapat R, Marwal A, Shaikh Z, Gaur RK (2012) *Geminivirus* database (GVDB): first database of family *Geminiviridae* and its genera *Begomovirus*. Pak J Biol Sci 15(14):702–706

Prajapat R, Marwal A, Gaur RK (2013) *Begomovirus* associated with alternative host weeds: a critical appraisal. Arch Phytopathol Plant Protect 47(2):157–170

Prajapat R, Marwal A, Gaur RK (2014) Recognition of errors in the refinement and validation of threedimensional structures of AC1 proteins of Begomovirus strains by using ProSA-web. J Viruses. https://doi.org/10.1155/2014/752656

Sahu AK, Marwal A, Nehra C, Shahid MS, Gaur RK (2014) First report of a begomovirus and associated betasatellite in Rosa indica and in India. Aust Plant Dis Notes 9(1):147

Sahu AK, Nehra C, Marwal A, Gaur RK (2015) First report of a begomovirus associated with betasatellites infecting new host spinach (*Spinacia oleracea*) in India. J Gen Plant Pathol 81 (2):146–150

Sahu PP, Prasad M (2015) Application of molecular antiviral compounds: novel approach for durable resistance against *Geminiviruses*. Mol Biol Rep 42:1157–1162

Saxena S, Kesharwani RK, Singh V, Singh S (2013) Designing of putative SiRNA against geminiviral suppressors of RNAi to develop *Geminivirus*-resistant papaya crop. Int J Bioinforma Res Appl 9(1):3

Schnippenkoetter WH, Martin DP, Hughes FL, Fyvie M et al (2001) The relative infectivities and genomic characterisation of three distinct *Mastreviruses* from South Africa. Arch Virol 146:1075–1088

Shepherd DN, Martin DP, McGivern DR, Boulton MI et al (2005) A three-nucleotide mutation altering the *Maize streak virus* Rep pRBR-interaction motif reduces symptom severity in maize and partially reverts at high frequency without restoring pRBR-Rep binding. J Gen Virol 86:803–813

Shepherd DN, Varsani A, Windram OP, Lefeuvre P et al (2008) Novel sugarcane streak and *Sugarcane streak Reunion Mastreviruses* from southern Africa and La Réunion. Arch Virol 153:605–609

Shepherd DN, Martin DP, Walt EV, Dent K et al (2010) *Maize streak virus*: an old and complex 'emerging' pathogen. Mol Plant Pathol 11(1):1–12

Storey HH (1925) The transmission of streak virus of maize by the leafhopper *Balclutha mbila* Naude. Ann Appl Biol 12:422–439

Thottappilly G, Bosque-Perez NA, Rossel HW et al (1993) Viruses and virus diseases of maize in tropical Africa. Ptant Pathol 42:494–509

Thottappily G, Bosque-Perez NA, Rossel HW (1993) Viruses and virus diseases of maize in tropical Africa. Plant Pathol 42:494–509

Tiendrébéogo F, Lefeuvre P, Hoareau M, Harimalala MA, De Bruyn A, Villemot J, Traoré VSE, Konaté G, Traoré AS, Barro N, Reynaud B, Traoré O, Lett JM (2012) Evolution of *African cassava mosaic virus* by recombination between bipartite and monopartite *Begomoviruses*. Virol J 9:67

van Antwerpen T, McFarlane SS, Buchanan GF, Shepherd DN, Martin DP, Rybicki EP, Varsani A (2008) First report of *Maize streak virus* infection of sugarcane in South Africa. Plant Dis 92:982

van der Walt E, Rybicki EP, Varsani A, Polston JE, Billharz R, Donaldson L, Monjane AL, Martin DP (2009) Rapid host adaptation by extensive recombination. J Gen Virol 90:734–746

Vanitharani R, Chellappan P, Pita JS, Fauquet CM (2004) Differential roles of AC2 and AC4 of cassava *Geminiviruses* in mediating synergism and suppression of posttranscriptional gene silencing. J Virol 78(17):9487–9498

Varsani A, Oluwafemi S, Windram OP, Shepherd DN et al (2008a) *Panicum streak virus* diversity is similar to that observed for *Maize streak virus*. Arch Virol 153:601–604

Varsani A, Shepherd DN, Monjane AL, Owor BE et al (2008b) Recombination, decreased host specificity and increased mobility may have driven the emergence of *Maize streak virus* as an agricultural pathogen. J Gen Virol 89:2063–2074

Varsani A, Monjane AL, Donaldson L, Oluwafemi S et al (2009a) Comparative analysis of *Panicum streak virus* and *Maize streak virus* diversity, recombination patterns and phylogeography. Virol J 6:194

Varsani A, Shepherd DN, Dent K, Monjane AL et al (2009b) A highly divergent South African *Geminivirus* species illuminates the ancient evolutionary history of this family. Virol J 6:36

Whitham SA, Hajimorad MR (2016) Plant genetic resistance to viruses. In: Wang A, Zhou X (eds) Current research topics in plant virology. Springer, Cham, pp 87–111

Willment JA, Martin DP, Rybicki EP (2001) Analysis of the diversity of African streak *Mastreviruses* using PCR-generated RFLPs and partial sequence data. J Virol Methods 93:75–87

Willment JA, Martin DP, Walt EV, Rybicki EP (2002) Biological and genomic sequence characterization of *Maize streak virus* isolates from wheat. Virology 92(1):81–86

Woolston CJ, Reynolds HV, Stacey NJ, Mullineaux PM (1989) Replication of *Wheat dwarf virus* DNA in protoplasts and analysis of coat protein mutants in protoplasts and plants. Nucleic Acids Res 17:6029–6041

Wright EA, Heckel T, Groenendijk J, Davies JW, Boulton MI (1997) Splicing features in *Maize streak virus* virion and complementary-sense gene expression. Plant J 12:1285–1297

Wu B, Melcher U, Guo X, Wang X et al (2008) Assessment of codivergence of *Mastreviruses* with their plant hosts. BMC Evol Biol 8:335

Yahaya A, Dangora DB, Alegbejo MD, Kumar PL et al (2016) Identification and molecular characterization of a novel sugarcane streak *Mastrevirus* and an isolate of the A-strain of maize streak virus from sugarcane in Nigeria. Arch Virol 162:597–602

Zaagueri T, Mnari-Hattab M, Zammouri S, Hajlaoui MR et al (2017) First report of *Chickpea chlorotic dwarf virus* in watermelon (*Citrullus lanatus*) in Tunisia. Plant Dis 101(2):392

Zhan X, Richardson KA, Haley A, Morris BAM (1993) The activity of the coat protein promoter of *Chloris striate mosaic virus* is enhanced by its own and C1-C2 gene products. Virology 193:498–502

Zvereva AS, Pooggin M (2012) Silencing and innate immunity in plant defense against viral and non-viral pathogens. Viruses 4(11):2578–2597

Global Weed-Infecting Geminiviruses

Poonam Roshan, Aditya Kulshreshtha, and Vipin Hallan

Abstract

Weeds are invasive species that grow along with cultivated plants due to their high phenotypic plasticity. They serve as reservoirs of geminiviruses during off-season for main crops and provide the source of virus inoculum during their plantation. Geminiviruses are single-stranded DNA viruses enclosed in icosahedral geminate particles. These viruses can be either monopartite or bipartite, depending upon the number of genomic circles present. The members of genus *Begomovirus* are responsible for huge economic crop losses and are transmitted through insect vector *Bemisia tabaci*. The majority of the weed-infecting monopartite begomoviruses are associated with *Betasatellite* genus of Tolecusatellitidae family and alphasatellites. Geminiviruses are reported to infect a variety of weeds in South-east Asia, Mediterranean region, Western Europe (mainly Spain and France), Africa, Latin America, Central America, Caribbean region, and Australia. Weeds harbor the mixed infection of viruses; therefore, these plants serve as melting pots for recombination and evolution of begomoviruses. This chapter presents the geminivirus infection on weeds, their recombination, and their spread to newer hosts.

P. Roshan · A. Kulshreshtha · V. Hallan (✉)
Academy of Scientific and Innovative Research (AcSIR), CSIR-Institute of Himalayan Bioresource Technology, Palampur, HP, India

Plant Virology Lab, CSIR-Institute of Himalayan Bioresource Technology, Palampur, HP, India
e-mail: hallan@ihbt.res.in

R. V. Kumar (ed.), *Geminiviruses*, https://doi.org/10.1007/978-3-030-18248-9_6

1 Introduction

The plant viruses of *Geminiviridae* family consist of circular single-stranded DNA as genetic material enclosed in icosahedral geminate capsid (Lazarowitz 1992). The family is classified into nine different genera: *Mastrevirus*, *Becurtovirus*, *Begomovirus*, *Capulavirus*, *Curtovirus*, *Eragrovirus*, *Grablovirus*, *Topocuvirus*, and *Turncurtovirus* on the basis of genome organization, insect vector, and host range. The members are transmitted by leafhoppers (*Mastrevirus*, *Becurtovirus*, and *Curtovirus*), treehoppers (*Curtovirus*, *Grablovirus*, and *Topocuvirus*), whitefly (*Begomovirus*), and aphid (*Capulavirus*) (Zerbini et al. 2017). Geminiviruses are able to infect both monocots and dicots; therefore, they are responsible for worldwide crop losses (Moffat 1999). The genome of these viruses can be monopartite or bipartite, based on the number of genomic components (Harrison and Robinson 1999). Monopartite geminiviruses possess single genomic component which encodes the proteins required for replication, encapsulation, movement, transcription, and suppression of gene silencing. Bipartite begomoviruses consist of two genomic components: DNA-A and DNA-B. DNA-A is homologous to DNA-A of monopartite geminiviruses and encodes proteins required for encapsulation, replication, transcription, and suppression of gene silencing. DNA-B encodes only two proteins, essential for intra- and intercellular movement (Hanley-Bowdoin et al. 2000). The viral genome replicates via rolling circle replication (RCR) mechanism that is initiated at stem-loop structure containing nonanucleotide sequence (TAATATT/AC). The *Begomovirus* is the largest genus of *Geminiviridae* family that infects dicots and is transmitted by whitefly (*Bemisia tabaci*) vector (Brown et al. 2015). Begomoviruses are distributed into two groups on the globe: old world (OW) and new world (NW) (Paximadis et al. 1999). The majority of monopartite begomoviruses are associated with members of *Betasatellite* genus and alphasatellite molecules which form disease complex in plants (Kulshreshtha et al. 2017; Sharma et al. 2019a; Zhou 2013).

Weeds are unwanted plants which grow along with cultivated crops and decrease their yield. They are responsible for 43% loss in crop yield and destroy the native habitats (Oerke 2006). Due to high environmental plasticity, weeds are widely distributed and are able to adapt in different ecological habitats (Holm et al. 1979). Weeds serve as reservoir host for the viruses and play important role in their persistence and spread (Hallan et al. 1998). During off-season of crops, they become main host for the virus population. The earliest record of geminivirus symptoms on a weed was described in a Japanese poem written 1265 years ago (Saunders et al. 2003). In this chapter, we described various geminiviruses-infecting weeds, their symptoms, and emergence.

2 Global Status of Weed-Infecting Geminiviruses

Begomovirus infection causes devastating disease and huge crop losses worldwide (Varma and Malathi 2003). The monopartite begomoviruses and some bipartite begomoviruses are present in OW. The majority of bipartite begomoviruses are present in the NW, which suggested their origin from OW begomoviruses (Rybicki 1994). The only exception is tomato leaf deformation virus (ToLDeV), a monopartite begomovirus present in NW (Melgarejo et al. 2013). The agricultural practices, trafficking of infected plant material by humans, and invasive polyphagous vectors are responsible for the global spread and diversification of geminiviruses. The best example is the spread of tomato yellow leaf curl virus (TYLCV) through infected seedlings from Israel to tomato-growing regions of NW during the early 1990s (Duffy and Holmes 2007). Similarly, the introduction of NW cotton to the Indian subcontinent resulted in incidence of leaf curl disease of cotton in Pakistan (Briddon and Markham 2000). The invasive polyphagous whitefly vector also (*Bemisia tabaci*) resulted in the transmission of native begomovirus to the new hosts and emergence of novel viruses. The yellow leaf curl-like symptoms were observed on tomato in the 1940s due to outbreaks of sweetpotato whitefly population and the infection was due to the presence of a geminivirus TYLCV (Cohen and Antignus 1994). Later on, TYLCV infection was reported on weeds such as *Euphorbia* sp., *Lamium amplexicaule*, *Malva parviflora*, and *Ageratum conyzoides* (Papayiannis et al. 2011; Kil et al. 2014). The weed species belonging to family Euphorbiaceae, Asteraceae, Fabaceae, Malvaceae, Solanaceae, Amaranthaceae, and Lamiaceae harbor virus inoculum in NW as well as OW. The weeds growing in South-east Asia are infected with both monopartite–satellite complex and bipartite begomoviruses. Therefore, South-east Asian region can be regarded as diversification center for weed-infecting begomoviruses (Fig. 1). Most of the weed species display chlorosis, yellow mosaic, vein yellowing symptoms, and stunting upon geminivirus infection (Fig. 2).

Ageratum conyzoides is a member of Asteraceae family and native of Central America. It is an annual invasive weed in tropical subtropical regions of the world and reported as natural host for ageratum yellow vein virus (Tan and Wong 1993; Saunders and Stanley 1999; Saunders et al. 2004), ageratum enation virus (Tahir et al. 2015), ageratum leaf curl virus (Huang and Zhou 2006a), chilli leaf curl virus (Iqbal et al. 2016), cotton leaf curl Rajasthan virus (Mubin et al. 2009), and malvastrum yellow vein virus (Jiang and Zhou 2004) in South-east Asia. *A. conyzoides* also serve as reservoir host for TYLCV in Tanzania and Mediterranean region as this weed grows along with tomato plantations (Papayiannis et al. 2011).

Datura stramonium commonly known as devil's snare and jimson weed is a member of Solanaceae family. It is a native of America and is now distributed to tropical and subtropical regions and parts of Europe. A novel begomovirus datura leaf distortion virus was found to infect this weed in Venezuela (Fiallo-Olive et al. 2013). Leaf curl disease of jimson weed was found to be associated with tomato yellow leaf curl China virus (TYLCChV) and a betasatellite (Ding et al. 2007). In France and Spain, this weed was infected with monopartite TYLCV (Bedford et al. 1998).

Croton bonplandianum is an annual weed found in Asia and is infected with croton yellow vein virus, croton yellow vein mosaic virus, and tomato leaf curl New Delhi virus (Hussain et al. 2011; Pramesh et al. 2013; Chowda-Reddy et al. 2005).

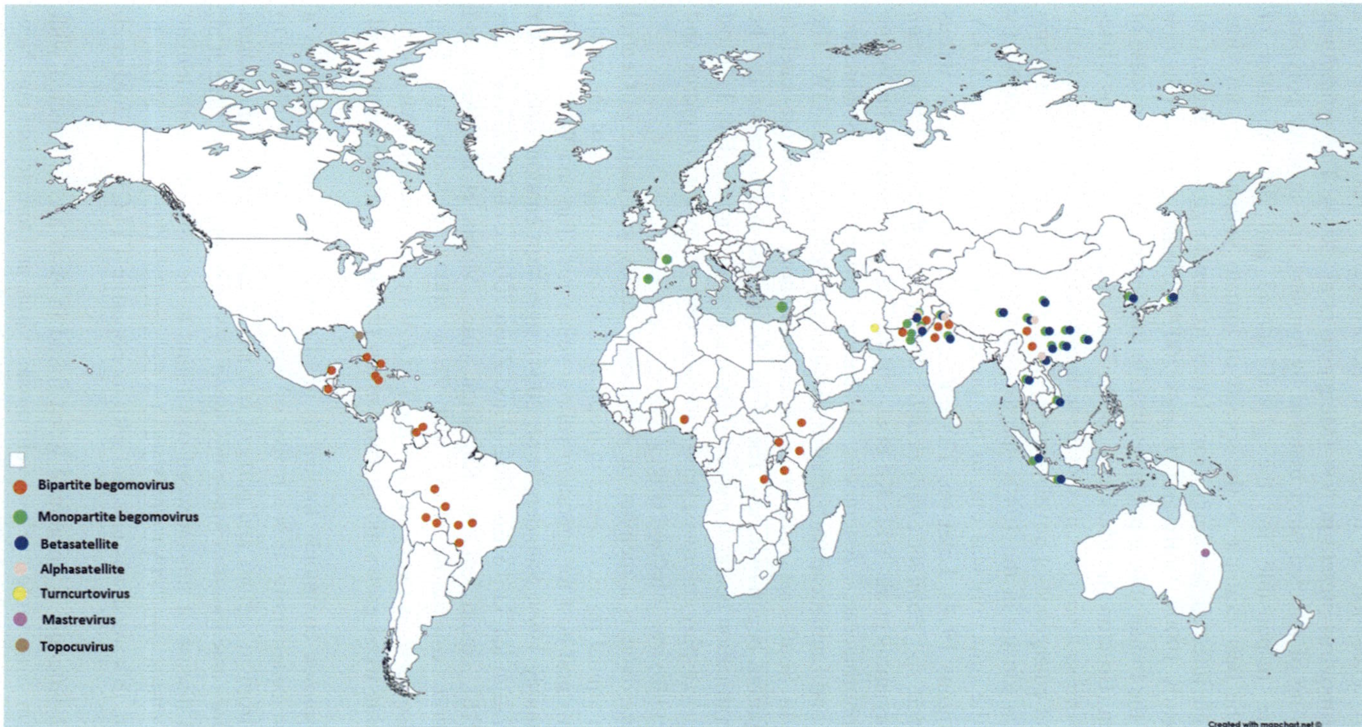

Fig. 1 Global distribution of geminiviruses-infecting weeds in NW and OW. Majority of bipartite begomoviruses are present in NW, whereas monopartite begomoviruses are found along with betasatellites and alphasatellites in South-east Asia

Fig. 2 Typical symptoms of vein yellowing, leaf curling, enation and chlorosis on weeds. (**a–l**) *Malvastrum coromandelianum*, *Eclipta* sp., *Synedrella* sp., *Sonchus asper*, *Ageratum conyzoides*, *Rumex* sp., *Urena lobata*, *Malvastrum* sp., *Urena lobata*, *Croton* sp., *Ipomea* sp. and *Croton* sp.

Malvastrum coromandelianum is an annual weed native to North America and now distributed in Africa, Asia, and South America. It acts as alternative host for monopartite begomoviruses such as malvastrum yellow mosaic virus, malvastrum leaf curl virus, malvastrum leaf curl Guangdong virus, and TYLCChV (Guo et al. 2007; Wu et al. 2007; Liu et al. 2011). In NW, it is reservoir host for sida golden yellow vein virus, sida golden mosaic Florida virus, and malvastrum yellow mosaic Jamaica virus (Fiallo-Olive et al. 2010, 2012; Graham et al. 2010).

Sida sp. is an invasive perennial weed found in tropical and subtropical areas of the world. *S. acuta* has been reported as natural host of sida golden mosaic virus and sida yellow mosaic China virus (Wong et al. 1993; Xiong et al. 2005). *S. micrantha* has been reported as reservoir host of abutilon mosaic virus, abutilon mosaic Bolivia virus, sida mosaic Bolivia virus 1, and sida golden mosaic buckup virus (Wyant et al. 2011; Stewart et al. 2014). In Brazil, *Sida* sp. served as reservoir host for sida micrantha virus and tomato mild mosaic virus which were found to infect tomatoes and beans (Castillo-Urquiza et al. 2010). The detailed information of geminivirus members infecting various weeds is given in Table 1.

Table 1 Geminivirus members and associated betasatellites infecting weeds

Name of the virus and genus	Geographical distribution	Weed host	Associated betasatellite	Associated satellite molecule/s	Symptom produced	References
Abutilon mosaic virus Bipartite begomovirus	Bolivia India	*Sida micrantha, Abutilon pictum*	–	–	Mosaic, bright yellow mosaic	Wyant et al. (2011) and Jyothsna et al. (2013)
Abutilon mosaic Bolivia virus Bipartite begomovirus	Bolivia	*S. micrantha, Abutilon* sp.	–	–	Yellow mosaic	Wyant et al. (2011)
African cassava mosaic virus Bipartite begomovirus	Nigeria	*Combretum confertum*	–	–	Chlorotic mosaic	Alabi et al. (2008)
Ageratum enation virus Monopartite begomovirus	India Pakistan	*Ageratum conyzoides, Cleome gynandra, Crassocephalum crepidioides, Sonchus oleraceus*	Ageratum yellow leaf curl betasatellite	Nanovirus-like DNA1	Vein enation, yellowing, stunting	Raj et al. (2010), Kumar et al. (2011) and Tahir et al. (2015)
Ageratum yellow vein virus Monopartite begomovirus	China Singapore	*A. conyzoides*	Ageratum yellow leaf curl betasatellite	Nanovirus-like DNA1	Vein yellowing, stunting	Tan and Wong (1993), Saunders and Stanley (1999) and Saunders et al. (2004)
Alternanthera yellow vein virus Monopartite begomovirus	China	*Alternanthera philoxeroides, Eclipta prostrata*	–	–	Yellow vein	Guo and Zhou (2005) and He et al. (2008)
Blainvillea yellow spot virus Bipartite begomovirus	Brazil	*Blainvillea rhomboidea*	–	–	Mosaic, yellowing and stunting	Castillo-Urquiza et al. (2008)

Blechum yellow vein virus Monopartite begomovirus	Philippines	*Blechum pyramidatum*	–	–	Yellow/ chlorotic leaf veins	Tsai et al. (2014)
Chilli leaf curl virus Monopartite begomovirus	Pakistan	*Urtica dioica*	Ageratum yellow leaf curl betasatellite	Ageratum yellow vein Pakistan alphasatellite, Bhendi yellow vein alphasatellite	Leaf curling, vein-yellowing	Iqbal et al. (2016)
Cotton leaf curl Rajasthan virus Monopartite begomovirus	Pakistan	*Digera arvensis*	Ageratum yellow leaf curl betasatellite, Tobacco leaf curl betasatellite	–	Yellow vein disease	Mubin et al. (2009)
Cotton leaf curl Burewala virus Monopartite begomovirus	Pakistan	*Xanthium strumarium*	Tomato yellow leaf curl Thailand betasatellite	Potato leaf curl alphasatellite	Leaf curling, vein thickening	Mubin et al. (2012)
Crassocephalum yellow vein virus Monopartite begomovirus	China	*C. crepidioides*	–	–	Vein yellowing	Dong et al. (2008)
Croton yellow vein mosaic virus Bipartite begomovirus	India	*A. conyzoides*, *Croton bonplandianum*, *Euphorbia geniculate*, *S. brachyotis*	–	–	Bright yellow vein, leaf curl	Pramesh et al. (2013)
Dalechampia chlorotic mosaic virus Bipartite begomovirus	Venezuela	*Boerhavia diffusa*, *Dalechampia* sp.	–	–	Vein yellowing	Fiallo-Olive et al. (2013)

(continued)

Table 1 (continued)

Name of the virus and genus	Geographical distribution	Weed host	Associated betasatellite	Associated satellite molecule/s	Symptom produced	References
Datura leaf distortion virus Bipartite begomovirus	Venezuela	*Datura stramonium*	–	–	Vein yellowing	Fiallo-Olive et al. (2013)
Deinbollia mosaic virus Bipartite begomovirus	East Africa	*Deinbollia borbonica*	–	–	Yellow mosaic	Kyallo et al. (2017)
Dicliptera yellow mottle virus Bipartite begomovirus	Cuba	*Dicliptera vahliana*	–	–	Yellow mottling	Echemendía et al. (2003)
East African cassava mosaic Cameroon virus Bipartite begomovirus	Nigeria	*C. confertum*		–	Chlorotic mosaic	Alabi et al. (2008)
Emilia yellow vein virus Bipartite begomovirus	China	*C. crepidioides*	–	–	Yellow vein	Yang et al. (2012)
Erectites yellow mosaic virus Monopartite begomovirus	Vietnam	*Erectites valerianifolia*	Erectites yellow mosaic betasatellite	–	Vein yellowing, leaf curling, chlorosis	Ha et al. (2008)

Euphorbia mosaic virus Bipartite begomovirus	Yucatan Peninsula	*Euphorbia heterophylla*	–	–	Golden mosaic	Hernandez-Zepeda et al. (2007)
Euphorbia yellow mosaic virus Bipartite begomovirus	Brazil	*E. heterophylla*	–	–	Bright yellow mosaic	Fernandes et al. (2011)
Honeysuckle yellow vein virus	Japan	*Lonicera japonica*			Vein yellowing	Ogawa et al. (2008)
Lindernia anagallis yellow vein virus Monopartite begomovirus	Vietnam	*Lindernia procumbens*	Lindernia anagallis yellow vein betasatellite	–	Vein yellowing and leaf curling	Ha et al. (2008)
Ludwigia yellow vein virus Monopartite begomovirus	China	*Ludwigia hyssopifolia*	Ludwigia yellow vein betasatellite	–	Vein yellowing	Huang et al. (2006)
Macroptilium yellow spot virus Bipartite begomovirus	Brazil	*Macroptilium lathyroides*	–	–	Mosaic, yellowing and stunting	Da Silva et al. (2011)
Malvastrum leaf curl virus Monopartite begomovirus	China	*Malvastrum coromandelianum*	Sida yellow vein China betasatellite	–	Leaf curling, vein thickening	Huang and Zhou (2006a, b)
Malvastrum yellow vein virus Monopartite begomovirus	China	*A. conyzoides*	Malvastrum yellow vein betasatellite	–	Yellow vein	Jiang and Zhou (2004)

(continued)

Table 1 (continued)

Name of the virus and genus	Geographical distribution	Weed host	Associated betasatellite	Associated satellite molecule/s	Symptom produced	References
Merremia mosaic virus Bipartite begomovirus	Belize	*E. heterophylla*, hot pepper, sweet pepper	–	–	Leaf curling, yellowing, mottling, mosaic	McLaughlin et al. (2008)
Papaya leaf curl China virus Monopartite begomovirus	China	*Corchoropsis timentosa*	–	–	Leaf curling	Huang and Zhou (2006a, b)
Rhynchosia yellow mosaic virus Bipartite begomovirus	Pakistan	*Rhynchosia minima*	–	–	Yellow mosaic	Ilyas et al. (2009)
Rhynchosia yellow mosaic India virus Bipartite begomovirus	India	*R. minima*	–	–	Yellow mosaic	Jyothsna et al. (2011)
Sida golden mosaic Buckup virus Bipartite begomovirus	Jamaica	*Sida* sp.	–	–	Yellow mosaic	Stewart et al. (2014)
Sida leaf curl virus Monopartite begomovirus	China	*S. cordifolia*	Ageratum leaf curl betasatellite	Sida leaf curl virus DNA1	Mild upward leaf curling	Guo and Zhou (2006)

Sida yellow mosaic China virus Monopartite begomovirus	China	*S. acuta*	Ageratum yellow vein betasatellite	–	Yellow mosaic	Xiong et al. (2005)
Sida mosaic Bolivia virus 1 Bipartite begomovirus	Bolivia	*S. micrantha*	–	–	Bright yellow mosaic	Wyant et al. (2011)
Tobacco leaf curl Cuba virus Bipartite begomovirus	Jamaica	*Malachra alceifolia*	–	–	Yellow mosaic	Hall et al. (2008)
Tomato leaf curl virus Monopartite begomovirus	India	*Parthenium hysterophorus*	Papaya leaf curl betasatellite	Ageratum yellow vein alphasatellite	Leaf curl, stunting	Kumar et al. (2016)
Tomato leaf curl New Delhi virus Bipartite begomovirus	Pakistan	*Eclipta prostrata*, *Calotropis procera*	–	–	Yellow vein, yellow mosaic	Haider et al. (2005) and Zaidi et al. (2017)
Tomato yellow leaf curl virus Monopartite begomovirus	Spain, France, Cyprus, Korea	*Euphorbia* sp., *Solanum nigrum*, *Datura stramonium*, *Malva* sp., *Lamium amplexicaule*	–	–	Leaf yellowing and curling	Bedford et al. (1998), Dalmon and Marchoux (2000), Papayiannis et al. (2011) and Kil et al. (2014)
Tobacco yellow dwarf virus Mastrevirus	Australia	*Rapistrum rugosum*	–	–	Necrosis and dwarfing	Schwinghamer et al. (2010)
Tomato pseudo curly top virus Topocuvirus	Florida	*Solanum nigrum*, *Datura stromonium*, *Stellaria media*	–	–	Leaf cupping and chlorosis	Tsai and Brown (1991)

(continued)

Table 1 (continued)

Name of the virus and genus	Geographical distribution	Weed host	Associated betasatellite	Associated satellite molecule/s	Symptom produced	References
Turnip curly top virus Turncurtovirus	Iran	*Solanum nigrum*, *Anchusa arvensis*	–	–	Inward rolling of the leaf margins	Razavinejad et al. (2013)
Wissadula golden mosaic virus Bipartite begomovirus	Jamaica	*Wissadula amplissima*	–	–	Leaf curling. Yellow mosaic	Collins and Roye (2006)

3 Weeds as Mixing Vessels for Recombination and Assortment of Viruses

Recombination, mutation, and pseudo-recombination between variants, species, and genera of the virus significantly contribute to genetic diversity, local adaptation, and emergence of new viruses (Pita et al. 2001; Martin et al. 2005; Graham et al. 2010). It has been demonstrated that recombination in geminiviruses is dependent on parental virus strain, host plant, and inoculum (Padidam et al. 1999). Weeds harbor mixed infections of geminiviruses which result in the evolution and emergence of new virus species or strains. The mixed infections result in the association of betasatellites with helper begomoviruses which led to emergence of more virulent strains or species. These satellite molecules are known to enhance the symptom severity and disease epidemics in new environments (Sharma et al. 2019b; Saunders et al. 2001; Briddon et al. 2004). The evolution of sida micrantha mosaic-associated viruses and alternanthera yellow vein virus is a result of recombination in the weed hosts (Jovel et al. 2007; Mubin et al. 2010). The recombinants of begomoviruses associated with cassava mosaic disease in Africa showed increased virulence in comparison to parental strains (Zhou et al. 1997). In Brazil, multiple recombination events among cleome leaf crumple virus isolates were reported in a single weed *Cleome affinis* (Da Silva et al. 2011). Furthermore, sida mottle virus and sida micrantha mosaic virus were originally characterized from weed species that were transmitted to crops by insect vector (Castillo-Urquiza et al. 2007, 2010). Weeds belonging to Euphorbiaceae and Fabaceae family are reported as reservoir host for cassava-infecting begomoviruses in Africa (Alabi et al. 2008). Therefore, weeds are designated as "mixing vessels" for genetic recombination between begomoviruses. Ageratum enation virus is a monopartite begomovirus associated with betasatellite and alphasatellite, is widely distributed in South-east Asia, and infects non-cultivated plants such as *Sonchus oleraceus* and *A. conyzoides* (Tahir et al. 2015). Two weeds *Chrozophora hierosolymitana* and *Herniaria* sp. were reported to harbor the TYLCV inoculums in Iran (Fazeli et al. 2009).

4 Relationship Between Weed, Virus Disease Complex, and Insect Vector

Weeds act as alternate host for both virus and insect vector during off-season of the main crops. As a result, these plants prevent the extinction of virus populations in the absence of annual crops (Seal et al. 2006). In such conditions, a dramatic increase in the whitefly-transmitted geminiviruses (WTG) population has been reported. The virus-infected plants have a greater tendency to attract the insect vector in comparison to healthy plants. Furthermore, virus infection alters the morphology and defense system which increases the infestation and fitness of the insect vector (Awmack and Leather 2002; Chen et al. 2013). The invasive polyphagous B-biotype *B. tabaci* is found to be responsible for the TYLCV disease epidemic in the Mediterranean region. The invasive whitefly species is reported to transmit about 200 species of the begomoviruses in both cultivated and non-cultivated plants (Delatte et al. 2005; Hogenhout et al. 2008). The mobile and polyphagous nature of this insect allows the dissemination of viral diversity into new crops (Lefeuvre et al. 2007). The association of betasatellites and alphasatellite with helper begomovirus offers a selective advantage for helper begomovirus to produce symptoms on the weed. The yellow vein disease of *A. conyzoides* was due to infection of ageratum yellow vein virus and a betasatellite. In the absence of betasatellite, the weed failed to develop the typical symptoms (Saunders et al. 2001). Similarly, the yellow vein disease of *C. bonplandianum* was associated with infection of a monopartite begomovirus and a betasatellite (Hussain et al. 2011). In case of tomato leaf curl virus infection on *Parthenium hysterophorus*, the betasatellite and alphasatellite complex developed typical leaf curling symptoms (Kumar et al. 2016). The begomovirus disease complex including multiple and recombinant betasatellites was reported from a common weed, *Digera arvensis*, in Pakistan (Mubin et al. 2009). The widespread distribution of weeds and polyphagous invasive whitefly vector in the warmer regions provides favorable platform for virus proliferation. The whitefly population prefers high temperature for reproduction, but is adversely affected by prolonged winters. Under such circumstances, weed species serve as reservoir inoculums for viral diversity and global warming in temperate areas offers great advantage for the virus spread.

5 Conclusion

The increasing incidence of geminivirus infection on economic important crops has become a major concern, and it has been found that weeds or non-cultivated hosts serve as source of virus infection. The presence of mixed infection renders the weeds as melting pots for begomovirus recombination and led to emergence of more fit variants of the virus. Additionally, the increasing population insect vector contributed to the spread of begomoviruses in new hosts. Therefore, reservoir host, insect vector, and virus constitute a cycle which seems to be the main reason for the outbreak and emergence of begomovirus disease complex in newer hosts. However, it is unclear whether weeds act as indigenous host of viruses or they get infected from the infected crop host? To address this question, further studies are required to demonstrate the role of weeds in primary source of virus inoculum and evolution of novel viruses. The outcome of such studies will reveal a potential way to combat the geminivirus infection in cultivated crops.

Acknowledgements Authors acknowledge Director CSIR-IHBT for providing the research facilities. PR is thankful to Council of Scientific and Innovative Research for providing Junior and Senior Research Fellowship. AK thanks the University Grants Commission (UGC) for providing Junior and Senior Research Fellowship. PR and AK duly acknowledge Academy of Scientific and Innovative Research (AcSIR), New Delhi, India.

References

Alabi OJ, Ogbe FO, Bandyopadhyay R, Kumar PL et al (2008) Alternate hosts of African cassava mosaic virus and East African cassava mosaic Cameroon virus in Nigeria. Arch Virol 153:1743

Awmack CS, Leather SR (2002) Host plant quality and fecundity in herbivorous insects. Annu Rev Entomol 47:817–844

Bedford I, Kelly A, Banks G, Briddon RW et al (1998) Solanum nigrum: an indigenous weed reservoir for a tomato yellow leaf curl geminivirus in southern Spain. Eur J Plant Pathol 104:221

Briddon RW, Markham PG (2000) Cotton leaf curl virus disease. Virus Res 71:151–159

Briddon RW, Bull SE, Amin I, Mansoor S et al (2004) Diversity of DNA 1: a satellite-like molecule associated with monopartite begomovirus-DNA b complexes. Virology 324:462–474

Brown JK, Zerbini FM, Navas-Castillo J, Moriones E et al (2015) Revision of begomovirus taxonomy based on pairwise sequence comparisons. Arch Virol 160:1593

Castillo-Urquiza GP, Beserra Junior JEA, Alfenas-Zerbini P et al (2007) Genetic diversity of begomoviruses infecting tomato in Paty do Alferes, Rio de Janeiro state, Brazil. Virus Rev Res 12:233

Castillo-Urquiza GP, Beserra JE, Bruckner FP, Lima GSA et al (2008) Six novel begomoviruses infecting tomato and associated weeds in Southeastern Brazil. Arch Virol 153(10):1985–1989

Castillo-Urquiza GP et al (2010) Genetic structure of tomato-infecting begomovirus populations in two tomato growing regions of Southeastern Brazil. In: International geminivirus symposium, 6.; international ssDNA comparative virology workshop, 4., 2010, Guanajuato, Mexico. Program and Abstracts in CD-ROM. Guanajuato, Mexico

Chen G, Pan H, Xie W, Wang S et al (2013) Virus infection of a weed increases vector attraction to and vector fitness on the weed. Sci Rep 3:2253

Chowda-Reddy R, Colvin J, Muniyappa V, Seal S (2005) Diversity and distribution of begomoviruses infecting tomato in India. Arch Virol 150:845

Cohen S, Antignus Y (1994) Tomato yellow leaf curl virus (TYLCV), a whitefly-borne geminivirus of tomatoes. Adv Dis Vector Res 10:259–288. New York: Springer-Verlag

Collins AM, Roye ME (2006) Two new bipartite begomoviruses infecting Wissadula amplissima in Jamaica. New Dis Rep 13:31

Da Silva SJC, Castillo-Urquiza GP, Hora Júnior BT, Assunçao IP et al (2011) High genetic variability and recombination in a begomovirus population infecting the ubiquitous weed Cleome affinis in northeastern Brazil. Arch Virol 156:2205

Dalmon A, Marchoux G (2000) Quelles plantes hotes pour le tomato yellow leaf curl virus? Phytoma 527:14–17

Delatte H, Martin DP, Naze F, Golbach RW et al (2005) South West Indian Ocean islands tomato begomovirus populations represent a new major monopartite begomovirus group. J Gen Virol 86:1533–1542

Ding M, Luo YQ, Dong JH et al (2007) First report of tomato yellow leaf curl China virus with DNA β infecting Datura stramonium in China. Aust Plant Dis Notes 2:63

Dong JH, Zhang ZK, Ding M, Fang Q et al (2008) Molecular characterization of a distinct begomovirus infecting Crassocephalum crepidioides in China. J Phytopathol 156:193–195

Duffy S, Holmes EC (2007) Multiple introductions of the Old World begomovirus tomato yellow leaf curl virus into the New World. Appl Environ Microbiol 73:7114–7117

Echemendía AL, Ramos PL, Peral R, Fuentes A et al (2003) First report of Dicliptera yellow mottle virus (DiYMoV) infecting Dicliptera vahliana in Cuba. Plant Pathol 52:787

Fazeli R, Heydarnejad J, Massumi H, Shaabanian M et al (2009) Genetic diversity and distribution of tomato-infecting begomoviruses in Iran. Virus Genes 38:311–319

Fernandes FR, Albuquerque LC, de Oliveira CL, Cruz AR et al (2011) Molecular and biological characterization of a new Brazilian begomovirus, euphorbia yellow mosaic virus (EuYMV), infecting Euphorbia heterophylla plants. Arch Virol 156(11):2063–2069

Fiallo-Olive E, Martinez-Zubiaur Y, Moriones E, Navas-Castillo J (2010) Complete nucleotide sequence of Sida golden mosaic Florida virus and phylogenetic relationships with other begomoviruses infecting malvaceous weeds in the Caribbean. Arch Virol 155(9):1535–1537

Fiallo-Olive E, Navas-Castillo J, Moriones E, Martínez-Zubiaur Y (2012) Begomoviruses infecting weeds in Cuba: increased host range and a novel virus infecting Sida rhombifolia. Arch Virol 157(1):141–146

Fiallo-Olive E, Chirinos DT, Geraud-Pouey F, Moriones E et al (2013) Complete genome sequences of two begomoviruses infecting weeds in Venezuela. Arch Virol 158:277

Graham AP, Martin DP, Roye ME (2010) Molecular characterization and phylogeny of two begomoviruses infecting Malvastrum americanum in Jamaica: evidence of the contribution of inter-species recombination to the evolution of malvaceous weed-associated begomoviruses from the northern Caribbean. Virus Genes 40(2):256–266

Guo X, Zhou X (2005) Molecular characterization of alternanthera yellow vein virus: a new begomovirus species infecting Alternanthera philoxeroides. J Phytopathol 153:694–696

Guo X, Zhou X (2006) Molecular characterization of a new begomovirus infecting Sida cordifolia and its associated satellite DNA molecules. Virus Genes 33(3):279–285

Guo X, Shi M, Zhou X (2007) Complete nucleotide sequences of Malvastrum yellow mosaic virus and its associated DNAβ molecule. Arch Virol 152(3):641–643

Ha C, Coombs S, Revill P, Harding R et al (2008) Molecular characterization of begomoviruses and DNA satellites from Vietnam: additional evidence that the New World geminiviruses were present in the Old World prior to continental separation. J Gen Virol 89:312–326

Haider MS, Tahir M, Latif S, Briddon RW (2005) First report of tomato leaf curl New Delhi virus infecting Eclipta prostrata in Pakistan. New Dis Rep 11:39

Hall GC, Graham AP, Roye ME (2008) Tobacco leaf curl Cuba virus infects the weed Malachra alceifolia in Jamaica. Plant Pathol 57:398

Hallan V, Saxena S, Singh B (1998) Ageratum, Croton and Malvastrum harbour geminiviruses: evidence through PCR amplification. World J Microbiol Biotechnol 14:931

Hanley-Bowdoin L, Settlage SB, Orozco BM, Nagar S, Robertson D (2000) Geminiviruses: models for plant DNA replication, transcription, and cell cycle regulation. Crit Rev Biochem Mol Biol 35:105–140

Harrison B, Robinson D (1999) Natural genomic and antigenic variation in whitefly-transmitted geminiviruses (begomoviruses). Annu Rev Phytopathol 37:369–398

He ZF, Mao MJ, Yu H, Wang XM et al (2008) First report of a strain of alternanthera yellow vein virus infecting Eclipta prostrate (L.) L. (Compositae) in China. J Phytopathol 156:496–498

Hernandez-Zepeda C, Idris AM, Carnevali G, Brow JK et al (2007) Molecular characterization and experimental host range of Euphorbia mosaic virus-Yucatan Peninsula, a begomovirus species in the Squash leaf curl virus clade. Plant Pathol 56:763–770

Hogenhout SA, Ammar ED, Whitfield AE, Redinbaugh MG (2008) Insect vector interactions with persistently transmitted viruses. Annu Rev Phytopathol 46:327–359

Holm L, Pancho JV, Herberger JP, Plucknett DL (1979) A geographical atlas of world weeds. Wiley, New York, NY

Huang JF, Zhou XP (2006a) Molecular characterization of two distinct begomoviruses from Ageratum conyzoides and Malvastrum coromandelianum in China. J Phytopathol 154:648–653

Huang JF, Zhou XP (2006b) First report of papaya leaf curl China virus infecting Corchoropsis timentosa in China. Plant Pathol 55:291

Huang JF, Jiang T, Zhou XP (2006) Molecular characterization of begomoviruses infecting Ludwigia hyssopifolia. J Phytopathol 88(1):83–88

Hussain K, Hussain M, Mansoor S, Briddon RW (2011) Complete nucleotide sequence of a begomovirus and associated betasatellite infecting croton (Croton bonplandianus) in Pakistan. Arch Virol 156:1101

Ilyas M, Qazi J, Mansoor S, Briddon RW (2009) Molecular characterisation and infectivity of a "Legumovirus" (genus Begomovirus: family Geminiviridae) infecting the leguminous weed Rhynchosia minima in Pakistan. Virus Res 145(2):279–284

Iqbal MJ, Hussain W, Zia-Ur-Rehman M, Hameed U et al (2016) First report of chilli leaf curl virus and associated alpha- and beta-satellite DNAs infecting nettle weed (Urtica dioica) in Pakistan. Plant Dis 100(4):870

Jiang T, Zhou XP (2004) First report of Malvastrum yellow vein virus infecting Ageratum conyzoides. Plant Pathol 53:799

Jovel J, Preiss W, Jeske H (2007) Characterization of DNA intermediates of an arising geminivirus. Virus Res 130:63–70

Jyothsna P, Rawat R, Malathi VG (2011) Molecular characterization of a new begomovirus infecting a leguminous weed Rhynchosia minima in India. Virus Genes 42(3):407–414

Jyothsna P, Haq QMI, Jayaprakash P, Malathi VG (2013) Molecular evidence for the occurrence of abutilon mosaic virus, a New World begomovirus in India. Indian J Virol 24(2):284–288

Kil EJ, Park J, Lee H, Kim J et al (2014) Lamium amplexicaule (Lamiaceae): a weed reservoir for tomato yellow leaf curl virus (TYLCV) in Korea. Arch Virol 159(6):1305–1311

Kulshreshtha A, Roshan P, Sharma D, Hallan V (2017) Molecular characterization of a new begomovirus infecting *Mirabilis jalapa* in northern India. Arch Virol 162(7):2163–2167

Kumar Y, Hallan V, Zaidi AA (2011) First report of ageratum enation virus infecting Crassocephalum crepidioides (Benth.) S. Moore and Ageratum conyzoides L. in India. J Gen Plant Pathol 77:214–216

Kumar S, Srivastava A, Jaidi A, Chauhan PS et al (2016) Molecular characterization of a begomovirus, α-satellite, and β-satellite associated with leaf curl disease of Parthenium hysterophorus in India. Plant Dis 100(11):2299–2305

Kyallo M, Ateka EM, Sseruwagi P, Ascencio-Ibanez JT et al (2017) Infectivity of Deinbollia mosaic virus, a novel weed-infecting begomovirus in East Africa. Arch Virol 162:3439–3445

Lazarowitz SG (1992) Geminiviruses: genome structure and gene function. Crit Rev Plant Sci 11:327–349

Lefeuvre P, Martin DP, Hoareau M, Naze F et al (2007) Begomovirus 'melting pot' in the south-west Indian Ocean islands: molecular diversity and evolution through recombination. J Gen Virol 88:3458–3468

Liu P, Xie Y, Zhou X (2011) Malvastrum coromandelianum is an alternative host of tomato yellow leaf curl China virus. New Dis Rep 17:30

Martin DP, van der Walt E, Posada D, Rybicki EP (2005) The evolutionary value of recombination is constrained by genome modularity. PLoS Genet 1(4):e51

McLaughlin PD, McLaughlin WA, Maxwell DP, Roye ME (2008) Identification of begomoviruses infecting crops and weeds in Belize. Plant Viruses 2(1):58–63

Melgarejo TA, Kon T, Rojas MR, Paz-Carrasco L et al (2013) Characterization of a New World monopartite begomovirus causing leaf curl disease of tomato in Ecuador and Peru reveals a new direction in geminivirus evolution. J Virol 87(10):5397–5413

Moffat AS (1999) Geminiviruses emerge as serious crop threat. Science 286:1835

Mubin M, Briddon RW, Mansoor S (2009) Diverse and recombinant DNA betasatellites are associated with a begomovirus disease complex of Digera arvensis, a weed host. Virus Res 142:208–212

Mubin M, Shahid MS, Tahir MN, Briddon RW et al (2010) Characterization of begomovirus components from a weed suggests that begomoviruses may associate with multiple distinct DNA satellites. Virus Genes 40(3):452–457

Mubin M, Akhtar S, Amin I et al (2012) Xanthium strumarium: a weed host of components of begomovirus–betasatellite complexes affecting crops. Virus Genes 44:112

Oerke EC (2006) Crop losses to pests. J Agric Sci 144(1):31–43

Ogawa T, Sharma P, Ikegami M et al (2008) First report of a strain of tobacco leaf curl Japan virus associated with a satellite DNA in honeysuckle in Japan. Plant Pathol 57:391

Padidam M, Sawyer S, Fauquet CM (1999) Possible emergence of new geminiviruses by frequent recombination. Virology 265:218–225

Papayiannis LC, Katis NI, Idris AM, Brown JK (2011) Identification of weed hosts of tomato yellow leaf curl virus in Cyprus. Plant Dis 95(2):120–125

Paximadis M, Idris AM, Torres-Jerez I, Villarreal A, Rey MEC, Brown JK (1999) Characterization of tobacco geminiviruses in the Old and New World. Arch Virol 144:703–717

Pita JS, Fondong VN, Sangare A, Otim-Nape GW, Ogwal S, Fauquet CM (2001) Recombination, pseudorecombination and synergism of geminiviruses are determinant keys to the epidemic of severe cassava mosaic disease in Uganda. J Gen Virol 82:655–665

Pramesh D, Mandal B, Phaneendra C et al (2013) Host range and genetic diversity of croton yellow vein mosaic virus, a weed-infecting monopartite begomovirus causing leaf curl disease in tomato. Arch Virol 158:531

Raj SK, Snehi SK, Khan MS, Tiwari AK et al (2010) Detection of ageratum enation virus from cat's whiskers (Cleome gynandra L.) with leaf curl symptoms in India. J Gen Plant Pathol 76:292–294

Razavinejad S, Heydarnejad J, Kamali M, Massumi H et al (2013) Genetic diversity and host range studies of turnip curly top virus. Virus Genes 46(2):345–353

Rybicki EP (1994) A phylogenetic and evolutionary justification for 3 genera of Geminiviridae. Arch Virol 139:49–77

Saunders K, Stanley J (1999) A nanovirus-like DNA component associated with yellow vein disease of Ageratum conyzoides: evidence for interfamilial recombination between plant DNA viruses. Virology 264(1):142–152

Saunders K, Bedford ID, Stanley J (2001) Pathogenicity of a natural recombinant associated with ageratum yellow vein disease: implications for geminivirus evolution and disease aetiology. Virology 282:38–47

Saunders K, Bedford ID, Yahara T, Stanley J (2003) Aetiology: the earliest recorded plant virus disease. Nature 422:831

Saunders K, Norman A, Gucciardo S, Stanley J (2004) The DNA β satellite component associated with ageratum yellow vein disease encodes an essential pathogenicity protein (βC1). Virology 324(1):37–47

Schwinghamer MW, Thomas JE, Schilg MA, Parry JN et al (2010) Mastreviruses in chickpea (Cicer arietinum) and other dicotyledonous crops and weeds in Queensland and northern New South Wales, Australia. Australas Plant Pathol 39(6):551–561

Seal SE, Van den Bosch F, Jeger MJ (2006) Factors influencing begomovirus evolution and their increasing global significance: implications for sustainable control. Crit Rev Plant Sci 25:23–46

Sharma D, Kulshreshtha A, Kumar R, Hallan V (2019a) First report of natural infection of alternanthera yellow vein virus and cotton leaf curl Multan betasatellite on a new host *Picrorhiza kurroa*, an important endangered medicinal herb. J Plant Pathol 101(1):149–153

Sharma D, Kulshreshtha A, Roshan P, Hallan V (2019b) Molecular characterization and infectivity analysis of a bipartite begomovirus associated with cotton leaf curl Multan betasatellite naturally infecting Rumex nepalensis in northern India. J Plant Pathol: 1–7

Stewart C, Kon T, Rojas M, Graham A et al (2014) The molecular characterisation of a Sida-infecting begomovirus from Jamaica. Arch Virol 159(2):375–378

Tahir M, Amin I, Haider MS, Mansoor S et al (2015) Ageratum enation virus—a begomovirus of weeds with the potential to infect crops. Viruses 7(2):647–665

Tan HNP, Wong SM (1993) Some properties of Singapore ageratum yellow vein virus (SAYVV). J Phytopathol 139:165–176

Tsai JH, Brown LG (1991) Pseudo-curly top of tomato. Plant Pathol 344:190–191

Tsai WS, Shih SL, Lee LM (2014) First report of a novel begomovirus associated with yellow vein disease of Browne's blechum (Blechum pyramidatum). Plant Dis 98(5):701

Varma A, Malathi VG (2003) Emerging geminivirus problems: a serious threat to crop production. Ann Appl Biol 142(2):145–164

Wong SM, Swanson MM, Harrison BD (1993) A new geminivirus causing vein yellowing of Ageratum conyzoides in Singapore. Plant Pathol 42:137–139

Wu J, Mugiira R, Zhou X (2007) Malvastrum leaf curl Guangdong virus is a distinct monopartite begomovirus. Plant Pathol 56:771–776

Wyant PS, Gotthardt D, Schafer B, Krenz B et al (2011) The genomes of four novel begomoviruses and a new Sida micrantha mosaic virus strain from Bolivian weeds. Arch Virol 156:347
Xiong Q, Guo XJ, Che HY, Zhou XP (2005) Molecular characterization of a distinct begomovirus and its associated Satellite DNA Molecule infecting Sida acuta in China. J Phytopathol 153:264–268
Yang CX, Luo JS, Zheng LM, Wu ZJ et al (2012) First report of the occurrence of Emilia yellow vein virus in Crassocephalum crepidioides in China. J Plant Pathol 94(4):87
Zaidi SS, Shakir S, Malik HJ, Farooq M et al (2017) First report of tomato leaf curl New Delhi virus on Calotropis procera, a weed as potential reservoir begomovirus host in Pakistan. Plant Dis 101 (6):1071
Zerbini FM, Briddon RW, Idris A, Martin DP et al (2017) ICTV virus taxonomy profile: Geminiviridae. J Gen Virol 98:131–133
Zhou X (2013) Advances in understanding begomovirus satellites. Annu Rev Phytopathol 51:387–381
Zhou X, Liu Y, Calvert L, Munoz C et al (1997) Evidence that DNA-A of a geminivirus associated with severe cassava mosaic disease in Uganda has arisen by interspecific recombination. J Gen Virol 78:2101–2111

Evolutionary Factors in the Geminivirus Emergence

Amir Hameed, Sara Shakir, and Syed Shan-e-Ali Zaidi

Abstract

The viruses belonging to family *Geminiviridae* have a small genome of 2.8–5.0 kb and only four to six protein coding genes, and yet they have evolved as the largest family of single-stranded DNA viruses and one of the most important plant viruses that are threat to several economical crops. Most of the efforts to introduce resistance against geminiviruses have met with limited success, mainly due to rapid evolution of geminiviruses and, in turn, efficient evasion from plant-pathogen resistance mechanisms. Here, we discuss different evolutionary pathways that shaped geminiviruses' genome and assisted their global spread.

1 Introduction

Geminiviruses are among the most destructive viruses, reducing crop productivity worldwide (Mansoor et al. 2006), and impose serious threats to food security. Geminiviruses are characterized by distinct geminate morphology, i.e., twin quasi-icosahedral particles enveloping a single-stranded (ss) DNA genome (~2.8–5.0 kb). The family *Geminiviridae* has been classified into nine genera: *Becurtovirus*, *Begomovirus*, *Capulavirus*, *Curtovirus*, *Mastrevirus*, *Eragrovirus*, *Grablovirus*, *Topocuvirus*, and *Tungrovirus*, based on viral genome organization, phylogenies,

A. Hameed (✉)
Department of Bioinformatics and Biotechnology, Government College University (GCU), Faisalabad, Pakistan

S. Shakir
Boyce Thompson Institute, Ithaca, NY, USA

S. S.-A. Zaidi
National Institute for Biotechnology and Genetic Engineering (NIBGE), Faisalabad, Pakistan

Gembloux Agro-Bio Tech, University of Liège, Gembloux, Belgium

R. V. Kumar (ed.), *Geminiviruses*, https://doi.org/10.1007/978-3-030-18248-9_7

vector transmission, and infectious host range (Varsani et al. 2017; Zerbini et al. 2017). Among these genera, *Begomovirus* is the most economically important and widespread worldwide. In the context of genomic components, most of the geminiviruses are monopartite, i.e., monopartite begomoviruses contain a single genomic component (identical to the DNA-A of bipartite begomoviruses), which is fully capable of self-existence involving replication, systemic movement, and host infections. Exceptionally, the genome of begomoviruses is either monopartite or bipartite; i.e., it consists of DNA-A and a second genomic component (DNA-B) (Fondong 2013). The genome of nearly all geminiviruses shares a highly conserved nonanucleotide sequence, i.e., "TAATATTAC" and some other four to eight multi-functional proteins required for efficient viral proliferation. The regulatory functions of geminivirus-encoded proteins and their role in virus-host interactions have been comprehensively reviewed in a number of studies (Fondong 2013; Hanley-Bowdoin et al. 2013; Lozano-Durán 2016; Yang et al. 2016). In addition, majority of the monopartite begomoviruses are associated with DNA satellites, including betasatellite, alphasatellite, and deltasatellite (Adams et al. 2017). Except alphasatellites that encode their own replication protein, these satellites are nucleic acid agents and, in most of the cases, entirely depend on their helper virus for encapsidation, movement, transmission, and replication.

Geminiviruses' prevalence has increased in the last two decades with the emergence of new strains prevailing across a wide geographical range. Outbreaks of geminiviral infections in food crops like cassava (Rey and Vanderschuren 2017), maize (Shepherd et al. 2010), tomato (Lefeuvre et al. 2010), etc., have threatened the food security of many developing nations across sub-Saharan Africa and Southern Asia. Geminiviruses are transmitted through insect vectors such as whiteflies, treehopper, leafhopper, etc., which facilitate their dispersal *in planta*. After invading the plant cell, geminiviruses hostile in the nucleus and start replication through double-stranded (ds) DNA intermediates. This dsDNA serves as a template for viral transcription and replication either through rolling-circle amplification (RCA) and/or pseudo-recombination-dependent mechanisms (Fondong 2013). Notably, geminiviruses have precisely evolved their replication cycle by utilizing host cellular resources and have reprogrammed/redirected many of the host regulatory proteins and pathways for their self-proliferation, reviewed in (Hanley-Bowdoin et al. 2013). The incidence of geminiviruses on diverse plant hosts suggests their adaptive evolution to invade and overcome the limitations imposed by host defense systems. To produce an effective infection, some geminiviral-encoded proteins also counteract with host RNA-mediated defense and other cellular signaling pathways (Hanley-Bowdoin et al. 2013). Later replication, the synthesized viral genome accumulates to high levels and shuttles to neighboring cells to expand the infection circle.

Despite sharing some common features, geminiviruses exhibit considerable diversity in distinct genomic characters among different genera. The diversification of geminiviruses suggests their evolutionary history and capability to reemerge as new species through genomic evolution. Efficient genome organization of geminiviruses facilitates them to take more risks of producing recombinant progenies owing to multifunctional proteins, overlapping genes, and highly optimized virus-host interactions.

2 Genomic Changes Through Mutations

Genomic changes through mutations are one of the prime routes for the evolution of all viruses and, also, for geminiviruses. Generation of recombinants and pseudo-recombinants are other mechanisms involved in viral genomic variations. As mixed viral infections are frequent in nature, the chances of heterologous genetic exchange through pseudo-recombination arise, which provide the basis for genomic evolution. This might lead to the development of new species/strains/variants, which have the ability to evolve quickly and infect new hosts. In case of bipartite geminiviruses, the interdependence of DNA-A and DNA-B components of viral genomes has resulted in enhanced virulence as well as in increased infectious range (Seal et al. 2006; Idris et al. 2008). In addition, the interactions among begomoviruses and their associated satellites have resulted in significant evolutionary impacts, which have broadened their epidemiology worldwide (Fig. 1) (Nawaz-ul-Rehman and Fauquet 2009). Recombinations among geminiviruses have resulted in significant variations ending up in enhanced virulence and broad association with new satellite components (Padidam et al. 1999). Sequence comparison analysis of geographically diverse geminiviruses exhibited some significant recombination events among conserved regions (CRs) of DNA-A components (Padidam et al. 1999). Examples of recombinant geminiviruses having re-assorted CRs include *Potato yellow mosaic virus* (PYMV) and *Potato yellow mosaic Panama virus* (PYMPV) (Urbino et al. 2004). Recombinant betasatellite associated with *Tomato yellow leaf curl China virus* (TYLCCNV) has been identified infecting tobacco plants, having genomic components of both begomoviral DNA-A and betasatellite molecules (Tao and Zhou 2008). Sequence analysis of recombinant betasatellite revealed the stable integration of CR as well as *AC*1 gene sequences from TYLCCNV. Some compound microsatellite (cSSR) motifs in geminiviral genomes have also been characterized as potential hotspots for recombination, suggesting functional advantages for geminiviral evolution (George et al. 2015). In the context of alphasatellites, a recent study has reported the recombinations among different alphasatellites associated with various hosts (Kumar et al. 2017). Recombination analysis of Chilli leaf curl alphasatellite (ChiLCA) has identified a nucleotide substitution rate of 2.25×10^{-3} of replication-associated protein (Rep) gene that might be facilitative of alphasatellite increasing epidemiology.

The cross-continental emergence of different geminiviral strains is subject to point mutations and pseudo-recombination that gradually gave rise to the development of new strains across diverse geographical locations. For example, recombinations among mastreviruses (*Maize streak virus*, MSV) infecting monocot and dicot plants across Africa have resulted in the emergence of some new MSV strains with novel characters that facilitated the dispersal of pathogens and promoted their host infectious range (Varsani et al. 2008).

Case studies of point mutations have been reported for three different geminiviruses, TYLCCNV, *East African cassava mosaic virus* (EACMV), and MSV, and reported a mutation frequency of $\sim 10^{-4}$ point substitutions/site/year (Isnard et al. 1998; Ge et al. 2007; Duffy and Holmes 2009). Investigation of the

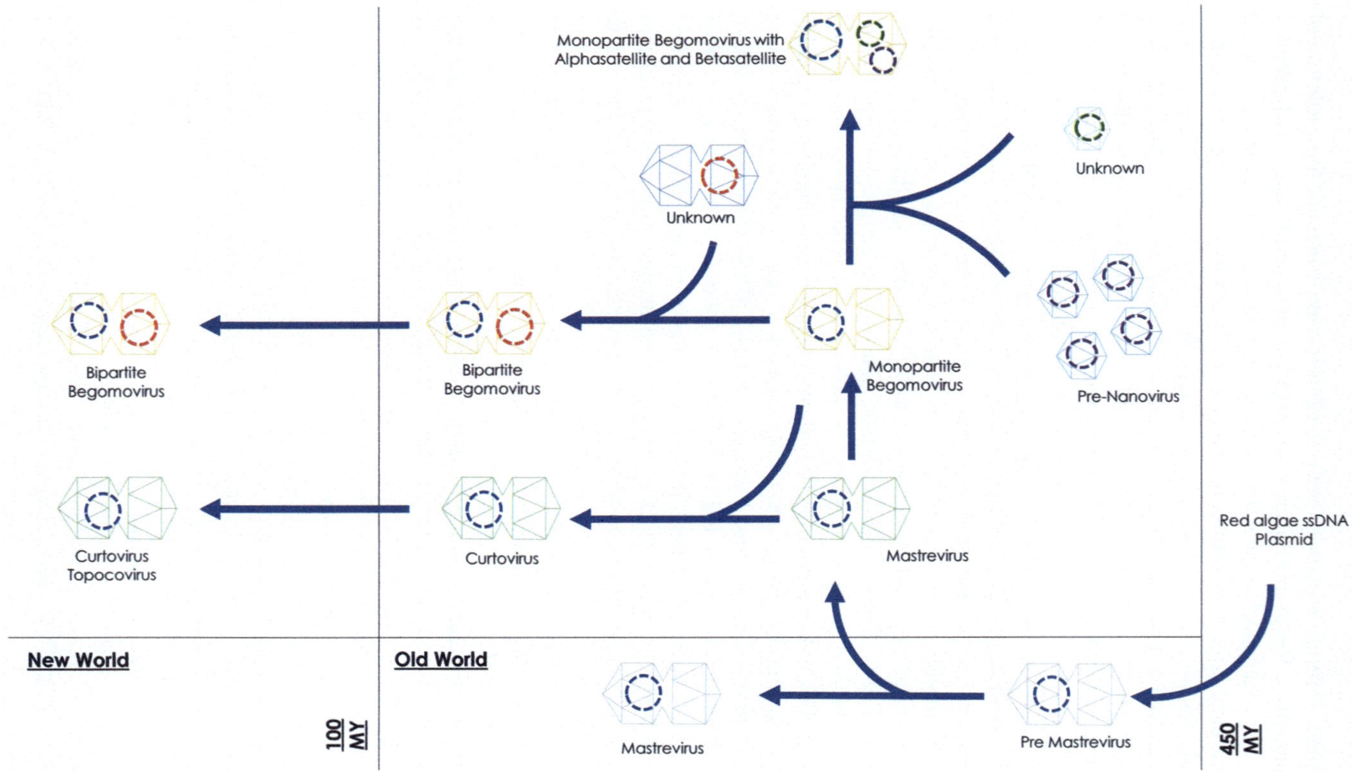

Fig. 1 A model for geminivirus evolution by component capture on space and time. Dotted circles represent single-stranded DNA (ssDNA) genomes where monopartite/DNA-A, DNA-B, alphasatellite and betasatellite ssDNA are represented by blue, red, purple and green colors, respectively

impacts of these mutations has predominantly shown transition point mutations in geminiviral genomes, as compared to insertion/deletion mutations. In addition, the viral mutations are dependent on several factors like viral strain, host type, growth stage of inoculated host plant, and inoculum consistency. Intrageneric recombinations among geminiviruses have facilitated the outbreaks of cotton leaf curl disease (CLCuD) in the Indian subcontinent. Phylogenetic and recombination detection analyses have indicated high recombination frequencies among two highly virulent components [(*Cotton leaf curl Multan virus* (CLCuMuV) and *Cotton leaf curl Kokhran virus* (CLCuKoV)] of CLCuD (Saleem et al. 2016). In several recombinants, the presence of coat protein (CP) from donor CLCuKoV and Rep gene from donor CLCuMuV, having nucleotide substitution rates of 2.706×10^{-4} and 4.96×10^{-4}, respectively, has been observed. Likewise, *Tomato leaf curl Yunnan virus* (TLCYnV), a recombinant begomovirus, acquired *C4* gene sequences from TYLCCNV, which systemically enhanced its virulence and assisted in overcoming host RNA defenses (Xie et al. 2013).

3 Mixed Infections

Mixed viral infections are common in nature and can result in unpredictable biological and epidemiological changes among co-infecting viruses. Importantly, an accumulative effect of viral disease, exhibiting severe infection, has been observed in most of the mixed infection cases. During these viral interactions, another form of assistance may be provided by one member in the ease of viral transmission through its vector species (Syller 2012). Co-infections of mastreviruses and begomoviruses infecting a single host (*Xanthium strumarium*) represent some other potential risks to the emergence of new species (Mubin et al. 2012). Likewise, a distinct strain of mastrevirus [(*Chickpea chlorotic dwarf virus* (CpCDV)] was found associated with CLCuD in Pakistan, co-infecting with whitefly-transmitted begomovirus [(*Cotton leaf curl Burewala virus* (CLCuBuV)] (Manzoor et al. 2014).

Intergenic recombination is another driving force assisting geminiviral evolution. Pseudo-recombination patterns among closely and/or distantly related geminiviruses have been creating variations leading to new variants or even new genera. For example, cutroviruses, inside *Geminiviridae*, are believed to be evolved through recombination among mastreviruses and begomoviruses (Varma and Malathi 2003). Detection of *Beet curly top Iran virus* (BCTIV) in Iran provides some phylogenetic evidence of becutroviruses evolution. The chimeric genome of BCTIV suggests that it may have evolved through intergeneric recombinations among cutrovirus and mastrevirus ancestors (Yazdi et al. 2008). *Spinach curly top Arizona virus* (SCTAV), an intergeneric recombinant, was reported from NW Arizona, which retained the CR/Rep genomic sequences of OW BCTIV and *Spinach curly top virus* (SCTV) (Hernández-Zepeda et al. 2013). Intergeneric recombinations among geminiviruses have been performed experimentally to produce recombinant chimeras. Recently, Khalid et al. (2017) reported the construction of a chimeric geminivirus (pGII0000MBC) that retained the infectious properties and was producing disease symptoms. Notably, they removed the CP part of a dicot-infecting mastrevirus (CpCDV) and molecularly replaced with a begomoviral-CP (*Cotton leaf curl Burewala virus*) and produced a recombinant mastrebegomo chimera.

4 Efficient Dependence on Insect Vectors

Indeed, insect vectors have contributed significantly in driving the geographical spread of phytopathogens worldwide. The efficient dependence of geminiviruses on insect vectors has developed co-evolutionary interactions between virus and vectors that facilitated their acquisition, retention, and systemic spread *in planta*. Vector transmissibility of viruses is a multifaceted interaction dependent on several factors including climatic season, alternate hosts/weeds, insect type, virion properties, host types, etc. Variability in insect vector causes a loss in viral transmission, but retains the virus inside the vector. For example, after exposure to an infected plant, mastrevirus survived for up to 28 days inside viruliferous leafhopper by circulative nonpropagative approach (Lett et al. 2002). The adaptation of vector-mediated transmission was a prerequisite for geminiviral evolution because of their limited natural capacity to be transmitted either by mechanical or seed-borne methods. Humans have also played a major role in enhancing viral epidemiology directly or indirectly by assisting the transmission of insect-vector populations. Introduction of heavy insecticide-based agricultural ecosystems has resulted in massive outbreaks of particular insects that later acquired insecticide resistance.

In case of begomovirus transmission, whitefly (*B. tabaci*) is responsible for a circulative and persistent mode of viral transmissibility. High reproduction rate, diverse phylogeny, alternative hosts, insecticide resistance, and global agricultural trades are some of the factors that proportionately accelerated the population of whiteflies worldwide. Explorations of begomovirus-whitefly complexes have revealed some unique interactions between vector and different viral components that facilitate the viral dispersal (Czosnek et al. 2002). The cross-continental spread of TYLCV and *Bean golden yellow mosaic virus* in the United States is hypothesized due to viruliferous whiteflies originated from the Caribbean through strong winds or hurricane waves (Rojas et al. 2005). The emergence of new world (NW) monopartite begomoviruses [*Tomato leaf deformation virus* (ToLDeV), *Tomato mottle leaf curl virus* (ToMoLCV), and *Tomato severe leaf curl virus* (ToSLCV)] infecting crops in Central and South America is suggestive of *B. tabaci*-transmitted old world (OW) geminiviruses (Gilbertson et al. 2015). Pandemic outbreaks of cassava mosaic diseases (CMDs) in Africa and in the Indian subcontinent are also accredited to the whitefly-transmitted geminiviruses. Infected seed/plant transmission was another huge factor for the geographical spread of CMD in Africa. Severe epidemics of curly top diseases in tomato (*Solanum lycopersicum*) were reported from western areas of the United States caused by beet leafhopper (*Circulifer tenellus*)-transmitted curtoviruses (Chen and Gilbertson 2016).

5 Introduction of New Hosts and Emergence of New Viruses

Rather following a rapid evolution, the global emergence of new strains of geminiviruses predominantly in the last two decades is related to their mobilizations to new hosts, transmission vectors, and/or movement of infectious plant materials. Transferring crops like tomato, cotton, cassava, etc., to new ecological habitats resulted in the invasion of local viral strains to encounter new resources that ultimately derived the viral evolution.

The introduction of NW cotton (*Gossypium hirsutum*) from Mexico to the Indian subcontinent elaborates the best example of geminiviral emergence on the new host. Due to favorable climate and geography, higher yield, and flourishing textile industry, cultivation of NW cotton dramatically increased in this region. Over a cultivated span of >100 years, there was no record of CLCuD in the Indian subcontinent (Briddon and Markham 2000; Sattar et al. 2013). The first report of CLCuD appeared in 1967 from Pakistan (Briddon and Markham 2000), and since then, its major outbreaks expended from Pakistan to neighboring countries (China and India). Since then, CLCuD is attributed as a disease complex comprising up to seven geminiviral species, which evolved overtime (Sattar et al. 2013). Recent studies report the reappearance of recombinant *Cotton leaf curl Kokhran virus* strain Burewala (CLCuKoV-Bur) in cotton breeding lines, which once, in 2001, was considered for resistance-breaking against CLCuD-resistant cotton (Hassan et al. 2017), and also for the reemergence of multiple begomoviruses found associated with CLCuD in Pakistan in the early 1990s (Zubair et al. 2017). This represents an indication of ongoing geminiviral evolution to sustain the threats of CLCuD complex.

During the sixteenth century, cross-continental movement of cassava from its origin (South America) to Africa is another example of geminiviral emergence on new hosts. After its cultivation in a new habitat, severe epidemics of CMD appeared and significantly reduced cassava production in sub-Saharan Africa (Rey and Vanderschuren 2017). CMD is associated with at least nine geminiviral species, which were believed to be natively prevailing in Africa as there were no reports of their incidence in South America (Ndunguru et al. 2005). The emergence of cassava geminiviruses in Africa is linked with their adaptation to exotic host (cassava) from some indigenous hosts (Ndunguru et al. 2005).

Tomato leaf curl New Delhi virus (ToLCNDV) is an emergent begomovirus infecting a number of important crops like tomato, potato, chili pepper, eggplant, cotton, etc., predominantly in the Indian subcontinent (Moriones et al. 2017). For the last few years, the diversity of ToLCNDV has extended to the Middle East, Western Mediterranean Basin, and North Africa infecting multiple crops (Zaidi et al. 2017e). Recently, recombinant strains of ToLCNDV are reported from Spain infecting new hosts which suggest the viral evolution to broaden its epidemiology (Fortes et al. 2016). Frequent occurrence of ToLCNDV has also been reported in cotton (Zaidi et al. 2016), either alone or in mixed infection with other bipartite begomoviruses (Zaidi et al. 2015). Movement of begomoviruses from cultivated crops to non-cultivated hosts is another driving force for viral evolution. Several new strains of begomoviruses (notably a new strain of TYLCV in Pakistan; Zaidi et al. 2017a) have been identified from weeds (such as *Calotropis procera* and *Eclipta prostrata* (Zaidi et al. 2017c, b), which exhibit the increasing diversity of geminiviruses to invade new hosts for their survival and recombinations (Ferro et al. 2017).

6 Agricultural Trades of Infectious Plant Materials

Trade of agricultural products is one of the major factors contributing to the spread of virus/insect pests worldwide. Several exogenous geminiviruses invaded new ecological zones of the world where they were introduced through human activity. The introduction of TYLCV in America is accredited to the movement of infected source material from Israel (Polston et al. 1999). Likewise, the prevalence of TYLCV in France and Spain is also believed due to infected plants coming elsewhere from Europe (Varma and Malathi 2003). Recently, ToLCNDV, a bipartite begomovirus, was reported from Pakistan infecting potato crop (Hameed et al. 2017). Earlier in 2004, the prevalence of ToLCNDV was predominantly reported from India causing severe leaf curl disease in potato (Usharani et al. 2004). The detection of ToLCNDV infecting potato in Pakistan is suggestive of viral movement from India through infectious potato materials.

Ongoing spread of geminiviruses in Oman, primarily a trading nation and recently establishing agriculture, is an evidence of viral emergence due to human activities (Khan et al. 2014). The emergence of various geographically distinct geminiviruses in Oman is potentially linked to the import of infectious plant materials like ornamental plants, which suggest the outward spread of viruses in the new ecological zone. Another example of ornamental import involved in geminiviral spread includes the emergence of CLCuD in China (Sattar et al. 2013). The geographical spread of *Tomato yellow leaf curl Thailand virus* (TYLCTHV) in China is traced back to the regions of Thailand-Myanmar from where it was infecting solanaceous crops (Kenyon et al. 2014).

7 Climatic Factors Enhancing Viral Dispersion

Environmental changes in temperature, wind, rainfall, etc., also impact viral distribution through favorable conditions for disease/vector proliferation. A little increase or decrease in temperature can generate considerable shifts in the insect-vector populations. In case of begomoviruses, the vector (*B. tabaci*) activity is greatly influenced by climatic conditions, which proportionally controls the begomoviral spread. Adult whiteflies prefer to reproduce on young leaves for oviposition sites that are more likely to disrupt during adverse climate (Van Lenteren and Noldus 1990). Heavy rainfall and low temperatures drastically reduce the whitefly population, whereas temperatures ranging within 25–33 °C and scanty rainfall build up an optimistic effect on their multiplication (Morales and Jones 2004). Strong winds and dust storms facilitate the long-distance movement of whiteflies, which also disperse the viruses within. The introduction of *Bean golden mosaic virus* (BGMV) in the Southern United States during the early 1990s was linked with exogenous whiteflies, which might have flown through Caribbean hurricanes (Varma and Malathi 2003). Rising threats of drought and salinity might result in enhanced viral infections due to increased crop vulnerability to counter back the combined stress conditions (Jones 2009). It might also reprogram the host-virus interactions directed towards increased viral epidemiology. Outward spreads of CLCuD complex in Asia are attributed to increased vector populations surviving in optimum climate, and this outbreak has initiated attempts to introduce dual geminivirus—*B. tabaci* resistance in cotton (Shukla et al. 2016; Zaidi et al. 2017d). This spread has also initiated higher incidence of a geminivirus-associated CLCuD, which is recorded in areas having temperature ranges of 33–45 °C with relatively fewer rainfalls (Mahatma et al. 2016).

8 Viral Dominance and Emergence by Suppressing Host Defense System

RNA interference (RNAi), an evolutionarily conserved mechanism in plants, provides defense against intruding viruses (Nicaise 2014). To counter back this defense system, viruses have co-evolved several pathways to interfere/block the host anti-viral silencing mechanism. Expression of some viral proteins is the most common approach that viruses employ against host defensive measures. These anti-silencing proteins/factors are termed as RNA silencing suppressors (RSS). Further expansions in plant virology have identified several RSS in different viral species and have investigated their modes of action. It was confirmed that viral-encoded RSS can interrupt host anti-viral silencing pathway at each step or could generate a multilevel interruption through interacting with individual RNA-induced silencing complex (RISC) components. Through involvement at the first phase of silencing, RSS try to block/interrupt the RNA interference (RNAi) trigger precursors, i.e., viral-derived short interfering RNA (siRNAs) formation as it could be more effective to stop silencing at an early stage. To conduct this process, RSS could either block the synthesis of siRNAs through targeting the cleavage of viral dsRNA transcripts or modify/reform the siRNAs before their release into RISC. Some viral-derived RSS have been reported to inhibit Dicer-like proteins (DCLs)' function such as p27 and p88 from *Red clover necrotic mosaic virus*, and P6 from *Cauliflower mosaic virus*. Another interruption in RNA silencing pathway is mediated through RSS binding with viral-derived long dsRNA precursors to prevent their cleavage/processing into siRNAs. A number of RSS are reported from different viral species having these affinity activities such as p38 from *Turnip crinkle virus*, p14 from *Pothos latent virus*, 2b from *Cucumber mosaic virus*, and p22 from *Tomato chlorosis virus*.

In case of geminiviruses, *V2* from *Tomato yellow leaf curl virus* (TYLCV) has been identified making interactions with host SGS3 effector proteins to interrupt the viral silencing pathway at the siRNA amplification phase. Several geminiviral-encoded RSS interact with host defense systems and block the viral targeting at transcriptional gene silencing (TGS) level. Examples include *AL2* from *Tomato golden mosaic virus*, *L2* from *Beet curly top virus*, *βC1* from *Tomato yellow leaf curl China virus*, *C2* from *Beet severe curly top virus*, and *TrAP* from *Cabbage leaf curl virus* that represent the RSS involved in disruption of DNA methylation during anti-viral TGS in host cells. Cassava-geminiviruses-encoded *AC2* and *AC4* proteins were investigated for their RSS activity in cassava and tobacco plants (Vanitharani et al. 2004). Transient expression of *AC4* of ACMV and *AC2* of *East African cassava mosaic Cameroon virus* (EACMCV) under constitutive *Cauliflower mosaic virus* 35S promoter resulted in a nearly eightfold increase in co-infecting viruses, suggestive of their role in suppressing the host PTGS (Vanitharani et al. 2004).

9 Conclusion

Geminiviruses have evolved as one of the most important plant viruses, and their rapid evolution poses a serious threat to the global agriculture. In context to their broad epidemiology, several factors have facilitated geminivirus emergence and their existence around the world. Factors like genomic evolution, efficient dependence on insect vectors, migrations to new hosts, climatic adaptability, agricultural trades, mixed infections, and suppression of host defense systems have proportionally influenced the ongoing geminiviral evolution. Overall, evolution, and specifically in case of geminiviruses, is an ongoing process. Emergence of new geminivirus strains, synergistic interactions among geminiviruses, new geminivirus-alpha/betasatellite complexes, and, consequently, their increasing host range and global spread pose a serious challenge for plant breeders and molecular biologists to design efficient and long-term resistance strategies in economically important crops. Meanwhile, rapid evolution of geminiviruses and their efficient evasion of resistance mechanisms must be considered while designing these resistance strategies.

References

Adams MJ, Lefkowitz EJ, King AMQ, Harrach B, Harrison RL, Knowles NJ, Kropinski AM, Krupovic M, Kuhn JH, Mushegian AR, Nibert M, Sabanadzovic S, Sanfacon H, Siddell SG, Simmonds P, Varsani A, Zerbini FM, Gorbalenya AE, Davison AJ (2017) Changes to taxonomy and the international code of virus classification and nomenclature ratified by the international committee on taxonomy of viruses (2017). Arch Virol 162:2505–2538

Briddon RW, Markham PG (2000) Cotton leaf curl virus disease. Virus Res 71:151–159

Chen LF, Gilbertson RL (2016) Transmission of curtoviruses (*beet curly top virus*) by the beet leafhopper (*Circulifer tenellus*). In: Vector-mediated transmission of plant pathogens. The American Phytopathological Society (APS), USA, pp 243–262

Czosnek H, Ghanim M, Ghanim M (2002) The circulative pathway of begomoviruses in the whitefly vector *Bemisia tabaci*—insights from studies with tomato yellow leaf curl virus. Ann Appl Biol 140:215–231

Duffy S, Holmes EC (2009) Validation of high rates of nucleotide substitution in geminiviruses: phylogenetic evidence from East African cassava mosaic viruses. J Gen Virol 90:1539–1547

Ferro C, Silva J, Xavier C, Godinho M, Lima A, Mar T, Lau D, Zerbini F (2017) The ever increasing diversity of begomoviruses infecting non-cultivated hosts: new species from Sida spp. and *Leonurus sibiricus*, plus two New World alphasatellites. Ann Appl Biol 170:204–218

Fondong VN (2013) Geminivirus protein structure and function. Mol Plant Pathol 14:635–649

Fortes IM, Sánchez-Campos S, Fiallo-Olivé E, Díaz-Pendón JA, Navas-Castillo J, Moriones E (2016) A novel strain of tomato leaf curl New Delhi virus has spread to the Mediterranean basin. Viruses 8:307

Ge L, Zhang J, Zhou X, Li H (2007) Genetic structure and population variability of tomato yellow leaf curl China virus. J Virol 81:5902–5907

George B, Alam CM, Kumar RV, Gnanasekaran P, Chakraborty S (2015) Potential linkage between compound microsatellites and recombination in geminiviruses: evidence from comparative analysis. Virology 482:41–50

Gilbertson RL, Batuman O, Webster CG, Adkins S (2015) Role of the insect supervectors Bemisia tabaci and *Frankliniella occidentalis* in the emergence and global spread of plant viruses. Annu Rev Virol 2:67–93

Hameed A, Tahir M, Amin I, Mansoor S (2017) First report of tomato leaf curl New Delhi virus and a tomato yellow leaf curl thailand betasatellite causing severe leaf curl disease of potato in Pakistan. Plant Dis 101:1065–1065

Hanley-Bowdoin L, Bejarano ER, Robertson D, Mansoor S (2013) Geminiviruses: masters at redirecting and reprogramming plant processes. Nat Rev Microbiol 11:777–788

Hassan I, Amin I, Mansoor S, Briddon RW (2017) Further changes in the cotton leaf curl disease complex: an indication of things to come? Virus Genes 53:759–761

Hernández-Zepeda C, Varsani A, Brown JK (2013) Intergeneric recombination between a new, spinach-infecting curtovirus and a new geminivirus belonging to the genus Becurtovirus: first New World exemplar. Arch Virol 158:2245–2254

Idris AM, Mills-Lujan K, Martin K, Brown JK (2008) Melon chlorotic leaf curl virus: characterization and differential reassortment with closest relatives reveal adaptive virulence in the squash leaf curl virus clade and host shifting by the host-restricted bean calico mosaic virus. J Virol 82:1959–1967

Isnard M, Granier M, Frutos R, Reynaud B, Peterschmitt M (1998) Quasispecies nature of three maize streak virus isolates obtained through different modes of selection from a population used to assess response to infection of maize cultivars. J Gen Virol 79:3091–3099

Jones RA (2009) Plant virus emergence and evolution: origins, new encounter scenarios, factors driving emergence, effects of changing world conditions, and prospects for control. Virus Res 141:113–130

Kenyon L, Tsai WS, Shih SL, Lee LM (2014) Emergence and diversity of begomoviruses infecting solanaceous crops in East and Southeast Asia. Virus Res 186:104–113

Khalid S, Zia-Ur-Rehman M, Ali SA, Hameed U, Khan F, Ahmed N, Farooq AM, Haider MS (2017) Construction of an infectious chimeric geminivirus by molecular cloning based on coinfection and recombination. Int J Agric Biol 19:629–634

Khan AJ, Mansoor S, Briddon RW (2014) Oman: a case for a sink of begomoviruses of geographically diverse origins. Trends Plant Sci 19:67–70

Kumar RV, Singh D, Singh AK, Chakraborty S (2017) Molecular diversity, recombination and population structure of alphasatellites associated with begomovirus disease complexes. Infect Genet Evol 49:39–47

Lefeuvre P, Martin DP, Harkins G, Lemey P, Gray AJ, Meredith S, Lakay F, Monjane A, Lett JM, Varsani A, Heydarnejad J (2010) The spread of tomato yellow leaf curl virus from the Middle East to the world. PLoS Pathog 6:e1001164

Lett JM, Granier M, Hippolyte I, Grondin M, Royer M, Blanc S, Reynaud B, Peterschmitt M (2002) Spatial and temporal distribution of geminiviruses in leafhoppers of the genus cicadulina monitored by conventional and quantitative polymerase chain reaction. Phytopathology 92:65–74

Lozano-Durán R (2016) Geminiviruses for biotechnology: the art of parasite taming. New Phytol 210:58–64

Mahatma L, Mahatma M, Pandya J, Solanki R, Solanki V (2016) Epidemiology of begomoviruses: a global perspective. In: Plant viruses: evolution and management. Springer, Singapore, pp 171–188
Mansoor S, Zafar Y, Briddon RW (2006) Geminivirus disease complexes: the threat is spreading. Trends Plant Sci 11:209–212
Manzoor MT, Ilyas M, Shafiq M, Haider MS, Shahid AA, Briddon RW (2014) A distinct strain of chickpea chlorotic dwarf virus (genus Mastrevirus, family Geminiviridae) identified in cotton plants affected by leaf curl disease. Arch Virol 159:1217–1221
Morales FJ, Jones PG (2004) The ecology and epidemiology of whitefly-transmitted viruses in Latin America. Virus Res 100:57–65
Moriones E, Praveen S, Chakraborty S (2017) Tomato leaf curl New Delhi virus: an emerging virus complex threatening vegetable and fiber crops. Viruses 9:264
Mubin M, Mansoor S, Briddon RW (2012) Letter to the editor: mastrevirus sequences in a begomovirus-infected plant. Virus Genes 44:536–538
Nawaz-Ul-Rehman MS, Fauquet CM (2009) Evolution of geminiviruses and their satellites. FEBS Lett 583:1825–1832
Ndunguru J, Legg JP, Aveling T, Thompson G, Fauquet CM (2005) Molecular biodiversity of cassava begomoviruses in Tanzania: evolution of cassava geminiviruses in Africa and evidence for East Africa being a center of diversity of cassava geminiviruses. Virol J 2:21
Nicaise V (2014) Crop immunity against viruses: outcomes and future challenges. Front Plant Sci 5:660
Padidam M, Sawyer S, Fauquet CM (1999) Possible emergence of new geminiviruses by frequent recombination. Virology 265:218–225
Polston J, Mcgovern R, Brown L (1999) Introduction of tomato yellow leaf curl virus in Florida and implications for the spread of this and other geminiviruses of tomato. Plant Dis 83:984–988
Rey C, Vanderschuren H (2017) Cassava mosaic and brown streak diseases: current perspectives and beyond. Annu Rev Virol 4:429–452
Rojas MR, Hagen C, Lucas WJ, Gilbertson RL (2005) Exploiting chinks in the plant's armor: evolution and emergence of geminiviruses. Annu Rev Phytopathol 43:361–394
Saleem H, Nahid N, Shakir S, Ijaz S, Murtaza G, Khan AA, Mubin M, Nawaz-Ul-Rehman MS (2016) Diversity, mutation and recombination analysis of cotton leaf curl geminiviruses. PLoS One 11:e0151161
Sattar MN, Kvarnheden A, Saeed M, Briddon RW (2013) Cotton leaf curl disease – an emerging threat to cotton production worldwide. J Gen Virol 94:695–710
Seal S, Vandenbosch F, Jeger M (2006) Factors influencing begomovirus evolution and their increasing global significance: implications for sustainable control. Crit Rev Plant Sci 25:23–46
Shepherd DN, Martin DP, Van Der Walt E, Dent K, Varsani A, Rybicki EP (2010) Maize streak virus: an old and complex 'emerging'pathogen. Mol Plant Pathol 11:1–12
Shukla AK, Upadhyay SK, Mishra M, Saurabh S, Singh R, Singh H, Thakur N, Rai P, Pandey P, Hans AL, Srivastava S, Rajapure V, Yadav SK, Singh MK, Kumar J, Chandrashekar K, Verma PC, Singh AP, Nair KN, Bhadauria S, Wahajuddin M, Singh S, Sharma S, Omkar, Upadhyay RS, Ranade SA, Tuli R, Singh PK (2016) Expression of an insecticidal fern protein in cotton protects against whitefly. Nat Biotechnol 34:1046–1051
Syller J (2012) Facilitative and antagonistic interactions between plant viruses in mixed infections. Mol Plant Pathol 13:204–216
Tao X, Zhou X (2008) Pathogenicity of a naturally occurring recombinant DNA satellite associated with tomato yellow leaf curl China virus. J Gen Virol 89:306–311
Urbino C, Polston JE, Patte CP, Caruana ML (2004) Characterization and genetic diversity of potato yellow mosaic virus from the Caribbean. Arch Virol 149:417–424
Usharani K, Surendranath B, Paul-Khurana S, Garg I, Malathi V (2004) Potato leaf curl – a new disease of potato in northern India caused by a strain of tomato leaf curl New Delhi virus. Plant Pathol 53:235–235

Van Lenteren JCV, Noldus L (1990) Whitefly-plant relationships: behavioural and ecological aspects. In: Gerling D (ed) Whiteflies: their bionomics, pest status and management. Intercept, Andove, pp 47–89

Vanitharani R, Chellappan P, Pita JS, Fauquet CM (2004) Differential roles of AC2 and AC4 of cassava geminiviruses in mediating synergism and suppression of posttranscriptional gene silencing. J Virol 78:9487–9498

Varma A, Malathi V (2003) Emerging geminivirus problems: a serious threat to crop production. Ann Appl Biol 142:145–164

Varsani A, Shepherd DN, Monjane AL, Owor BE, Erdmann JB, Rybicki EP, Peterschmitt M, Briddon RW, Markham PG, Oluwafemi S (2008) Recombination, decreased host specificity and increased mobility may have driven the emergence of maize streak virus as an agricultural pathogen. J Gen Virol 89:2063–2074

Varsani A, Roumagnac P, Fuchs M, Navas-Castillo J, Moriones E, Idris A, Briddon RW, Rivera-Bustamante R, Zerbini FM, Martin DP (2017) Capulavirus and Grablovirus: two new genera in the family Geminiviridae. Arch Virol 162:1819–1831

Xie Y, Zhao L, Jiao X, Jiang T, Gong H, Wang B, Briddon RW, Zhou X (2013) A recombinant begomovirus resulting from exchange of the C4 gene. J Gen Virol 94:1896–1907

Yang X, Wang B, Li F, Yang Q, Zhou X (2016) Research advances in geminiviruses. In: Current research topics in plant virology. Springer, Cham, pp 251–269

Yazdi HR, Heydarnejad J, Massumi H (2008) Genome characterization and genetic diversity of beet curly top Iran virus: a geminivirus with a novel nonanucleotide. Virus Genes 36:539–545

Zaidi S, Iqbal Z, Amin I, Mansoor S (2015) First report of tomato leaf curl Gujarat virus, a bipartite begomovirus on cotton showing leaf curl symptoms in Pakistan. Plant Dis 99:1655

Zaidi SS, Shafiq M, Amin I, Scheffler BE, Scheffler JA, Briddon RW, Mansoor S (2016) Frequent occurrence of tomato leaf curl New Delhi virus in cotton leaf curl disease affected cotton in Pakistan. PLoS One 11:e0155520

Zaidi S, Shakir S, Farooq M, Amin I, Mansoor S (2017a) First report of a novel strain of tomato yellow leaf curl virus causing yellow leaf curl disease on cluster bean in Pakistan. Plant Dis 101:1071–1071

Zaidi S, Shakir S, Farooq M, Amin I, Mansoor S (2017b) First report of alternanthera yellow vein virus from *Eclipta prostrata* in Pakistan. Plant Dis 101:266–266

Zaidi S, Shakir S, Malik H, Farooq M, Amin I, Mansoor S (2017c) First report of tomato leaf curl New Delhi virus on *Calotropis procera*, a weed as potential reservoir begomovirus host in Pakistan. Plant Dis 101:1071

Zaidi SS, Briddon RW, Mansoor S (2017d) Engineering dual begomovirus-*Bemisia tabaci* resistance in plants. Trends Plant Sci 22:6–8

Zaidi SS, Martin DP, Amin I, Farooq M, Mansoor S (2017e) Tomato leaf curl New Delhi virus: a widespread bipartite begomovirus in the territory of monopartite begomoviruses. Mol Plant Pathol 18:901–911

Zerbini FM, Briddon RW, Idris A, Martin DP, Moriones E, Navas-Castillo J, Rivera-Bustamante R, Roumagnac P, Varsani A, Ictv Report C (2017) ICTV virus taxonomy profile: Geminiviridae. J Gen Virol 98:131–133

Zubair M, Zaidi SS-E-A, Shakir S, Farooq M, Amin I, Scheffler JA, Scheffler BE, Mansoor S (2017) Multiple begomoviruses found associated with cotton leaf curl disease in Pakistan in early 1990 are back in cultivated cotton. Sci Rep 7:680

Geminivirus–Vector Relationship

Nicolas Bejerman

Abstract

Geminiviruses are the most abundant plant viruses. This group of ssDNA viruses infects a wide range of hosts including weeds, ornamentals, as well as economically important crops and is widely distributed on the planet Earth. Geminiviruses cause some of the most damaging and economically important diseases of crop plants. This chapter summarizes biological and molecular aspects of the relationships between geminiviruses and their vectors.

1 General Considerations

Geminiviruses are nonenveloped plant-infecting viruses, which comprise either two circular DNA components (bipartite: DNA A and DNA B) or a single circular DNA component (monopartite) of 2.5–5.2 kb in length encapsidated in twinned icosahedral particles. The Geminiviridae comprises nine different genera: Becurtovirus, Begomovirus, Capulovirus, Curtovirus, Eragrovirus, Grablovirus, Mastrevirus, Topucovirus, and Turncurtovirus (Zerbini et al. 2017). Geminiviruses are the most abundant plant viruses, and their insect vectors play significant roles in geminivirus spread and evolution in nature (Yang et al. 2017). This group of single-stranded (ssDNA) viruses infects a wide range of hosts including weeds as well as economically important crops and is widely distributed on the planet Earth. Geminiviruses cause significant economic losses in food, feed, and fiber crops affecting food and nutritional

N. Bejerman (✉)
Consejo Nacional de Investigaciones Científicas y Técnicas (CONICET), Buenos Aires, Argentina

Instituto de Patología Vegetal – Centro de Investigaciones Agropecuarias – Instituto Nacional de Tecnología Agropecuaria (IPAVE-CIAP-INTA), Córdoba, Argentina
e-mail: bejerman.nicolas@inta.gob.ar

R. V. Kumar (ed.), *Geminiviruses*, https://doi.org/10.1007/978-3-030-18248-9_8

security worldwide. For instance, severe losses in cotton in Asia and cassava and maize in Africa are attributed to geminivirus infection. Legumes in India are infected with begomoviruses that cause annual yield losses estimated at $300 million. Tomato leaf curl viruses (ToLCVs) cause annual yield losses of $140 million in Florida, USA, and continue to be a constraint on tomato production worldwide (Ramesh et al. 2017). The geminivirus–vector–host interactions have important influences on the population dynamics of vectors and the epidemiology of these economically important viruses. Therefore, the understanding of the geminivirus transmission by their vector, regarding both biological and molecular aspects, is crucial for the development of effective disease control strategies against these viruses. This chapter summarizes the current knowledge about biological and molecular aspects of insect–geminivirus relationships.

2 Biological Aspects

The process of virus transmission by an insect vector varies based on how the virus is acquired, retained, and inoculated into plants and has been classified into four categories: nonpersistent, semi-persistent, persistent circulative, and persistent propagative (Hogenhout et al. 2008). Geminiviruses are transmitted in a persistent circulative manner. These viruses enter the insect body and disseminate to various tissue systems prior to their transmission to plant hosts but do not replicate in the body of the insect (Whitfield et al. 2015).

Members of the family Geminiviridae are transmitted by hemipteran insects, while members of the largest genus Begomovirus are transmitted by whiteflies (*Bemisia tabaci*, family Aleyrodidae) (Gray et al. 2014). Mastreviruses, Turncurtoviruses, Becutoviruses, Eragroviruses, and Curtoviruses are transmitted by specific leafhoppers (Cicadellidae). *Psammotettix Alienus* and *Cicadulina mbila* transmit the mastreviruses *wheat dwarf virus* (WDF) and *maize streak virus* (MSV), respectively (Wang et al. 2014), whereas the Curtovirus *beet curly top virus* is transmitted by *Circulifer tenellus* (Wang et al. 2014), and the Becurtovirus *beet curly top iran virus* is transmitted by *Circulifer haematoceps* (Varsani et al. 2014), while *Circulifer haematoceps* transmits the turncurtovirus *turnip curly top virus* (Varsani et al. 2014). Grablovirus and Topocuvirus members are transmitted by the specific treehoppers (Membracidae) *Spissistilus festinus* and *Micrutalis malleifera*, respectively (Bahder et al. 2016; Briddon et al. 1996), whereas the Capulavirus members are transmitted by specifics aphids, like *Aphis craccivora*, which transmits the alfalfa leaf curl virus (Roumagnac et al. 2015) (Table 1).

Begomoviruses are the most numerous and widespread viruses within the *Geminiviridae* family, and the emergence of begomoviruses as important pathogens is closely related to the increasing prevalence of whiteflies worldwide (Wei et al. 2017) (Fig. 1). Begomoviruses and whiteflies have been co-evolving for millions of years (Ghanim 2014).

As was mentioned above, begomoviruses are transmitted exclusively by the sweetpotato whitefly *Bemisia tabaci*. Whiteflies have a wide host range and feed on many crops such as tobacco, tomato, pepper, cucumber, potato, and some weeds. Hot and dry conditions in tropical and subtropical regions favor whitefly feeding and

Table 1 Geminivirus genera vector specificity

Genera	Insect vector
Mastrevirus	Leafhopper
Curtovirus	Leafhopper
Topocuvirus	Treehopper
Begomovirus	Whitefly
Capulavirus	Aphid
Eragrovirus	Leafhopper
Grablovirus	Treehopper
Turncurtovirus	Leafhopper
Becurtovirus	Leafhopper

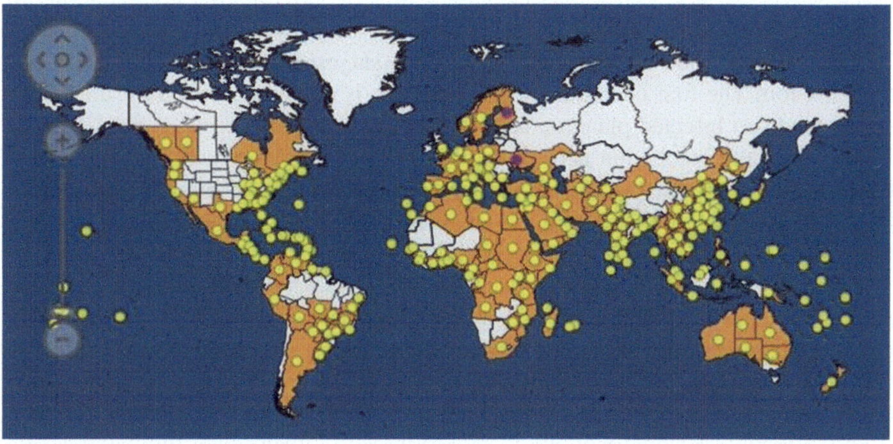

Fig. 1 Worldwide distribution of *Bemisia tabaci*. Adapted from https://gd.eppo.int/taxon/BEMITA/distribution

high reproductive rates and help the spread of the geminiviruses. Winds and temperature deviations have a big impact on the spread of whitefly-transmitted infections (Navas-Castillo et al. 2014).

Bemisia tabaci is a complex of cryptic species that comprises at least 39 species (Vyskočilová et al. 2018). Divergence among *B. tabaci* populations was shown to be high and the species within the complex can be distinguished by sequencing the mitochondrial *COI* gene. Data on the sequence of *mtCOI* of worldwide collected whitefly strains are stored in databases (www.ebi.ac.uk). Analysis of these sequences has shown that the genetic differentiation of *B. tabaci* populations corresponded to the geographical origin, except for two species that are found worldwide. The strong geographical association of populations with hardly any gene flow between them supports the view that *B. tabaci* is a species complex consisting of numerous morphologically cryptic species (Firdaus et al. 2013). Two species are especially of significant economic importance as highly invasive pests: Mediterranean (MED) and Middle East-Asia Minor 1 (MEAM1), known formerly as the Q and B biotypes, respectively, which are found in tropical and subtropical countries around the world (Vyskočilová et al. 2018). Other groups are found in a specific continent or region such as Asia I, Asia IV, Australia, New World, SubSahAf1, and SubSahAf5 that are restricted to specific countries or areas such as Asia, China, Australia, America, Africa, and Uganda, respectively, which suggests that geographical barriers are an important factor affecting differentiation and speciation of *B. tabaci* (Firdaus et al. 2013). The species in this complex differ in host range, resistance to insecticide, virus transmission, and their ability to induce plant disorders (Rosen et al. 2015).

Several studies have shown that most of the *B. tabaci* species complex members may transmit most, if not all, begomoviruses; however, the transmission efficiencies vary significantly among different *B. tabaci* species (Polston et al. 2014). Variation in transmission efficiency is even observed among different populations of the same species (Kollenberg et al. 2014). Variation in virus acquisition and transmission efficiency is the rule rather than exception in different tripartite combinations of whiteflies, begomoviruses, and plants (Polston et al. 2014).

Virus transmission is characterized for three stages: acquisition, incubation, and transmission. The first one is the acquisition that is the stage needed for the vector which feeds on infected plants to acquire the virus. Once the virus is acquired, the next stage is the incubation period, named also as the latency period, which is the time interval between the acquisition and the beginning of infectivity; once the latency period is completed, the vector injects the viruses in a healthy plant and this stage is named as transmission (Hull 2014).

An infection cycle of a geminivirus starts by acquiring virus particles from the plant phloem through the insect stylet (Fig. 2). It has been reported for begomoviruses that the minimum acquisition period (AP) length ranges from 5 to 60 min (Rosen et al. 2015). For example, *tomato yellow leaf curl virus* (TYLCV) and the bipartite begomovirus *Squash leaf curl virus* (SLCuV) transmission by *B. tabaci* is very efficient. A single whitefly is able to infect a tomato plant following a 24 h AP, and the efficiency of transmission reaches 100% when 5–15 insects are used (Czosnek et al. 2001). However, for leafhopper-transmitted geminiviruses, acquisition times range from a few seconds to an hour (Hull 2014). Following acquisition, begomoviruses such as TYLCV and SLCuV are retained in the whitefly vector for its entire life, while *tomato yellow leaf curl Sardinia virus* (TYLCSV) is undetectable after approximately 20 days (Rosen et al. 2015). Longer feeding times give higher transmission rates and longer persistence in the vector (Brown and Czosnek 2002; Hull 2014). The efficiency of the transmission of several begomoviruses tested decreases with an increase in whitefly age, and the sex of the whitefly may also influence virus transmission efficiency as males have been shown to be less efficient vectors (Rosen et al. 2015). Once ingested, there is a latency period before a Begomovirus can be transmitted. During the latency period, the virus first translocates through the midgut to the hemolymph (insect blood) and the salivary glands (the final organ) before it is secreted with saliva during feeding (Fig. 2). The average latency periods vary among different begomoviruses, but the minimum latent period is approximately 19 h (Rosen et al. 2015); for leafhopper-transmitted geminiviruses, this period is 23 ± 4.1 h (Hull 2014). A crucial, yet unknown, number of virions must accumulate in the salivary glands before a successful inoculation of a new host plant for each species of virus (Rosen et al. 2015).

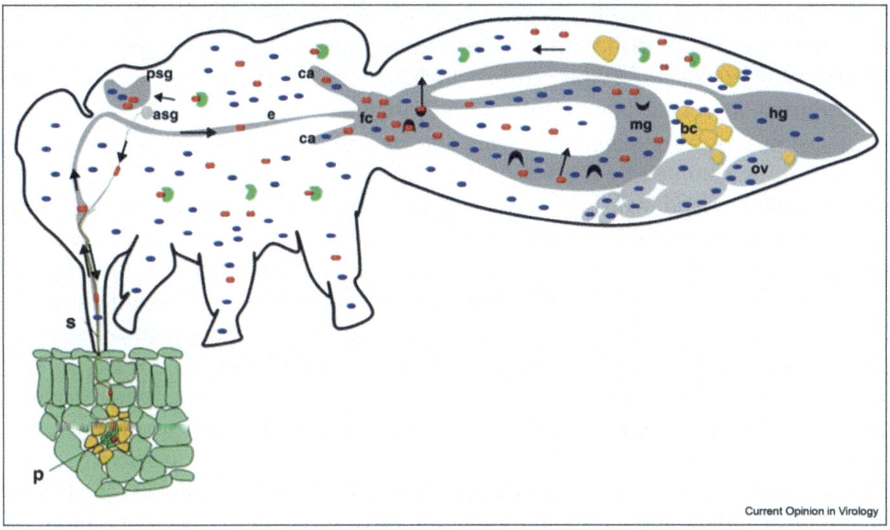

Fig. 2 The circulative transmission pathway for geminiviruses. After acquisition into the insect, virions interact with HSP70 (black particles) in the midgut (mg) and cross the hemolymph. In the hemolymph, virions interact with the GroEL protein (green particles) and cross the insect primary salivary glands (PSG) and then are spit into a host plant with salivary secretions. *s* stylet, *ov* ovary, *ca* caeca, and *asg* accessory salivary gland. Adapted from Rosen et al. (2015)

3 Molecular Aspects

As was stated above, virus transmission has been classified into four categories: nonpersistent, semi-persistent, persistent circulative, and persistent propagative (Hogenhout et al. 2008). In the former two categories, also called noncirculative transmission, the interactions between vector and virus are transient, with the virus only associated with the mouthparts or foregut of the insect vector (Whitfield et al. 2015). In contrast, in the other two categories, the virus develops intimate interactions with internal organs of the vector(s), such as the transmission of geminiviruses (Whitfield et al. 2015). Although geminiviruses are transmitted by diverse vector species, they use a similar route of dissemination, which is the sequential path head-midgut-hemolymph-salivary gland, in the vector, and the viral CP is the viral determinant of this process (Whitfield et al. 2015). The geminivirus CP is the protein encoded by the virus that determines vector specificity; therefore, phylogenetic analysis based on the sequence of this protein is useful to shed light on the likely vector involved in the transmission of the different Geminiviridae members. It is likely that the CP protein interacts with a receptor protein(s) lining the insect gut and salivary gland and thus determines the specificity of insect vector–virus interactions (Bahder et al. 2016; Wang et al. 2014).

Begomoviruses accumulate in vesicle-like structures in whitefly midgut cells (Xia et al. 2018). During circulation, begomovirus virions translocate from the insect midgut epithelial cells into the hemolymph possibly via receptor-mediated endocytosis (Rosen et al. 2015). While in the hemolymph, virions reach and enter the primary salivary glands (PSGs) from which they are egested into the plant with saliva during insect feeding (Czosnek and Ghanim 2012). During the circulative transmission process, geminiviruses have to overcome at least four barriers to be successfully transmitted by the vector: midgut invasion barrier, midgut penetrating barrier, salivary gland invasion barrier, and salivary gland penetrating barrier (Gray et al. 2014); the travel from the gut lumen into the hemolymph of the insect vector is likely the most important step in the circulative transmission; therefore, the crossing of the midgut walls is a significant barrier to virus transmission (Whitfield et al. 2015; Xia et al. 2018). Passage of viruses through these barriers requires specific interactions between virus and vector components (Rosen et al. 2015). Several proteins have been implicated in the circulation of begomoviruses, which include two heat shock proteins. The GroEL protein secreted by endosymbionts, Cyclophilin B, knottin-1, and clathrin have been shown to be involved in the begomovirus circulative transmission (Ghanim 2014; Gotz et al. 2012; Hariton Shalev et al. 2016; Kanakala and Ghanim 2016; Kliot et al. 2014; Luan et al. 2011; Pan et al. 2017; Rana et al. 2016; Wang et al. 2016). Knowledge of the factors involved in begomovirus transmission is not only important to our general understanding of the virus–vector relationship, but also essential to the development of new strategies and techniques for the management of these virus diseases in plants.

Recently, it was reported for the first time that begomoviruses could have a transovarial transmission from female whiteflies to offspring (Wei et al. 2017). It was found that specific interaction between viral coat protein and vector vitellogenin determines transovarial transmissibility of some begomoviruses, which have caused great damage to agricultural production and are generally believed not to be transovarially transmitted by insect vectors; therefore, this provides insights into the evolution and has great significance to their epidemiology Transovarial transmission may have contributed significantly to the global spread of some begomoviruses, such as TYLCV (Wei et al. 2017). Therefore, identification of vector and virus components involved in transovarial transmission can lead to new strategies to combat virus spread.

Geminiviruses depend on their vectors for transmission to the host; therefore, the vector behavior has overwhelming ecological and evolutionary significance for the pathogens that they carry and transmit. Consequently, the ability of a pathogen to alter the behavior of its vector in a manner that facilitates its own transmission would be highly adaptive (Li et al. 2014). Geminiviruses have been shown mainly to modify vector behavior via their shared host plant to achieve an indirect mutualistic relationship between pathogen and vector which could result in an increased probability of transmission to new hosts (Luan et al. 2014). Indirect mutualistic relationships in pathogen–vector–plant interactions have two main aspects. First, the pathogen causes nutritional changes in infected plants, resulting in improved fitness of the vectors. Second, the pathogen increases plant attractiveness and suitability to the vectors by overcoming plant defenses against the vector species, thereby promoting vector performance and increasing pathogen spread (Luan et al.

2014). Pathogens may modulate plant volatile production to influence vector behavior. For instance, volatile terpenoids mediate direct defense against whiteflies (Luan et al. 2013). Infection of tobacco by TYLCCNV and its betasatellite complex reduces the synthesis of the sesquiterpene cedrene. This reduction in turn benefits its vector resulting in a vector–virus mutualism (Luan et al. 2013).

The plant jasmonic acid (JA) signaling pathway plays an important role in whitefly resistance (Li et al. 2014). Begomovirus infection leads to reduced transcription of some JA-responsive genes (Li et al. 2014), and the impairment of JA signaling enhances vector performance (Luan et al. 2013). It was shown that *Arabidopsis thaliana* ASYMMETRIC LEAVES1 (AS1) is a molecular target of the TYLCCNV pathogenicity factor βC1 to explain whitefly–geminivirus mutualism. βC1 directly binds to AS1 and depresses the AS1-mediated suppression of leaf development; by contrast, βC1 promotes the repressive role of AS1 in regulating JA signaling (Li et al. 2014). Furthermore, it was reported that βC1 protein is a key viral genetic factor for the suppression of terpene synthesis to achieve indirect vector–virus mutualism (Li et al. 2014). Also, the transcription factor MYC2 was identified as an additional interaction partner of βC1. MYC2 is a key component in the JA pathway that regulates genes involved in terpene synthesis. So, begomoviruses establish mutualistic relationships with their whitefly vectors by targeting the activity of the plant MYC2 protein (Li et al. 2014) (Fig. 3). These strategies that are based on alterations in JA signaling are employed by begomoviruses for persistent transmission.

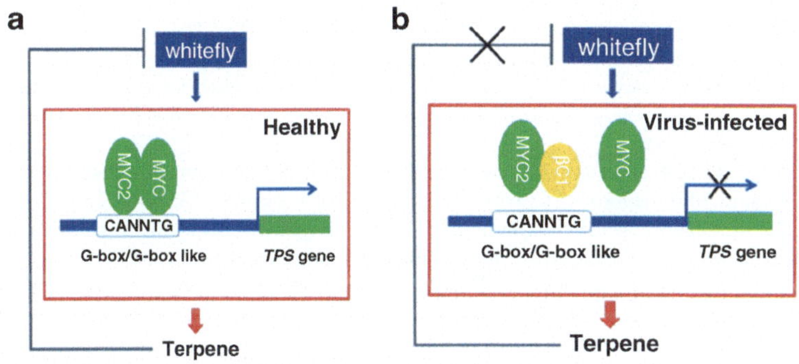

Fig. 3 Working model of plant MYC2 and MYC2-like transcription factors in begomovirus–whitefly–plant tripartite interactions. (**a**) Plant MYC2 mediates the transcription activation of *TPS* genes by direct binding to G-box/G-box-like elements of the promoter region. Whitefly feeding activates the transcription of *MYC2* and *TPS* genes. Monoterpenes or sesquiterpenes are released from plants to defend against whitefly. (**b**) In begomovirus-infected plants, however, βC1 interacts with MYC2, interfering with MYC2 dimerization, which is necessary for the activation of JA-mediated plant resistance. This interaction decreases the DNA binding activity of MYC2 and suppresses transcript levels of *TPS*, leading to reduced release of terpenes. Therefore, begomovirus-infected plants become more susceptible to whiteflies. Adapted from Li et al. (2014)

4 Conclusions

Geminiviruses are the most abundant plant viruses. These viruses share host plants with their insect vectors, and these viruses manipulate host defense to indirectly influence the behavior and performance of their vectors. These manipulations are crucial for geminiviruses in order be transmitted with high efficiency, which is linked not only with their global distribution but also with their wide host range. Understanding both the biological and molecular mechanisms of geminivirus transmission is essential to develop strategies for a sustainable management of geminiviruses and their invasive vectors in order to get an increase in the production of every crop affected with these viruses.

References

Bahder BW, Zalom FG, Jayanth M, Sudarshana MR (2016) Phylogeny of geminivirus coat protein sequences and digital PCR aid in identifying *Spissistilus festinus* as a vector of Grapevine red blotch-associated virus. Phytopathology 106(10):1223–1230

Briddon RW, Bedford ID, Tsai JH, Markham PG (1996) Analysis of the nucleotide sequence of the treehopper-transmitted geminivirus, tomato pseudo-curly top virus, suggests a recombinant origin. Virology 219:3387–3394

Brown JK, Czosnek H (2002) Whitefly transmission of plant viruses. Adv Bot Res 36:65–100

Czosnek H, Ghanim M (2012) Back to basics: are begomoviruses whitefly pathogens? J Integr Agric 11:225–234

Czosnek H, Ghanim M, Rubinstein G, Morin S, Fridman V, Zeidan M (2001) Whiteflies: vectors, and victims (?), of geminiviruses. Adv Virus Res 57:291–322

Firdaus S, Vosman B, Hidayati N, Jaya Supena ED et al (2013) The *Bemisia tabaci* species complex: additions from different parts of the world. Insect Sci 20:723–733

Ghanim M (2014) A review of the mechanisms and components that determine the transmission efficiency of tomato yellow leaf curl virus (Geminiviridae; *Begomovirus*) by its whitefly vector. Virus Res 186:47–54

Gotz M, Popovski S, Kollenberg M, Gorovits R et al (2012) Implication of *Bemisia tabaci* heat shock protein 70 in begomovirus-whitefly interactions. J Virol 86:13241–13252

Gray S, Cilia M, Ghanim M (2014) Circulative, "nonpropagative" virus transmission: an orchestra of virus-, insect-, and plant-derived instruments. Adv Virus Res 89:141–199

Hariton Shalev A, Sobol I, Ghanim M, Liu SS, Czosnek H (2016) The whitefly *Bemisia tabaci* knottin-1 gene is implicated in regulating the quantity of tomato yellow leaf curl virus ingested and transmitted by the insect. Viruses 8:205

Hogenhout SA, Ammar ED, Whitfield AE, Redinbaugh MG (2008) Insect vector interactions with persistently transmitted viruses. Annu Rev Phytopathol 46:327–359

Hull R (2014) Plant to plant movement. In: Hull R (ed) Plant virology. Academic, London, pp 669–751

Kanakala S, Ghanim M (2016) Implication of the whitefly *Bemisia tabaci* cyclophilin B protein in the transmission of tomato yellow leaf curl virus. Front Plant Sci 7:1702

Kliot A, Cilia M, Czosnek H, Ghanim M (2014) Implication of the bacterial endosymbiont Rickettsia spp. in interactions of the whitefly *Bemisia tabaci* with tomato yellow leaf curl virus. J Virol 88:5652–5660

Kollenberg M, Winter S, Götz M (2014) Quantification and localization of Watermelon chlorotic stunt virus and tomato yellow leaf curl virus (Geminiviridae) in populations of *Bemisia tabaci* (Hemiptera, Aleyrodidae) with differential virus transmission characteristics. PLoS One 9: e111968

Li R, Weldegergis BT, Li J, Jung C et al (2014) Virulence factors of geminivirus interact with MYC2 to subvert plant resistance and promote vector performance. Plant Cell 26:4991–5008

Luan JB, Li JM, Varela N, Wang YL et al (2011) Global analysis of the transcriptional response of whitefly to tomato yellow leaf curl China virus reveals their relationship of coevolved adaptations. J Virol 85:3330–3340

Luan JB, Yao DM, Zhang T, Walling LL et al (2013) Suppression of terpenoid synthesis in plants by a virus promotes its mutualism with vectors. Ecol Lett 16:390–398

Luan JB, Wang XW, Colvin J, Liu SS (2014) Plant-mediated whitefly-begomovirus interactions: research progress and future prospects. Bull Entomol Res 104:267–276

Navas-Castillo J, Lopez-Moya JJ, Aranda MA (2014) Whitefly-transmitted RNA viruses that affect intensive vegetable production. Ann Appl Biol 165:155–171

Pan L, Chen Q, Zhao J, Guo T et al (2017) Clathrin-mediated endocytosis is involved in tomato yellow leaf curl virus transport across the midgut barrier of its whitefly vector. Virology 502:152–159

Polston JE, De Barro P, Boykin LM (2014) Transmission specificities of plant viruses with the newly identified species of the *Bemisia tabaci* species complex. Pest Manag Sci 70:1547–1552

Ramesh SV, Sahu PP, Prasad M, Praveen S, Pappu HR (2017) Geminiviruses and plant hosts: a closer examination of the molecular arms race. Viruses 9:256

Rana VP, Popli S, Saurav GK, Raina HS et al (2016) *Bemisia tabaci* midgut protein interacts with begomoviruses and plays a role in virus transmission. Cell Microbiol 18:663–678

Rosen R, Kanakala S, Kliot A et al (2015) Persistent, circulative transmission of begomoviruses by whitefly vectors. Curr Opin Virol 15:1–8

Roumagnac P, Granier M, Bernardo P, Deshoux M, Ferdinand R, Galzi S et al (2015) Alfalfa leaf curl virus: an aphid-transmitted geminivirus. J Virol 89:9683–9688

Varsani A, Navas-Castillo J, Moriones E, Hernandez-Zepeda C, Idris A, Brown JK et al (2014) Establishment of three new genera in the family Geminiviridae: becurtovirus, eragrovirus and turncurtovirus. Arch Virol 159:2193–2203

Vyskočilová S, Tek Tay W, van Brunschot S, Seal S, Colvin J (2018) An integrative approach to discovering cryptic species within the *Bemisia tabaci* whitefly species complex. Sci Rep 8:10886

Wang YJ, Mao QZ, Liu WW, Mar TT et al (2014) Localization and distribution of wheat dwarf virus in its vector leafhopper, *Psammotettix alienus*. Phytopathology 104:897–904

Wang Z-Z, Shi M, Huang Y-C, Wang X-W, Stanley D, Chen X-X (2016) A peptidoglycan recognition protein acts in whitefly (*Bemisia tabaci*) immunity and involves in Begomovirus acquisition. Sci Rep 6:37806

Wei J, He YZ, Guo Q, Guo T et al (2017) Vector development and vitellogenin determine the transovarial transmission of begomoviruses. Proc Natl Acad Sci U S A 114:201701720

Whitfield AE, Falk BW, Rotemberg D (2015) Insect vector-mediated transmission of plant viruses. Virology 479–480:278–289

Xia W, Liang Y, Chi Y, Pan LL et al (2018) Intracellular trafficking of begomoviruses in the midgut cells of their insect vector. PLoS Pathog 14:e1006866

Yang Q, Ding B, Zhou X (2017) Geminiviruses and their application in biotechnology. J Integr Agric 16:2761–2771

Zerbini FM, Briddon RW, Idris A, Martin DP, Moriones E, Navas-Castillo J et al (2017) ICTV virus taxonomy profile: geminiviridae. J Gen Virol 98:131–133

Replication of DNA Satellites and Their Role in Viral Pathogenesis

Muhammad N. Sattar, Zafar Iqbal, and Amir Hameed

Abstract

The white-fly borne begomoviruses (family *Geminiviridae*) have circular single-stranded (css) DNA genome, which is encapsidated as monopartite (DNA-A) or bipartite (DNA-A and DNA-B) in the twinned icosahedrons. During the course of their evolution and to escape host defense machinery, begomoviruses adopt small cssDNA satellites called alpha-, beta-, and deltasatellites. Alphasatellties are found to be associated with begomovirus–betasatellite complexes and encode their own replication-associated protein (Rep), thus capable of autonomous replication. These satellite-like molecules are not well known to serve any critical function for their helper begomovirus except for few reports about attenuation of helper-virus accumulation and/or occasionally suppression of the host defense. Most of the monopartite begomoviruses in the Old World (OW) are found to be associated with betasatellites; however, none of the New World (NW) begomoviruses are known to be associated with betasatellites. Begomoviruses replicate their genome through rolling circle replication (RCR), which requires the virus-encoded Rep to recognize and bind to the iterated sequences (iterons) in the origin of replication (*ori*) region. Betasatellites lack such iterated sequences; however, they can be transreplicated by a diverse range of begomoviruses, following a similar pattern for replication. Betasatellites play a significant role in viral pathogenesis by interacting with certain host factors, attenuation of disease symptoms, suppression of host defense, and sometimes

M. N. Sattar (✉)
Department of Biotechnology, College of Agriculture and Food Science, King Faisal University, Al-Hasa, Kingdom of Saudi Arabia

Z. Iqbal
Central Laboratories, King Faisal University, Al-Hasa, Kingdom of Saudi Arabia

A. Hameed
Department of Bioinformatics & Biotechnology, Government College University (GCU), Faisalabad, Pakistan

R. V. Kumar (ed.), *Geminiviruses*, https://doi.org/10.1007/978-3-030-18248-9_9

inter- or intracellular shuttling of begomovirus genome. Likewise, the noncoding molecules deltasatellites depend upon their helper virus for their replication. However, their precise role in viral pathogenesis still needs to be explored.

1 Introduction

Many plant viruses coexist with certain nucleic acid, either DNA or RNA molecules, termed as "satellites." Satellites lack the ability for an independent existence and are entirely dependent on their helper virus for their replication, encapsidation, movement, and proliferation (Briddon and Stanley 2006). Satellite molecules have a very simple genomic organization and usually encode few or no genes and have little or no sequence homology to their helper virus. The word "satellite" coins two major classes of nucleic acid agents: satellite viruses, which are capable of self-encapsidation by producing capsid protein, and the virus-associated satellites, which lack their own capsid protein and therefore utilize helper-virus proteins for their encapsidation (Mayo et al. 2005). Satellite virus was described for the first time in 1962 when scientists discovered some exogenous nucleic acid agents being associated with few strains of *Tobacco necrosis virus* (TNV), a *Necrovirus*, laterally recognized as *Tobacco necrosis satellite virus* (TNSV) (Kassanis 1962). Until now, several satellite molecules have been found to be associated with different classes of plant viruses, particularly RNA viruses such as *Rice yellow mottle virus* satellite (RYMV-sat) and *Cucumber mosaic virus* satellites (CMV-sat) (Adams et al. 2017; Mayo et al. 2005). Majority of the satellites have single-stranded (ss) RNA genome; however, double-stranded (ds) RNA genome is also present in few members. RNA genome of few ssRNA satellites encodes some proteins that may or may not assist in replication process (Palukaitis et al. 2008). The first begomovirus-associated DNA satellite [Tomato leaf curl satellite (ToLCV-sat)] was reported in 1997 from tomato plants infected with *Tomato leaf curl virus* (ToLCV) (Dry et al. 1997). In most cases, these satellites overload the resources of helper virus for their own replication/survival and interfere with viral infectivity (Brown et al. 2012). However, few members of DNA satellites have been identified causing coinfections with helper viruses and result in severe disease symptoms as compared to single viral infection (Nawaz-ul-Rehman and Fauquet 2009).

2 History and Current Status of ssDNA Satellites Associated with Begomoviruses

Most of the begomoviruses (family: *Geminiviridae*) in the Old World (OW) and few in the New World (NW) are being associated with circular ssDNA satellites (Fig. 1). Till now, DNA satellites associated with majority of the OW monopartite begomoviruses include more frequently found betasatellites and occasionally

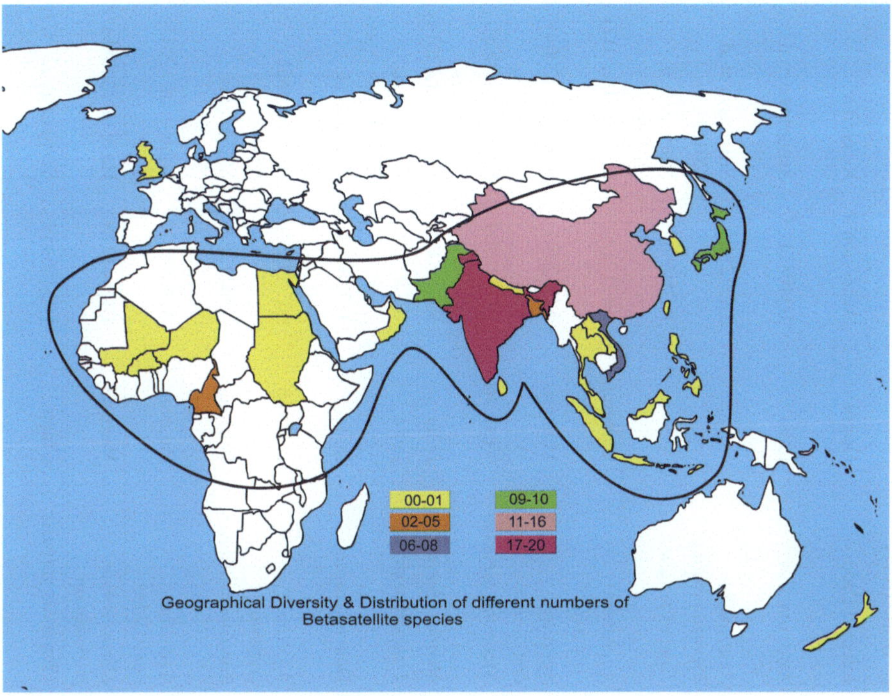

Fig. 1 Genome organization of cssDNA satellites. Schematic position and orientation of genes is shown with colored arrows. (**a**) Alphasatellites encode Rep protein and contain an adenine rich (A-rich) sequence. (**b**) Betasatellites encode single protein, βC1, have a satellite conserved region (SCR) and an A-rich sequence. The purple and green parts of deltasatellite SCR represent the predicted conserved stem-loop and secondary stem-loop structures, respectively. For all satellites, the intergenic region (IR) contains a predicted hairpin-loop structure, which contains the nonanucleotide sequence (TAA/GTATTAC) as part of the loop

alphasatellites, whereas newly characterized deltasatellites are most frequently reported in association with the NW begomoviruses (Adams et al. 2017) (Table 1). Recently, betasatellites and deltasatellites have been assigned new genera *Betasatellite* and *Deltasatellite* in the sub-viral family, *Tolecusatellitidae*, respectively (Adams et al. 2017).

2.1 Alphasatellites

After their first discovery in 1999 from *Ageratum conyzoides* (Saunders and Stanley 1999), approximately 200 complete alphasatellite (earlier described as DNA-1) sequences have been deposited in the GenBank database. Alphasatellites are circular and ssDNA (css) molecules associated with OW monopartite begomoviruses, begomovirus–betasatellite complexes, and/or NW bipartite begomoviruses

Table 1 List of major DNA satellites prevailing in the Old World (OW) and the New World (NW)

Satellite	Acronym	Associated Begomovirus	Infected Crop	Origin	Accession #	Reference
Betasatellites						
Ageratum leaf curl betasatellite	ALCB	*Papaya leaf curl virus*	*Aster amellus*	India	JQ408217	Srivastava et al. (2013)
		Ageratum enation virus	*Amaranthus hypochondriacus*		JX512904	Srivastava et al. (2015)
		Ageratum enation virus	*Tagetes patula*		KC589700	Marwal et al. (2013b)
Ageratum leaf curl Cameroon betasatellite	ALCCMB	*Ageratum leaf curl Cameroon virus*	*Ageratum conyzoides*	Cameroon	FR717141	Leke et al. (2012)
Ageratum yellow vein betasatellite	AYVB	*Ageratum yellow vein virus*	*Ageratum conyzoides*	Malaysia	AJ542497	Bull et al. (2004)
		Papaya leaf curl China virus	*Nicotiana tabacum*	Vietnam	DQ641709	Ha et al. (2008a)
		Chili leaf curl India virus	*Mentha piperita*	India	KF364485	Saeed et al. (2014)
Ageratum yellow leaf curl betasatellite	AYLCB	*Wheat dwarf India virus*	*Triticum aestivum*	India	KC305092	Kumar et al. (2014)
		Ageratum enation virus	*Daucus carota*	India	JF728869	Kumar et al. (2013)
		Tobacco curly shoot virus	*Ageratum conyzoides*	Pakistan	NC_005046	Briddon et al. (2003)
		Alternanthera yellow vein virus	*Sonchus oleraceus*	Pakistan	AM412239	Mubin et al. (2010)
		Ageratum enation virus	*Ageratum conyzoides*	Pakistan	AM698010	Tahir et al. (2015)
Alternanthera yellow vein betasatellite	AlYVB	*Alternanthera yellow vein mosaic virus*	*Chrysogonum peruvianum*	Vietnam	DQ641716	Ha et al. (2008b)
Bean leaf curl China betasatellite	BLCCNB	*Tomato yellow leaf curl China virus*	*Phaseolus vulgaris*	China	DQ256459	Dong et al. (2007)

Bhendi yellow vein betasatellite	BYVB	*Bhendi yellow vein mosaic virus*	*Abelmoschus esculentus*	India	AJ308425	Jose and Usha (2003)
Chili leaf curl betasatellite	ChiLCB	*Chili leaf curl virus*	*Petunia hybrida*	India	KJ700655	Nehra and Gaur (2014)
		Pepper leaf curl Lahore virus	*Capsicum sp.*	Pakistan	AM849549	Tahir et al. (2010)
Chili leaf curl Sri Lanka betasatellite	ChiLCSiB	*Chili leaf curl Sri Lanka virus*	*Capsicum sp.*	Sri Lanka	JN555600	Senanayake et al. (2013)
Cotton leaf curl Gezira betasatellite	CLCuGB	*Cotton leaf curl Gezira virus*	Gossypium sp.	Sudan	AY669329	Fauquet et al. (2005)
			Abelmoschus esculentus	Niger	FJ469629	Shih et al. (2009)
Croton yellow vein mosaic betasatellite	CroYVMB	*Sweet potato leaf curl virus*	*Ipomoea purpurea*	India	JX050198	Geetanjali et al. (2013)
Erechtites yellow mosaic betasatellite	ErYMB	*Erechtites yellow mosaic virus*	*Daucus carota*	Vietnam	DQ641713	Ha et al. (2008b)
Honeysuckle yellow vein betasatellite	HYVB	*Honeysuckle yellow vein mosaic virus*	*Lonicera japonica*	Japan	NC_009571	Ogawa et al. (2008)
Malvastrum leaf curl betasatellite	MaLCB	*Malvastrum leaf curl virus*	*Ageratum malvastrum*	China	AJ971264	Huang and Zhou (2006)
Okra leaf curl betasatellite	OLCuB	*Okra leaf curl Oman virus*	*Abelmoschus esculentus*	Oman	KF267444	Akhtar et al. (2014)
Papaya leaf curl betasatellite	PaLCuB	*Papaya leaf curl China virus*	*Carica papaya*	Vietnam	NC_009555	Ha et al. (2008b)
Radish leaf curl betasatellite	RaLCB	*Radish leaf curl virus*	*Daucus carota*	India	NC_010239	Singh et al. (2012)
Sida leaf curl betasatellite	SiLCuB	*Sida leaf curl virus*	*Sida sp.*	China	AM050733	Guo and Zhou (2006)
Sida yellow mosaic China betasatellite	SiYMCNB	*Sida yellow mosaic China virus*	*Sida sp.*	China	AJ810095	Xiong et al. (2005)
Tobacco curly shoot betasatellite	TbCSB	*Tobacco leaf curl virus*	*Tobacco sp.*	China	AJ457822	Zhou et al. (2003)

(continued)

Table 1 (continued)

Satellite	Acronym	Associated Begomovirus	Infected Crop	Origin	Accession #	Reference
Tomato leaf curl Bangalore betasatellite	ToLCBB	*Tomato leaf curl Bangalore virus*	*Solanum lycopersicum*	India	GU984046	Tiwari et al. (2010)
Tomato leaf curl Bangladesh betasatellite	ToLCBDB	*Tomato leaf curl Bangalore virus*	*Solanum lycopersicum*	Bangladesh	AJ542489	Bull et al. (2004)
Tomato leaf curl Gandhinagar betasatellite	ToLCGaB	*Tomato leaf curl Gandhinagar virus*	*Solanum lycopersicum*	India	NC_023038	Rathore et al. (2014)
Tomato leaf curl Philippines betasatellite	ToLCPHB	*Tomato leaf curl Philippines virus*	*Solanum lycopersicum*	Philippines	NC_009570	Sharma et al. (2011)
Tomato yellow leaf curl China betasatellite	TYLCCNB	*Tomato yellow leaf curl China virus*	*Solanum lycopersicum*	China	AJ781301	Tao and Zhou (2008)
Tomato leaf curl betasatellite	TLCB	*Tomato yellow leaf curl Vietnam virus*	*Solanum lycopersicum*	Nepal	AJ542492	Bull et al. (2004)
Vernonia yellow vein betasatellite	VeYVVB	*Vernonia yellow vein virus*	*Vernonia cinerea*	India	NC_013423	Packialakshmi et al. (2010)
Zinnia leaf curl betasatellite	ZLCuB	*Zinnia yellow leaf curl virus*	*Zinnia sp.*	Pakistan	AJ316028	Briddon et al. (2003)
Tomato yellow leaf curl Thailand betasatellite	TYLCThB	*Tomato leaf curl New Delhi virus*	*Solanum tuberosum*	Pakistan	LK933548	Hameed et al. (2017)
Alphasatellites						
Ageratum leaf curl Cameroon alphasatellite	ALCCMA	*Ageratum leaf curl Cameroon virus*	*Ageratum conyzoides*	Cameroon	NC_014744	Leke et al. (2012)
Ageratum yellow vein China alphasatellite	AYVCHA	*Ageratum yellow vein China virus*	*Syndrella nodiflora*	Philippines	KF785752	She et al. (2015)
Ageratum yellow vein alphasatellite	AYVA	*Ageratum yellow vein virus*	*Ageratum conyzoides*	Singapore	AJ416153	Saunders et al. (2002)
Ageratum yellow vein Pakistan alphasatellite	AYVPKA	*Ageratum yellow vein virus*	*Ageratum conyzoides*	Pakistan	AJ512949	Briddon et al. (2004)
Ageratum yellow vein Singapore alphasatellite	AYVSGA	*Tomato yellow leaf curl virus*	*Solanum lycopersicum*	Oman	FJ956707	Idris et al. (2011)

Cassava mosaic Madagascar alphasatellite	CMMaA	*Cassava mosaic virus*	*Manihot esculenta*	Madagascar	HE984148	Harimalala et al. (2013)
Chili leaf curl Multan alphasatellite	ChiLCMA	*Chili leaf curl virus*	*Solanum tuberosum*	Pakistan	NC_013103	Mubin et al. (2009)
Cotton leaf curl alphasatellite	CLCuA	*Cotton leaf curl virus*	*Gossypium sp.*	Pakistan	AJ132344	Mansoor et al. (1999)
Cotton leaf curl Gezira alphasatellite	CLCuGeA	*Cotton leaf curl Gezira virus*	*Solanum lycopersicum*	Sudan	KC763634	Fiallo-Olivé et al. (2013)
Cyamopsis tetragonoloba leaf curl alphasatellite	CyTLCA	*Gaur leaf curl virus*	*Guar sp.*	India	GU385877	Kumar et al. (2010)
Lantana yellow vein alphasatellite	LYVA	*Lantana yellow vein mosaic virus*	*Lantana sp.*	India	KC206075	Marwal et al. (2013a)
Malvastrum yellow mosaic Cameroon alphasatellit	MYMCA	*Tomato leaf curl Cameroon virus*	*Solanum lycopersicum*	Cameroon	FN675298	Leke et al. (2011)
Melon chlorotic mosaic alphasatellite	MeCMA	*Melon chlorotic mosaic virus*	*Melon*	Venezuela	KF670682	Romay et al. (2014)
Mimosa yellow leaf curl alphasatellite	MiYLCA	*Mimosa yellow leaf curl virus*	*Mimosa*	Vietnam	DQ641719	Ha et al. (2008a)
Okra leaf curl Oman alphasatellite	OLCOMA	*Okra leaf curl virus*	*Abelmoschus esculentus*	Oman	KF267445	Akhtar et al. (2014)
Okra yellow crinkle Cameroon alphasatellite	OYCrCA	*Okra yellow crinkle virus*	*Abelmoschus esculentus*	Cameroon	FN675288	Leke et al. (2011)
Sida yellow vein China alphasatellite	SYVCA	*Tomato yellow leaf curl virus (TYLCV)*	*Solanum lycopersicum*	Pakistan	KC677736	Shahid et al. (2014)
Sida yellow vein Vietnam alphasatellite	SYVVA	*Side yellow vein Vietnam virus*	*Sida rhombifolia*	Vietnam	DQ641718	Ha et al. (2008a)
Tobacco curly shoot alphasatellite	TCSA	*Tobacco curly shoot virus*	*Nicotiana benthamiana*	China	NC_005057	Xie et al. (2004)
Tobacco leaf curl PUSA alphasatellite	TLCPA	*Tobacco leaf curl Pusa virus*	*Nicotiana tabacum*	India	NC_014597	Singh et al. (2011)

(continued)

Table 1 (continued)

Satellite	Acronym	Associated Begomovirus	Infected Crop	Origin	Accession #	Reference
Tomato yellow leaf curl China alphasatellite	TYLCChA	*Tomato leaf curl China virus*	*Duranta sp.*	Pakistan	AM749494	Unpublished
Vernonia yellow vein Fujian alphasatellite	VYVFA	*Vernonia yellow vein Fujian virus*	*Vernonia cinerea*	China	JF265670	Zulfiqar et al. (2012)
Deltasatellites						
Croton yellow vein deltasatellite	CrYVD	*Croton yellow vein mosaic virus*	*Croton bonplandianus*	India	AJ968684	Unpublished
Malvastrum leaf curl deltasatellite	MaLCuD	*Malvastrum leaf curl virus*	*Malvastrum coromandelianum*	China	KF433066	Unpublished
Sida golden yellow vein deltasatellite 1	SiGYVD1	*Sida golden yellow vein virus*	*Malvastrum coromandelianum*	Cuba	JN986808	Fiallo-Olivé et al. (2012)
Sida golden yellow vein deltasatellite 2	SiGYVD2	*Sida golden yellow vein viru*	*Malvastrum coromandelianum*	Cuba	JN819490	Fiallo-Olivé et al. (2012)
Sida golden yellow vein deltasatellite 3	SiGYVD3	*Sida golden yellow vein viru*	*Malvastrum coromandelianum*	Cuba	JN819498	Fiallo-Olivé et al. (2012)
Sweet potato leaf curl deltasatellite 1	SPLCD1	*Sweet potato leaf curl virus*	*Sweet potato*	Spain	FJ914390	Unpublished
Sweet potato leaf curl deltasatellite 2	SPLCD2	*Sweet potato leaf curl virus*	*Merremia dissecta*	Venezuela	KF716173	Unpublished
Sweet potato leaf curl deltasatellite 3	SPLCD3	*Sweet potato leaf curl virus*	Unidentified host	Puerto Rico	KT099179	Rosario et al. (2016)
Tomato leaf curl deltasatellite	ToLCD	*Tomato leaf curl virus*	*Solanum lycopersicum*	Australia	U74627	Dry et al. (1997)
Tomato yellow leaf distortion deltasatellite 1	ToYLDD1	*Tomato yellow leaf distortion virus*	*Sidastrum micranthum*	Cuba	JN819495	Fiallo-Olivé et al. (2012)
Tomato yellow leaf distortion deltasatellite 2	ToYLDD2	*Tomato yellow leaf distortion virus*	*Sidastrum micranthum*	Cuba	KU232893	Fiallo-Olivé et al. (2012)

(Paprotka et al. 2010). Alphasatellites do not truly represent satellite molecules because of their self-encoded replication-associated protein (Rep) and autonomous replication ability (Briddon et al. 2004) and could survive in permissive hosts (Mansoor et al. 1999). However, for encapsidation, transmission by the insect vector, and *in planta* movement, they are reliant on helper viruses.

The genome of alphasatellites comprised of ~1380 nucleotides (nt) that encode a single open reading frame (ORF): *Rep* (36 kDa) subsiding in virion-sense strand (coding strand), a highly conserved A-rich genomic sequence (~200 nt), and an origin of replication (*Ori*) containing a conserved nonanucleotide sequence (TAGTATT/AC) present in a predicted hairpin structure (Fig. 1a) (Briddon et al. 2004). The nonanucleotide sequence and the Rep-encoding segments of alphasatellite genome resemble the nanoviruses (another family of ssDNA viruses) (Brown et al. 2012), which suggests their possible capture by a begomovirus during mixed infections (Briddon and Stanley 2006). It is presumed that the captured Rep-encoded component (~1000 nt) of nanoviruses was reorganized through embedding A-rich sequences to gain a ~1400 nt size (half the size of begomovirus, i.e., ~2800 nt) in order to encode a structurally stabilized Rep required for self-encapsidation (Briddon and Stanley 2006; Mansoor et al. 2003). Three different subclasses of alphasatellites, DNA-1-type, DNA-2-type, and DNA-3-type, are frequently reported. The most commonly occurring alphasatellites are DNA-1-type that predominantly occur in the Indian subcontinent (Paprotka et al. 2010). The DNA-2-type alphasatellites are far rare, found in Singapore and Oman having low nt sequence identity with DNA-1-type, while DNA-3-type are novel alphasatellites detected from Guatemala, Brazil, and Puerto Rico (Rosario et al. 2016). According to Rosario et al. (2016), the DNA-3-type alphasatellites share 51–55% nt sequence identity with the DNA-1-type. Moreover, they help to increase the symptom severity in the host plants and form a separate monophyletic group when analyzed through phylogenetic studies (Rosario et al. 2016).

Although predominantly alphasatellites are found to be associated with begomoviruses, quite recently, an alphasatellite has been found associated with a mastrevirus, *Wheat dwarf India virus* (WDIV), in a natural field infection, which shows that alphasatellites have fewer constraints for their helper virus, host plant, or the insect vector. Their enigmatic role in virus pathogenesis has not been clearly answered yet. However, in few studies, the Rep proteins of alphasatellites have been described as the post-transcriptional gene silencing (PTGS) suppressors (Nawaz-ul-Rehman et al. 2010). Alphasatellites still need extensive explorations in plant virology as there are no consolidated reports available that describe their precise function and association mechanism with begomoviruses.

2.2 Betasatellites

Betasatellites (formerly described as DNAβ) are cssDNA-satellite molecules, which have recently been classified into the sub-viral family *Tolecusatellitidae* genus *Betasatellite* (Adams et al. 2017) (Fig. 1b). Betasatellites are predominantly found to be associated with monopartite begomoviruses in the OW; however, since the last few years, these molecules have also been found in association with bipartite begomoviruses (Hameed et al. 2017; Jyothsna et al. 2013) and recently with a mastrevirus (Kumar et al. 2014). Unlike alphasatellites, betasatellites are true satellite molecules as they entirely depend on the helper virus for their encapsidation, replication, and systemic dispersal (Briddon and Stanley 2006). Betasatellites variably interact with their helper-virus component and result in multiple types of coinfections. In some cases, the betasatellites synergistically infect the host plants by increasing their helper-virus accumulation and are essential for symptom induction (Chandel et al. 2016). For example, the interactions of *Cotton leaf curl Multan virus* (CLCuMuV) with *Cotton leaf curl Multan betasatellite* (CLCuMuB) and *Ageratum yellow vein virus* (AYVV) with *Ageratum yellow vein betasatellite* (AYVB) result in severe disease symptoms and enhanced virus titer as compared to their helper begomovirus alone in cotton and *A. conyzoides*, respectively (Briddon et al. 2001; Saunders et al. 2000). In other cases, a facultative interaction has also been observed in begomovirus:betasatellite complex, where the begomovirus component could infect alone and does not necessarily require betasatellite for symptom induction and/or enhanced viral titer (Chandel et al. 2016). For example, *Tobacco curly shoot virus* (TbCSV):(TbCSB) could make coinfections but TbCSV could infect alone; however, the presence of betasatellite induces more severe symptoms (Li et al. 2005). Another promising role played by betasatellite is the substitution of DNA-B component. Coinoculation of CLCuMuB and *Tomato leaf curl New Delhi virus* (ToLCNDV) DNA-A induced leaf curl disease phenotype in the model plant *Nicotiana benthamiana* in the absence of DNA-B component (Saeed et al. 2007). Since their first description in 2000, genome sequences of more than 1000 betasatellite isolates have been submitted to GenBank, depicting their ongoing diversity and evolution (Adams et al. 2017) (Fig. 2).

The betasatellite genome (~1350 nt) exhibits three conserved features: a complementary-sense single *βC1* gene (Briddon et al. 2003), a highly conserved A-rich region (~150-200 nt), and a ~100 nt satellite conserved region (SCR) (Nawaz-ul-Rehman and Fauquet 2009). The betasatellite genome shares no sequence homology with their cognate viruses except for a similar nonanucleotide sequence (TAATATTAC) present in the SCR (Briddon et al. 2003). Betasatellite-encoded βC1 (13-14 kDa) is a multifunctional protein involved in pathogenesis, enhancing viral DNA accumulation in the nucleus and suppression of the host antiviral defense response (Saunders et al. 2004). The other important ability of βC1 is self-interaction and localization at cell periphery, thus presumably having a role in viral movement (Cheng et al. 2011). Additionally, βC1 interacts with numerous host factors like ubiquitin-conjugating enzymes (UBC) (Eini et al. 2009) and asymmetric leaves 1 (AS1) factor, etc. (Yang et al. 2008).

The studies on geographical occurrence and diversity of betasatellites showed that the major center of their diversity lies in the Indian subcontinent and Southeast Asia (Fig. 2). On the basis of phylogeny, betasatellites may be broadly categorized into two major groups: the first group constitutes all the betasatellites isolated from the plant family *Malvaceae* (hibiscus, cotton, okra, hollyhock, etc.) while the

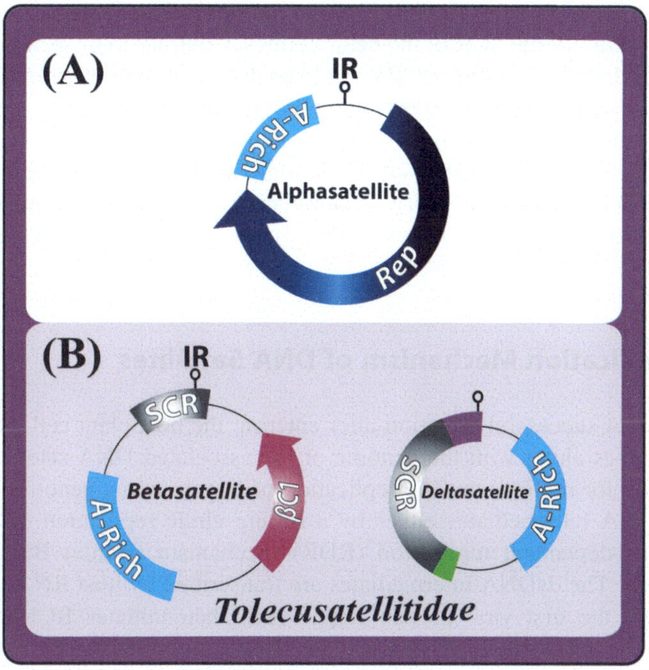

Fig. 2 Geographical genetic diversity of betasatellites. The map shows the diversification of betasatellites in colored parts, and the number of species identified in each country is labeled in respective colored boxes

betasatellites belonging to the second group were isolated from non-malvaceous plants (tomato, chillies, ageratum, zinnia, etc.). These findings are suggestive of the important role of host plants in the evolution of betasatellites.

2.3 Deltasatellites

Recently, another class of cssDNA satellites (approx. quarter the size of helper begomoviruses) has been classified as a new genus "*Deltasatellite*" (family *Tolecusatellitidae*) (Fiallo-Olivé et al. 2012; Lozano et al. 2016). Deltasatellites have been further categorized into three types of noncoding DNA satellites, i.e., *Tomato leaf curl virus*–satellite (ToLCV-sat) identified from Australia (Dry et al. 1997), DNA satellites associated with sweepoviruses in Venezuela and Spain (Lozano et al. 2016), and those isolated from malvaceous hosts in the Caribbean (Fiallo-Olivé et al. 2012). Besides their structural resemblance (Fig. 1b), phylogenetically deltasatellites are not closely related to each other; however, they entirely depend upon the helper virus for their vital functions. Their genome contains a stem-loop structure with a nonanucleotide (TAATATT/AC) sequence, an A-rich region, and does not encode any putative ORF. These satellites have a second putative stem-loop structure situated close to the iteron-like sequences, and a short region that resembles the SCR of the betasatellites. Contrary to the betasatellites, the emergence of deltasatellites in the NW might be due to agricultural trades of infected plants, like sweet potato, from OW to the NW (Lozano et al. 2016).

To date, the precise function of deltasatellites in context to begomovirus-deltasatellite complex is unclear, although some studies reported the influence of deltasatellite in lowering helper-virus accumulation in host plant that might facilitate the sequential movement of viruses to other plant parts (Fiallo-Olivé et al. 2016; Hassan et al. 2016).

3 Replication Mechanism of DNA Satellites

In the course of successful infection after entering the host plant cell, begomoviral ssDNA genomes along with the genome of the associated DNA satellite(s) access the cell nuclei for replication. The replication of begomovirus genome is achieved through dsDNA intermediates either by a rolling circle replication (RCR) and/or recombination-dependent replication (RDR) mechanism (Hanley-Bowdoin et al. 2013) (Fig. 3). The dsDNA intermediates are transcribed by host RNA-polymerase II to translate the first viral protein Rep, which then initiates RCR of both the begomovirus and the satellite DNA (Hanley-Bowdoin et al. 2013). These circular dsDNAs are assembled into transcriptionally active viral mini-chromosomes with the help of host histone proteins (Pilartz and Jeske 2003). The viral Rep protein creates a conducive environment to commence the replication. The Rep protein of begomoviruses drives the initiation and termination of RCR by nicking the dsDNA intermediates and rejoining the circular DNA at the specific site in the nonanucleotide sequences (TAATATT/AC) (Hanley-Bowdoin et al. 2013; Laufs et al. 1995). The successful commencement of RCR specifically requires high-affinity interactions between the begomovirus Rep protein and *ori* in the intergenic region (IR). The synthesis of complementary strand is initiated with a nick by Rep protein in the nonanucleotide sequence of the ssDNA, and an RNA primer is synthesized by the host DNA primase to initiate this phenomenon. The host plant replisome machinery (DNA polymerase and associated factors) is hijacked and reprogrammed during the elongation step to accomplish the viral dsDNA synthesis (Bagewadi et al. 2004; Kaliappan et al. 2011). The newly synthesized strand is displaced and released by Rep as a circular ssDNA. The synthesized dsDNA is used as a template to start next replication cycle. At the end of optimum replication cycles, the Rep protein downregulates its own synthesis and ultimately activates the expression of TrAP, which leads to the production of CP to start the virus and DNA satellites assembly (Fig. 3).

The DNA satellites employ a similar mechanism of DNA replication as their helper begomovirus (Alberter et al. 2005). However, beta- and deltasatellites are devoid of Rep protein (unlike begomovirus and alphasatellites) and hence depend exclusively upon the helper virus to commence their replication (Zhou 2013). Alphasatellites are capable of autonomous replication mechanism through RCR

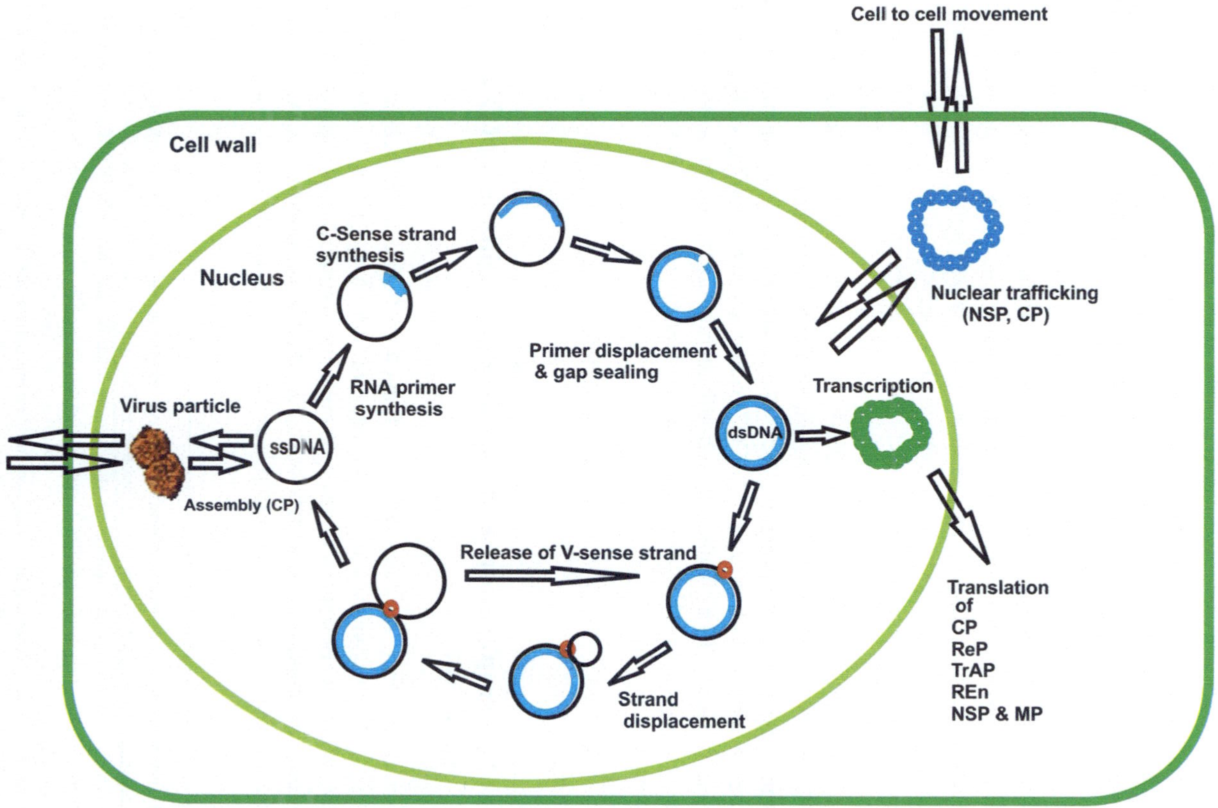

Fig. 3 Pictorial model of replication mechanism of begomoviruses/satellites. Once the virion DNA is delivered to the nucleus, host machinery initiates the synthesis of the complementary strand. Host-derived polymerases convert the single-stranded virion DNA into a double-stranded intermediate, which performs

directed by their own protein, alpha-Rep. As betasatellites are devoid of self-replication, the replication strategy of the betasatellites is determined by the helper begomovirus (Alberter et al. 2005). The replication model for begomoviruses suggests the imperative binding of Rep protein to the iterative sequences upstream of the nonanucleotide sequences followed by the recognition of the *ori*. Apparently, betasatellites frequently lack the iteron sequences, which is suggestive of some other mechanism involved in the *ori* recognition of betasatellites (Leke et al. 2012). The nicking site for Rep in betasatellites is expected to present in the nonanucleotide stem-loop sequences adjacent to the SCR. The stem sequences and the adjacent hairpin structures of betasatellites are remarkably similar in all betasatellites, and thus, it reaffirms that they participate in the *ori* recognition by helper Rep (Zhou 2013). The position of the highly conserved SCR present immediately upstream of the stem-loop sequences is also analogous to the relative position of IR of the helper begomoviruses. Furthermore, the conservation of SCRs in the defective forms of betasatellites further supports that the SCR region is involved in the replication of the betasatellites (Zhang et al. 2016). However, the sequence between the downstream of SCR and βC1 is required for efficient replication of betasatellites (Eini and Behjatnia 2016). The betasatellites lack iteron-like Rep-binding motifs (RBM); thus, the presence of G-box motif (CACGTG) may serve the binding of Rep protein (Eini and Behjatnia 2016). Thus, the high-affinity binding of Rep has a critical role in betasatellite replication. However, the exact mechanism of Rep binding is needed to be explored yet.

4 Transreplication and Pseudo-Recombination of DNA Satellites

As discussed earlier, the only sequence homology between the betasatellites and their helper begomoviruses is the presence of stem-loop structure (Briddon and Stanley 2006). In contrast to DNA-B of bipartite begomoviruses, betasatellites do not necessarily require cognate DNA-A of a bipartite or monopartite begomovirus genome. Instead, they are transreplicated by a diverse range of non-cognate

Fig. 3 (continued) as a transcriptionally active mini-chromosome. This mini-chromosome mediates the cell-to-cell movement and nuclear trafficking. Virion derived replication-associated protein (Rep) then binds to the iterons, produces a site-specific nick in the origin of replication (*ori*), and becomes covalently linked to the 5′ end of the nicked DNA via a tyrosine residue. The 3'OH end acts as a primer for synthesis of new virion-sense DNA by host-encoded factors, using the complementary-sense as a template. The nicking-joining activity of Rep releases unit length virion-sense ssDNA molecules. The newly synthesized ssDNA either continues the replication cycle (acting as a template for complementary-strand synthesis), is moved from cell-to-cell (possibly as virions), or is packaged by the coat protein (CP) for onward transmission by insect vectors. Image was reproduced from Briddon and Stanley (2006)

begomoviruses for their transreplication. For example, *Cotton leaf curl Gezira betasatellite* (CLCuGeB) is known to be a cognate associate of *Okra yellow crinkle virus* (OYCrV) and *Cotton leaf curl Gezira virus* (CLCuGeV) causing okra leaf curl disease (Leke et al. 2013). However, it can also be transreplicated with three other distinct begomoviruses *Tomato yellow leaf curl virus* (TYLCV), *Tomato leaf curl Mali virus* (ToLCMLV), and *Tomato yellow leaf crumple virus* (TYLCrV), each from a diverse geographic origin (Saunders 2008). Similarly, CLCuMuB is a cognate member of CLCuD-associated begomoviruses (CABs) in Asia (Iqbal et al. 2012; Sattar et al. 2017). The non-cognate associations between CLCuMuB with ToLCNDV and TbCSB with *Clerodendrum golden mosaic China virus* (CGMCV) are significant examples of its transreplication by a bipartite begomovirus (Li and Zhou 2010; Saeed 2010). Moreover, CLCuMuB can also be transreplicated by other non-cognate viruses such as TYLCV, *Tomato leaf curl Karnatka virus* (ToLCKnV), and *Tomato leaf curl virus* (ToLCV) (Kharazmi et al. 2012). Apart from begomoviruses, betasatellites are also known to be transreplicated by the members of other genera of the family *Geminiviridae*. *Beet curly top virus* (BCTV; genus *Becurtovirus*) successfully transreplicates *Ageratum yellow vein betasatellite* (AYVB) and *Tomato yellow leaf curl China betasatellite* (TYLCCNB) (Yang et al. 2011a). In another study, a *Curtovirus, Beet severe curly top virus* (BSCTV), successfully supported the transreplication of CLCuMuB (Kharazmi et al. 2012). Likewise, association of Cotton leaf curl Multan alphasatelliite (CLCuMuA), Guar leaf curl alphasatellite (GLCuA), and *Ageratum yellow leaf curl betasatellite* (AYLCB) with the WDIV (genus *Mastrevirus*) highlights the natural transreplication of DNA satellites by the member of a different genus (Kumar et al. 2014). Such associations are quite surprising because the functional betasatellites are mostly known to be associated with monopartite begomoviruses during natural infections. Moreover, the DNA-B component of bipartite begomoviruses has specific interactions with the Rep of the cognate DNA-A only. Such indistinguishable replication of betasatellites depicts that these molecules are quite flexible for their transreplication as compared to the specificity of recognition between Rep protein and DNA-B component of a bipartite begomovirus.

Apart from the fact that betasatellites have quite a promiscuous mode of replication, two distinct betasatellite species rarely coexist with a single helper virus within the same host plant. Apparently, this is because betasatellites are adapted to their cognate helper virus for replication during the course of evolution (Zhou et al. 2003). Thus, the cognate betasatellites are shown to accumulate to higher levels than the non-cognate betasatellites within the same host (Qing and Zhou 2009). For example, the coinoculation of TYLCCNB and TbCSB with one helper virus creates a competition, which causes cognate betasatellite dominant over non-cognate betasatellite. However, switching their sequence elements also switched the preferential replication of the respective cognate helper virus (Zhang et al. 2016).

Under natural environmental conditions, although betasatllites may coexist with the alphasatellites, the binding of alpha-Rep with the Rep protein of the helper virus may obstruct betasatellite replication. The alphasatellites can ameliorate begomovirus symptoms and hinder high accumulation of betasatellites during coinfections (Idris et al. 2011).

The deltasatellites contain a stem-loop with nonanucleotide, TATA box, and a second predicted stem-loop with iteron-like sequences. Moreover, their A-rich region and a short region also share high homology with betasatellite SCRs (Lozano et al. 2016). However, further investigations are needed to decipher their mode of replication and roles in viral pathogenesis. Most probably, these molecules are also transreplicated by the helper begomovirus Rep due to the presence of begomovirus iteron-like sequences upstream of the second stem loop (Fiallo-Olivé et al. 2016). The deltasatellite, Tomato leaf curl virus-satellite (ToLCV-sat), has been shown to be transreplicated by ToLCV as well as geographically distinct geminiviruses like *Tomato yellow leaf curl Sardinia virus* (TYLCSV), *African cassava mosaic virus* (ACMV), and a becurtovirus BCTV. In another study, the deltasatellites sat-177 and sat-603 could only be transreplicated by the cognate begomoviruses, *Sida golden yellow vein virus* (SiGYVV), and a monopartite begomovirus *Tomato leaf deformation virus* (ToLDV) from the NW. However, the OW TYLCV, TYLCSV, and ACMV could not support their transreplication (Fiallo-Olivé et al. 2016).

5 Deciphering the Role of DNA Satellites in the Begomovirus Pathogenesis

All the functions of betasatellites are accredited to their single gene product βC1. This protein, when expressed transiently through *Potato virus X* (PVX) or through a stable transformation in model host plants (*N. benthamiana* and *N. tabacum*) cells, induces typical begomovirus disease symptoms of leaf curling, vein thickening, and enations (Kon et al. 2007; Qazi et al. 2007). The βC1 protein regulates the expression of several different miRNAs involved in the developmental processes when expressed through PVX in *N. benthamiana* plants. The accumulation of miR159 and miR160 was significantly enhanced, while the accumulation of miR164, miR165/166, miR169, and miR170 was reduced when the βC1 gene was transiently expressed in the inoculated plants (Amin et al. 2011). The βC1 accumulates primarily in the nucleus, localizes at the periphery of the infected host cells, and colocalizes along the endoplasmic reticulum. These localization patterns and presence of both nuclear import and export signals point toward the putative role of βC1 in intracellular transport and movement (Cheng et al. 2011). Moreover, βC1 forms punctate bodies, both in vivo and in vitro, by self-interaction, which presumably has a role in symptom induction. A deletion mutagenesis study shows that amino acids spanning two α-helices at C-terminal are important in self-interaction.

The βC1 interacts with a variety of host-encoded factors such as with AS1 and AS2 factors. Self-interaction of these two factors is required for leaf development; βC1 mimics the function of AS2 by interacting with AS1 and thus affects the leaf development (Yang et al. 2008). The CLCuMuB-encoded βC1 interacts with a UBC to induce betasatellite-specific symptoms in the host plant (Eini et al. 2009). It is speculated that βC1 interaction with UBC perturbs the ubiquitin-proteasome pathway to enhance βC1 accumulation, which ultimately led to the development of viral symptoms.

Plants have developed a fine-tuned defense mechanism, which is operated through PTGS and TGS, against invading pathogens. To counter the host defense response, βC1 has the ability to suppress the PTGS-mediated host defense by interacting with one of the important host defense components, Argonaute-1 (AGO-1), which binds to the siRNAs and represses the target RNAs (Eini 2017). The βC1 protein can bind to ss- as well as dsDNA, dsRNA, and both long and short RNAs in a sequence-independent manner to suppress the host defense. This binding activity is mediated by the nuclear localization signals (NLS) present in the βC1. TYLCCNB-encoded βC1 has the ability to suppress PTGS by upregulating the *N. benthamiana* calmodulin-related protein (Nbrgs-CaM), which can repress the expression of *N. benthamiana* RNA-dependent RNA polymerase 6 (Zhou 2013). Besides PTGS, the βC1 also has the ability to reduce the TGS or can even reverse the established TGS (Yang et al. 2011b) in the host plants. This suppression of TGS is mediated by interacting with S-adenosyl homocysteine hydrolase (SAHH), an enzyme generally required for the generation of S-adenosyl methionine (SAM), through a NLS (49KKK51) present in βC1 (Yang et al. 2011b). The CLCuMuB-β C1 can suppress the host defense by downregulating the jasmonic acid (JA)-responsive genes such as CORI3, PR4, NbPHAN, and PDF1 (Yang et al. 2008) and can interact with certain host-encoded factors involved in metabolic and defense pathways (Tiwari et al. 2013). The expression of βC1 can differentially regulate the genes involved in electron carrier for photosynthesis, respiration, and ATP synthesis (Andleeb et al. 2010). The CLCuMuB-βC1 also interacts with ATG8 protein, a ubiquitin-like protein having a role in the biogenesis of autophagosomes (Shelly et al. 2009). It is thus speculated that the interaction of βC1 with ATG8 may likely be an antiviral defense mechanism. Besides, βC1 can interact with tomato UBC, an enzyme required for ubiquitination and ultimately the degradation of the target protein (Eini et al. 2009). This interaction interferes with ubiquitin-proteasome pathway that could enhance the βC1 accumulation.

To counter the βC1 pathogenesis, host plants have developed a sophisticated counterattack mechanism. Tomato plants employed Sucrosenonfermenting1-related kinase (SISnRK1)-mediated defense against the betasatellites. Hyperexpression of SISnRK1 leads to the reduction in betasatellite accumulation and delayed onset of the symptoms. It has been showed that SISnRK1 phosphorylates TYLCCNB-βC1 at the amino acid positions 33 (serine residue) and 78 (threonine), thus negatively regulating the βC1 functions (Cui et al. 2004; Yang et al. 2008).

6 Role of Rep-A of Alphasatellite in Viral Pathogenesis

To date, the interactions of the alphasatellite-encoded Rep protein with the host-encoded factors and its role in successful begomovirus infection have not been fully explored. Only a few studies are available, which reported that the Rep protein encoded by few alphasatellites have PTGS suppressor activity (Nawaz-ul-Rehman et al. 2010), suggesting the role of alphasatellite in overcoming RNAi-mediated host defense. The type-2 alphasatellites are known to have a role in the symptom attenuation by reducing the accumulation of begomovirus (Nawaz-ul-Rehman et al. 2010) and/or betasatellites (Idris et al. 2011; Wu and Zhou 2005) in the begomovirus–betasatellite complexes. This attenuation in symptoms may likely increase the chance of host survival and virus transmission.

7 Conclusion

Host–pathogen interactions are like arms race with typical zero-sum game, which ultimately leads to the disease development or the host recovery. In this subtle type of intimate relationship, both counterparts continuously deploy different strategies to take advantage over each other. The acquisition of DNA satellites by begomoviruses is the continuity of this process. DNA satellites have equipped their helper begomoviruses to suppress the host defense (both TGS and PTGS) and/or help in the symptom attenuation, which ultimately helps the virus to evade the host defense. During the acquisition process, begomoviruses resized these DNA satellites precisely in a mathematical way, alpha- and betasatellites are almost half, while deltasatellites are one-fourth of the helper-virus genome, to support their replication.

The maintenance of these DNA satellites by the helper-virus replication machinery is dependent upon dynamic, mainly undefined, interactions between begomovirus, DNA satellite, and host-encoded factors. However, *ori* is the only common feature between DNA satellites and the helper viruses, so interaction between the geminivirus Rep and DNA satellites is principally dependent on this region. Likewise, *ori* region (particularly nonanucleotide) determines the successful commencement of RCR. The importance of *ori* in replication has been proven experimentally where switching of the *ori* sequence has switched the preferential transreplication of betasatellite by helper begomovirus (Zhang et al. 2016). Although, no strong selection mechanism is present between DNA satellites and their helper virus, the interaction between DNA satellites and their helper virus is not merely a transreplication but stacking of a multilayer interaction (Iqbal et al. 2012, 2017).

References

Adams MJ, Lefkowitz EJ, King AMQ, Harrach B et al (2017) Changes to taxonomy and the international code of virus classification and nomenclature ratified by the International Committee on Taxonomy of Viruses. Arch Virol 162:2505–2538

Akhtar S, Khan AJ, Singh AS, Briddon RW (2014) Identification of a disease complex involving a novel monopartite begomovirus with beta-and alphasatellites associated with okra leaf curl disease in Oman. Arch Virol 159:1199–1205

Alberter B, Rezaian MA, Jeske H (2005) Replicative intermediates of *Tomato leaf curl virus* and its satellite DNAs. Virology 331:441–448

Amin I, Patil BL, Briddon RW, Mansoor S et al (2011) A common set of developmental miRNAs are upregulated in *Nicotiana benthamiana* by diverse begomoviruses. Virol J 8:143

Andleeb S, Amin I, Bashir A, Briddon RW et al (2010) Transient expression of βC1 protein differentially regulates host genes related to stress response, chloroplast and mitochondrial functions. Virol J 7:373

Bagewadi B, Chen S, Lal SK, Choudhury NR et al (2004) PCNA interacts with Indian mung bean yellow mosaic virus rep and downregulates rep activity. J Virol 78:11890–11903

Briddon RW, Stanley J (2006) Sub-viral agents associated with plant single-stranded DNA viruses. Virology 344:198–210

Briddon RW, Mansoor S, Bedford ID, Pinner MS et al (2001) Identification of DNA components required for induction of cotton leaf curl disease. Virology 285:234–243

Briddon RW, Bull SE, Amin I, Idris AM et al (2003) Diversity of DNA β, a satellite molecule associated with some monopartite begomoviruses. Virology 312:106–121

Briddon RW, Bull SE, Amin I, Mansoor S et al (2004) Diversity of DNA 1; a satellite-like molecule associated with monopartite begomovirus-DNA β complexes. Virology 324:462–474

Brown JK, Fauquet CM, Briddon RW, Zerbini M et al (2012) *Geminiviridae*. Virus taxonomy – Ninth report of the International Committee on Taxonomy of Viruses. Associated Press, Elsevier Inc., London, Waltham, San Diego, pp 351–373

Bull SE, Tsai W-S, Briddon RW, Markham PG et al (2004) Diversity of begomovirus DNA β satellites of non-malvaceous plants in east and south East Asia. Arch Virol 149:1193–1200

Chandel V, Singh MK, Jangid A, Dhatwalia S (2016) Emerging satellites associated with begomoviruses: world scenario. In: Gaur RK, Petrov NM, Patil BL, Stoyanova MI (eds) Plant viruses: evolution and management. Springer, Singapore, pp 145–169

Cheng X, Wang X, Wu J, Briddon RW et al (2011) βC1 encoded by tomato yellow leaf curl China betasatellite forms multimeric complexes *in vitro* and *in vivo*. Virology 409:156–162

Cui XF, Tao XR, Xie Y, Fauquet CM et al (2004) A DNAβ associated with tomato yellow leaf curl China virus is required for symptom induction. J Virol 78:13966–13974

Dong JH, Luo YQ, Ding M, Zhang ZK et al (2007) First report of tomato yellow leaf curl China virus infecting kidney bean in China. Plant Pathol 56:342

Dry I, Krake LR, Rigden JE, Rezaian MA (1997) A novel subviral agent associated with a geminivirus: the first report of a DNA satellite. Proc Natl Acad Sci USA 94:7088–7093

Eini O (2017) A betasatellite-encoded protein regulates key components of gene silencing system in plants. Mol Biol 51:579–585

Eini O, Behjatnia SAA (2016) The minimal sequence essential for replication and movement of Cotton leaf curl multan betasatellite DNA by a helper virus in plant cells. Virus Genes 52:679–687

Eini O, Dogra S, Selth LA, Dry IB et al (2009) Interaction with a host ubiquitin-conjugating enzyme is required for the pathogenicity of a geminiviral DNA b satellite. Mol Plant-Microbe Interact 22:737–746

Fauquet CM, Sawyer S, Idris AM, Brown JK (2005) Sequence analysis and classification of apparent recombinant begomoviruses infecting tomato in the Nile and Mediterranean basins. Phytopathology 95:549–555

Fiallo-Olivé E, Martínez-Zubiaur Y, Moriones E, Navas-Castillo J (2012) A novel class of DNA satellites associated with New World begomoviruses. Virology 426:1–6

Fiallo-Olivé E, Hamed A, Navas-Castillo J, Moriones E (2013) Cotton leaf curl Gezira alphasatellite associated with tomato leaf curl Sudan virus approaches the expected upper size limit in the evolution of alphasatellites. Virus Res 178:506–510

Fiallo-Olivé E, Tovar R, Navas-Castillo J (2016) Deciphering the biology of deltasatellites from the New World: maintenance by New World begomoviruses and whitefly transmission. New Phytol 212:680–692

Geetanjali SA, Shilpi S, Mandal B (2013) Natural association of two different betasatellites with sweet potato leaf curl virus in wild morning glory (*Ipomoea purpurea*) in India. Virus Genes 47:1–5

Guo XJ, Zhou XP (2006) Molecular characterization of a new begomovirus infecting *Sida cordifolia* and its associated satellite DNA molecules. Virus Genes 33:279–285

Ha C, Coombs S, Revill P, Harding R et al (2008a) Molecular characterization of begomoviruses and DNA satellites from Vietnam: additional evidence that the New World geminiviruses were present in the Old World prior to continental separation. J Gen Virol 89:312–326

Ha C, Revill P, Harding RM, Vu M et al (2008b) Identification and sequence analysis of potyviruses infecting crops in Vietnam. Arch Virol 153:45–60

Hameed A, Tahir MN, Amin I, Mansoor S (2017) First report of tomato leaf curl New Delhi virus and a tomato yellow leaf curl Thailand betasatellite causing severe leaf curl disease of potato in Pakistan. Plant Dis 101:1065

Hanley-Bowdoin L, Bejarano ER, Robertson D, Mansoor S (2013) Geminiviruses: masters at redirecting and reprogramming plant processes. Nat Rev Microbiol 11:777–788

Harimalala M, Bruyn A, Hoareau M, Andrianjaka A et al (2013) Molecular characterization of a new alphasatellite associated with a cassava mosaic geminivirus in Madagascar. Arch Virol 158 (8):1–4

Hassan I, Orílio AF, Fiallo-Olivé E, Briddon RW et al (2016) Infectivity, effects on helper viruses and whitefly transmission of the deltasatellites associated with sweepoviruses (genus *Begomovirus*, family *Geminiviridae*). Sci Rep 6:30204

Huang JF, Zhou XP (2006) Molecular characterization of two distinct begomoviruses from *Ageratum conyzoides* and *Malvastrum coromandelianum* in China. J Phytopathol 154:648–653

Idris AM, Shahid MS, Briddon RW, Khan AJ et al (2011) An unusual alphasatellite associated with monopartite begomoviruses attenuates symptoms and reduces betasatellite accumulation. J Gen Virol 92:706–717

Iqbal Z, Sattar MN, Kvarnheden A, Mansoor S et al (2012) Effects of the mutation of selected genes of *Cotton leaf curl Kokhran virus* on infectivity, symptoms and the maintenance of Cotton leaf curl Multan betasatellite. Virus Res 169:107–116

Iqbal Z, Shafiq M, Ali I, Mansoor S et al (2017) Maintenance of Cotton leaf curl Multan betasatellite by tomato leaf curl New Delhi virus—analysis by mutation. Front Plant Sci 8:2208

Jose J, Usha R (2003) Bhendi yellow vein mosaic disease in India is caused by association of a DNA b satellite with a begomovirus. Virology 305:310–317

Jyothsna P, Haq QMI, Singh P, Sumiya KV et al (2013) Infection of tomato leaf curl New Delhi virus (ToLCNDV), a bipartite begomovirus with betasatellites, results in enhanced level of helper virus components and antagonistic interaction between DNA B and betasatellites. Appl Microbiol Biotechnol 97:5457–5471

Kaliappan K, Choudhury NR, Suyal G, Mukherjee SK (2011) A novel role for RAD54: this host protein modulates geminiviral DNA replication. FASEB J 26(3):1142–1160

Kassanis B (1962) Properties and behaviour of a virus depending for its multiplication on another. Microbiology 27:477–488

Kharazmi S, Behjatnia SAA, Hamzehzarghani H, Niazi A (2012) Cotton leaf curl Multan betasatellite as a plant gene delivery vector trans-activated by taxonomically diverse geminiviruses. Arch Virol 157:1269–1279

Kon T, Sharma P, Ikegami M (2007) Suppressor of RNA silencing encoded by the monopartite tomato leaf curl Java begomovirus. Arch Virol 152:1273–1282

Kumar J, Kumar A, Roy J, Tuli R et al (2010) Identification and molecular characterization of begomovirus and associated satellite DNA molecules infecting *Cyamopsis tetragonoloba*. Virus Genes 41:118–125

Kumar J, Singh S, Kumar A, Khan J et al (2013) Detection and characterization of a new betasatellite: variation in disease symptoms of tomato leaf curl Pakistan virus-India due to associated betasatellite. Arch Virol 158:257–261

Kumar J, Kumar J, Singh SP, Tuli R (2014) Association of satellites with a mastrevirus in natural infection: complexity of *Wheat dwarf India virus* disease. J Virol 88:7093–7104

Laufs J, Schumacher S, Geisler N, Jupin I et al (1995) Identification of the nicking tyrosine of geminivirus rep protein. FEBS Lett 377:258–262

Leke W, Kvarnheden A, Ngane E, Titanji V et al (2011) Molecular characterization of a new begomovirus and divergent alphasatellite from tomato in Cameroon. Arch Virol 156:925–928

Leke WN, Brown JK, Ligthart ME, Sattar N et al (2012) *Ageratum conyzoides*: a host to a unique begomovirus disease complex in Cameroon. Virus Res 163:229–237

Leke WN, Sattar MN, Ngane EB, Ngeve JM et al (2013) Molecular characterization of begomoviruses and DNA satellites associated with okra leaf curl disease in Cameroon. Virus Res 174:116–125

Li J, Zhou X (2010) Molecular characterization and experimental host-range of two begomoviruses infecting *Clerodendrum cyrtophyllum* in China. Virus Genes 41:1–10

Li ZH, Xie Y, Zhou XP (2005) *Tobacco curly shoot virus* DNA b is not necessary for infection but intensifies symptoms in a host-dependent manner. Phytopathology 95:902–908

Lozano G, Trenado HP, Fiallo-Olivé E, Chirinos D et al (2016) Characterization of non-coding DNA satellites associated with sweepoviruses (genus *Begomovirus*, *Geminiviridae*) – definition of a distinct class of begomovirus-associated satellites. Front Microbiol 7:162

Mansoor S, Khan SH, Bashir A, Saeed M et al (1999) Identification of a novel circular single-stranded DNA associated with cotton leaf curl disease in Pakistan. Virology 259:190–199

Mansoor S, Briddon RW, Zafar Y, Stanley J (2003) Geminivirus disease complexes: an emerging threat. Trends Plant Sci 8:128–134

Marwal A, Kumar Sahu A, Gaur RK (2013a) Molecular characterization of begomoviruses and DNA satellites associated with a new host Spanish flag (*Lantana camara*) in India. ISRN Virol 2013:5

Marwal A, Sahu A, Choudhary D, Gaur RK (2013b) Complete nucleotide sequence of a begomovirus associated with satellites molecules infecting a new host *Tagetes patula* in India. Virus Genes 47:1–5

Mayo MA, Leibowitz MJ, Palukaitis P, Scholthof K-BG et al (2005) Satellites. In: Fauquet CM, Mayo MA, Maniloff J, Desselberger U, Ball LA (eds) VIIIth report of the International Committee on Taxonomy of Viruses. Virus taxonomy. Elsevier/Academic Press, London, pp 1163–1169

Mubin M, Briddon RW, Mansoor S (2009) Complete nucleotide sequence of chili leaf curl virus and its associated satellites naturally infecting potato in Pakistan. Arch Virol 154:365–368

Mubin M, Shahid MS, Tahir MN, Briddon RW et al (2010) Characterization of begomovirus components from a weed suggests that begomoviruses may associate with multiple distinct DNA satellites. Virus Genes 40:452–457

Nawaz-ul-Rehman MS, Fauquet CM (2009) Evolution of geminiviruses and their satellites. FEBS Lett 583:1825–1832

Nawaz-ul-Rehman MS, Nahid N, Mansoor S, Briddon RW et al (2010) Post-transcriptional gene silencing suppressor activity of the alpha-rep of non-pathogenic alphasatellites associated with begomoviruses. Virology 405:300–308

Nehra C, Gaur RK (2014) Molecular characterization of Chilli leaf curl viruses infecting new host plant Petunia hybrida in India. Virus Genes 50:1–5

Ogawa T, Sharma P, Ikegami M (2008) The begomoviruses Honeysuckle yellow vein mosaic virus and tobacco leaf curl Japan virus with DNAb satellites cause yellow dwarf disease of tomato. Virus Res 137:235–244

Packialakshmi R, Srivastava N, Girish K, Usha R (2010) Molecular characterization of a distinct begomovirus species from *Vernonia cinerea* and its associated DNA-β using the bacteriophage Φ29 DNA polymerase. Virus Genes 41:135–143

Palukaitis P, Rezaian A, García-Arenal F (2008) Satellite nucleic acids and viruses. In: Mahy BWJ, van Regenmortel MHV (eds) Encyclopedia of virology. Academic Press, Oxford, pp 526–535

Paprotka T, Metzler V, Jeske H (2010) The first DNA 1-like a satellites in association with New World begomoviruses in natural infections. Virology 404:148–157

Pilartz M, Jeske H (2003) Mapping of abutilon mosaic geminivirus minichromosomes. J Virol 77:10808–10818

Qazi J, Amin I, Mansoor S, Iqbal J et al (2007) Contribution of the satellite encoded gene βC1 to cotton leaf curl disease symptoms. Virus Res 128:135–139

Qing L, Zhou X (2009) Trans-replication of, and competition between, DNA β satellites in plants inoculated with tomato yellow leaf curl China virus and tobacco curly shoot virus. Phytopathology 99:716–720

Rathore S, Bhatt B, Yadav B, Kale R et al (2014) A new begomovirus species in association with betasatellite causing tomato leaf curl disease in Gandhinagar, India. Plant Dis 98:428–428

Romay G, Lecoq H, Desbiez C (2014) Melon chlorotic mosaic virus and associated alphasatellite from Venezuela: genetic variation and sap transmission of a begomovirus-satellite complex. Plant Pathol 64(5):1224–1234

Rosario K, Marr C, Varsani A, Kraberger S et al (2016) Begomovirus-associated satellite DNA diversity captured through vector-enabled metagenomic (VEM) surveys using whiteflies (Aleyrodidae). Viruses 8:36

Saeed M (2010) Tomato leaf curl New Delhi virus DNA a component and Cotton leaf curl Multan betasatellite can cause mild transient symptoms in cotton. Acta Virol 54:317–318

Saeed M, Zafar Y, Randles JW, Rezaian MA (2007) A monopartite begomovirus-associated DNA b satellite substitutes for the DNA B of a bipartite begomovirus to permit systemic infection. J Gen Virol 88:2881–2889

Saeed ST, Khan A, Kumar B, Ajayakumar PV et al (2014) First report of Chilli leaf curl India virus infecting *Mentha spicata* (Neera) in India. Plant Dis 98:164–164

Sattar MN, Iqbal Z, Tahir MN, Ullah S (2017) The prediction of a new CLCuD epidemic in the Old World. Front Microbiol 8:631

Saunders K (2008) Analysis of geminivirus DNA replication by 2-D gel. In: Gary D, Foster IEJ, Hong Y, Nagy PD (eds) Plant virology protocols, pp 135–143

Saunders K, Stanley J (1999) A nanovirus-like component associated with yellow vein disease of *Ageratum conyzoides*: evidence for interfamilial recombination between plant DNA viruses. Virology 264:142–152

Saunders K, Bedford ID, Briddon RW, Markham PG et al (2000) A unique virus complex causes *Ageratum* yellow vein disease. Proc Natl Acad Sci USA 97:6890–6895

Saunders K, Bedford ID, Stanley J (2002) Adaptation from whitefly to leafhopper transmission of an autonomously-replicating nanovirus-like DNA component associated with ageratum yellow vein disease. J Gen Virol 83:909–915

Saunders K, Norman A, Gucciardo S, Stanley J (2004) The DNA β satellite component associated with ageratum yellow vein disease encodes an essential pathogenicity protein (βC1). Virology 324:37–47

Senanayake DMJB, Jayasinghe JEARM, Shilpi S, Wasala SK et al (2013) A new begomovirus–betasatellite complex is associated with chilli leaf curl disease in Sri Lanka. Virus Genes 46:128–139

Shahid M, Ikegami M, Waheed A, Briddon R et al (2014) Association of an alphasatellite with tomato yellow leaf curl virus and ageratum yellow vein virus in Japan is suggestive of a recent introduction. Viruses 6:189–200

Sharma P, Matsuda N, Bajet NB, Ikegami M (2011) Molecular analysis of new isolates of tomato leaf curl Philippines virus and an associated betasatellite occurring in the Philippines. Arch Virol 156:305–312

She X, He Z, Yin G, Du Z et al (2015) A new alphasatellite molecule associated with Ageratum yellow vein China virus in the Philippines. J Phytopathol 163:54–57

Shelly S, Lukinova N, Bambina S, Berman A et al (2009) Autophagy plays an essential anti-viral role in Drosophila against vesicular stomatitis virus. Immunity 30:588–598

Shih S, Kumar S, Tsai W, Lee L et al (2009) Complete nucleotide sequences of okra isolates of Cotton leaf curl Gezira virus and their associated DNA-β from Niger. Arch Virol 154:369–372

Singh M, Singh K, Haq Q, Mandal B et al (2011) Molecular characterization of tobacco leaf curl Pusa virus, a new monopartite *Begomovirus* associated with tobacco leaf curl disease in India. Virus Genes 43:1–11

Singh M, Haq QMR, Mandal B, Varma A (2012) Evidence of the association of radish leaf curl virus with tobacco yellow leaf curl disease in Bihar, India. Indian J Virol 23:64–69

Srivastava A, Raj S, Kumar S, Snehi S (2013) New record of Papaya leaf curl virus and Ageratum leaf curl beta satellite associated with yellow vein disease of aster in India. New Dis Rep 28(6). https://doi.org/10.5197/j.2044-0588.2013.028.006

Srivastava A, Jaidi M, Kumar S, Raj SK et al (2015) Association of Papaya leaf curl virus with the leaf curl disease of grain amaranth (*Amaranthus cruentus* L.) in India. Phytoparasitica 43:97–101

Tahir M, Haider MS, Briddon RW (2010) Chili leaf curl betasatellite is associated with a distinct recombinant begomovirus, pepper leaf curl Lahore virus, in *Capsicum* in Pakistan. Virus Res 149:109–114

Tahir M, Amin I, Haider S, Mansoor S et al (2015) *Ageratum enation virus* – a begomovirus of weeds with the potential to infect crops. Viruses 7:647–665

Tao X, Zhou X (2008) Pathogenicity of a naturally occurring recombinant DNA satellite associated with tomato yellow leaf curl China virus. J Gen Virol 89:306–311

Tiwari N, Padmalatha K, Singh V, Haq Q et al (2010) Tomato leaf curl Bangalore virus (ToLCBV): infectivity and enhanced pathogenicity with diverse betasatellites. Arch Virol 155:1343–1347

Tiwari N, Sharma PK, Malathi VG (2013) Functional characterization of βC1 gene of Cotton leaf curl Multan betasatellite. Virus Genes 46:111–119

Wu P-J, Zhou X-P (2005) Interaction between a nanovirus-like component and the *Tobacco curly shoot virus*/satellite complex. Acta Biochim Biophys Sin 37:25–31

Xie Y, Wu P, Tao X, Zhou X (2004) Identification of a nanovirus-like DNA molecule associated with *Tobacco curly shoot virus* isolates containing satellite DNA. Prog Nat Sci 14:689–693

Xiong Q, Guo XJ, Che HY, Zhou XP (2005) Molecular characterization of a distinct begomovirus species and its associated satellite DNA molecule infecting *Sida acuta*. J Phytopathol 153:264–268

Yang J-Y, Iwasaki M, Machida C, Machida Y et al (2008) βC1, the pathogenicity factor of TYLCCNV, interacts with AS1 to alter leaf development and suppress selective jasmonic acid responses. Genes Dev 22:2564–2577

Yang X, Guo W, Ma X, An Q et al (2011a) Molecular characterization of *Tomato leaf curl China virus*, infecting tomato plants in China, and functional analyses of its associated betasatellite. Appl Environ Microbiol 77:3092–3101

Yang X, Xie Y, Raja P, Li S et al (2011b) Suppression of methylation-mediated transcriptional gene silencing by βC1-SAHH protein interaction during geminivirus-betasatellite infection. PLoS Pathog 7:e1002329

Zhang T, Xu X, Huang C, Qian Y et al (2016) A novel DNA motif contributes to selective replication of a geminivirus-associated betasatellite by a helper virus-encoded replication-related protein. J Virol 90:2077–2089

Zhou X (2013) Advances in understanding begomovirus satellites. Annu Rev Phytopathol 51:357–381

Zhou X, Xie Y, Tao X, Zhang Z et al (2003) Characterization of DNA b associated with begomoviruses in China and evidence for co-evolution with their cognate viral DNA-A. J Gen Virol 84:237–247

Zulfiqar A, Zhang J, Cui X, Qian Y et al (2012) A new begomovirus associated with alpha- and betasatellite molecules isolated from *Vernonia cinerea* in China. Arch Virol 157:189–191

Geminiviruses Versus Host's Gene Silencing Mechanism

Omid Eini

Abstract

RNA silencing is a well-known antiviral pathway that also controls geminiviruses at two levels: inhibition of viral transcription (TGS) and degradation of viral transcripts (PTGS). Plant viruses encode proteins to suppress this antiviral system. In this chapter, the gene silencing pathway and the steps in which the geminiviral suppressors act have been reviewed. More specifically, the type of viral small RNAs and their role in local and systemic silencing have been described. In addition, this chapter provides an overview of latest researches and findings in geminivirus–host interaction in gene silencing pathway.

1 Introduction

Geminiviruses (family *Geminiviridae*) are single-stranded DNA viruses with a monopartite or bipartite genome (DNA A and DNA B) encapsidated in twinned icosahedral particles. This family contains nine different genera including *Begomovirus*, *Becurtovirus*, *Curtovirus*, *Capulovirus*, *Eragrovirus*, *Grablovirus*, *Mastrevirus*, *Topucovirus*, and *Turncurtovirus* (Brown et al. 2012; Varsani et al. 2017). They replicate in the host cell nucleus (Arguello-Astorga et al. 1994).

In bipartite geminiviruses, the DNA A component encodes proteins responsible for viral DNA replication, vector transmission, encapsidation, and suppression of gene silencing, whereas the DNA B encodes proteins that are required for the movement of virus (Duan et al. 1997; Sanderfoot and Lazarowitz 1995). In monopartite viruses, the genome is homologous to DNA A of bipartite viruses and

O. Eini (✉)
Department of Plant Protection, School of Agriculture, University of Zanjan, Zanjan, Iran
e-mail: omid.eini@znu.ac.ir

R. V. Kumar (ed.), *Geminiviruses*, https://doi.org/10.1007/978-3-030-18248-9_10

the movement function is provided by V2 gene or the coat protein gene (Poornima Priyadarshini et al. 2011). Most of the monopartite begomoviruses are associated with additional ssDNA components (size ~1.3–1.4 kb) referred to as alpha- or betasatellites (Briddon et al. 2003; Paprotka et al. 2010). The satellite DNA components are partly or entirely dependent on the helper virus for their replication, movement, and encapsidation functions (Briddon et al. 2001; Saunders et al. 2000). Alpha-satellites are self-replicating ssDNAs and are related to nanoviral DNA component. They contain a typical genome organization with one open reading frame (ORF) that encodes replication-associated protein (Rep), a conserved hairpin structure, and an A-rich region (Paprotka et al. 2010). Unlike the alphasatellites, betasatellites encode a single gene, which is a pathogenicity determinant (Briddon et al. 2001; Saunders et al. 2000; Saeed et al. 2005) and was shown to suppress gene silencing (Cui et al. 2005; Eini et al. 2012).

2 Gene Silencing and Plant Viruses

RNA silencing pathway is a cellular defense against viruses. This pathway also plays a role in controlling transposon mobility, development of the organism via microRNAs, histone, and DNA methylation, and in the establishment of heterochromatin (Baulcombe 2004; Voinnet 2005). In plants, the gene silencing phenomenon based on co-suppression was first discovered in transgenic petunia, when an attempt to overexpress chalcone synthase by introducing a chimeric petunia CHS gene resulted in the loss of expression of both the transgene and the homologous endogenous gene (Napoli et al. 1990). This phenomenon is known as quelling in fungi (Cogoni and Giuseppe 1997) and RNA interference (RNAi) in animals (Fire et al. 1998).

Three silencing pathways have been described in plants including PTGS, cytoplasmic gene silencing, which is important for virus protection; transcriptional gene silencing (TGS), which is associated with DNA methylation and suppression of transcription that may protect the genome from transposons; and microRNA (miRNA) pathway, which regulates gene expression by silencing of endogenous messenger RNAs (mRNAs) and has a key role in plant development (Baulcombe 2004).

2.1 Post-transcriptional Gene Silencing Pathway

Gene silencing has common steps in plants, animals, insects, and fungi (Pickford and Cogoni 2003; Roth et al. 2004). First, double-stranded RNAs (dsRNAs), produced from different sources such as intermediate replication forms of viruses, are diced by RNaseIII-like proteins, known as Dicer, into small interfering RNA (siRNA) with 2 nt overhangs at 3′ ends. Four Dicer-like (DCL) proteins have been described in *Arabidopsis*. They possibly have been specialized to cleave dsRNA of different origins (Deleris et al. 2006; Xie et al. 2005a). For example, DCL2 and DCL4 are implicated in antiviral defense (Bouché et al. 2006; Deleris et al. 2006; Fusaro et al. 2006). DCL2, DCL4, and DCL3 have a role in dicing dsRNAs that are produced from the inverted-repeat transcripts. DCL4 has a role in sense-PTGS (S-PTGS) in which the target gene is transcribed to high levels and is copied into a duplex by a host RNA-dependent RNA polymerase (RdRP) (Brodersen and Voinnet 2006; Dunoyer and Voinnet 2005; Fusaro et al. 2006). More details for the biochemical properties of plant DCL proteins have been reviewed (Fukudome and Fukuhara 2017).

The generated siRNAs vary in size depending on the DCL that cleaves the dsRNA. The action of DCL1 and DCL4 produces 21 nt siRNA, whereas DCL2 generates 22 nt and DCL3 produces 24 nt siRNA (Dunoyer and Voinnet 2005; Xie et al. 2005b). The siRNAs are methylated at their 3′ termini by a small RNA-specific methyltransferase, HUA Enhancer1 (HEN1) (Boutet et al. 2003), which probably protects siRNAs from degradation and polyuridylation.

In another ATP-dependent step, siRNAs are denatured and then join into an endonuclease silencing complex called RNA-induced silencing complex (RISC). In the activated RISC, single-stranded siRNAs act as guides to bring the complex into contact with complementary mRNAs and thereby cause their degradation (Mlotshwa et al. 2002; Roth et al. 2004). The target RNA is cleaved by a RNA-binding and slicer protein, ARGONAUTE1 (AGO1), which cuts the target RNA at the position between the 10 and 11th nt of the siRNA (Bartel 2004; Elbashir et al. 2001). In Arabidopsis, there are 10 members for the *AGO* family. *AGO1* acts in various RNA silencing pathways (Csorba et al. 2007; Fagard et al. 2000; Hutvagner and Simard 2008; Morel et al. 2002; Zhang et al. 2006a). The antiviral roles of plant AGO and the strategies that viruses have evolved to modulate, attenuate, or suppress AGO antiviral functions have been reviewed (Carbonell and Carrington 2015).

2.2 Transcriptional Gene Silencing Pathway

In TGS, the promoter and sometimes the coding region of the target gene are methylated. Methylation or methylation-associated chromatin remodeling of promoter sequences is thought to prevent binding of factors required for gene transcription. The pattern of DNA methylation is inherited and maintained across the next generations in plants (Gehring and Henikoff 2007).

In the RNA-mediated silencing pathway, RNA may induce methylation of DNA, known as RNA-directed DNA methylation (RdDM). This type of methylation was first reported from viroid-infected tobacco plants, where the genome of a viroid, which was integrated as a transgene into the tobacco genome, became methylated only during replication of the homologous viroid (Wassenegger et al. 1994). RdDM depends on the presence of dsRNA. Thus, endogenous, sense transgene- or invert repeat transgene-derived siRNAs can guide DNA methylation in homologous DNA sequences (Aufsatz et al. 2002). The siRNA-directed DNA methylation in plants is also linked to histone methylation (Zilberman et al. 2003). Cell factors implicated in the RdDM pathway are referred to as the RNA-induced transcriptional silencing complex (Finnegan and Matzke 2003).

The mechanism of methylation directed by siRNA and how the siRNA interacts with its homologous genomic DNA is unknown. The nascent transcripts and/or the DNA itself are possible targets of siRNAs (Brodersen and Voinnet 2006). The role of small RNAs in regulating cytosine methylation of DNA has been reviewed (Hardcastle and Lewsey 2016). In addition to silencing elements, methylation requires other components such as methyltransferases. There are three known classes of methyltransferase, which add methyl groups to cytosine, reviewed in previously (Bender 2004; Gehring and Henikoff 2007).

3 Systemic Silencing and Transitivity

Most plant viruses are systemic in their hosts (Hull 2002). It seems that in the co-evolution process, the host plant initiates RNA silencing against the viral RNA and produces mobile silencing signals that can spread with or ahead of the virus in infected plants (Kalantidis et al. 2008; Roth et al. 2004). In fact, the sequence-specific signal of gene silencing was shown to spread from cells that had received the ectopic DNA to other cells and tissues through plasmodesmata and phloem channels (Voinnet et al. 1998).

Silencing signals act at a highly specific level. In addition, transmission of silencing requires the active transcriptional state of both the trigger and target transgenes. Therefore, this nucleic acid must have a RNA structure. There are two types of movement of silencing signals in plants: short-range movement (10–15 cells) in which 21 nt siRNAs are involved and long-range movement in which 24 nt siRNAs are necessary (Lecellier and Voinnet 2004). In the second type, primary siRNAs derived from viral dsRNA recruit RdRP to produce complementary RNA and subsequently to produce secondary siRNAs. Therefore, the long-range movement requires RdRP6, SDE3 (RNA helicase), and possibly SDE5 to produce new dsRNA by using short-range signals (Voinnet 2005).

Secondary siRNAs are derived from either initiator regions or adjacent regions on both the 5′ and the 3′ side of the initial target sequence (Sijen et al. 2001; Vaistij et al. 2002). Therefore, a primary siRNA molecule could generate many dsRNAs, which would then trigger silencing of even more target molecules. The change from production of primary siRNAs, which correspond to a specific sequence of a targeted RNA, to that of secondary siRNAs, which target regions outside the initial target sequence, is known as transitivity. Transitivity can lead to methylation of a target DNA or cleavage of its transcript (Vaistij et al. 2002). It should be noted that transitivity is important in virus defense, since this process allows the defense system to counteract the replicating viral RNAs. In addition, this amplification step ensures that a few molecules of transposon RNA could activate the chromatin silencing pathway sufficiently to suppress all copies of a transposable element (Baulcombe 2004; Baulcombe 2007).

4 MicroRNAs

MicroRNAs are endogenous small RNAs (20–24 nt in length) that are processed by the action of DCL proteins from imperfectly paired hairpin precursor RNAs, and typically they target a single site in their target mRNA (Reinhart et al. 2002), whereas *siRNAs* are derived from perfectly paired double-stranded RNA molecules that can be endogenous or derived from introduced RNAs, viruses, or transgenes. SiRNAs target multiple sites on the cognate RNA (Bartel 2004). A similar mechanism directs miRNA processing in both plants and animals (Reinhart et al. 2002). MiRNAs play a role in plant growth and development, signal transduction, and response to biotic and abiotic stresses (Bologna and Voinnet 2014; Gu et al. 2014; Voinnet 2009). They also play key roles in plant–virus interactions (Ramesh et al. 2017; Scaria et al. 2006).

Based on the miRNA pathway, two types of antiviral defenses have been reported: indirectly, by regulation of genes that have a role in virus resistance (Ghanbari et al. 2016; Naqvi et al. 2010) or directly by *targeting viral RNAs* in which plant miRNAs can facilitate viral mRNA cleavage (Ramesh et al. 2017). *Virus-encoded miRNAs* were first identified from Epstein-Barr virus (EBV) in infected human B cells (Pfeffer et al. 2004). Subsequently, hundreds of animal virus-encoded miRNAs were reported from various groups of viruses such as polyomaviruses, herpesviruses, and adenoviruses (Gottwein and Cullen 2008). In animals, some virus-encoded miRNAs can regulate both viral and host gene expression (Nair and Zavolan 2006; Pfeffer et al. 2004). This type of viral gene regulation facilitates infection and enhances virulence (Lu et al. 2008; Pumplin and Voinnet 2013). Therefore, animal viruses can employ their own miRNAs to regulate the cellular environment to support the viral life cycle (Roberts et al. 2011). In plants, numerous virus-derived siRNAs (vsiRNAs) or viroid-related siRNAs have been identified. These siRNAs play diverse functions in plant–virus interactions (Huang et al. 2016; Shimura et al. 2011; Smith et al. 2011). However, virus-encoded miRNAs have been reported from only a few numbers of plant RNA viruses (Gao et al. 2012; Viswanathan et al. 2014). The possible reason for the small number of virus-encoded miRNAs from plant viruses as compared to the animal viruses is not clear, but can be depended on their differences in mode of action, genome size, or life cycle (Ramesh et al. 2014a). In fact, a larger DNA genome in animal viruses that are known to encode miRNAs as compared to the small genome in most plant viruses may explain the abundance of virus-encoded miRNAs in animal viruses (Wang et al. 2012). Based on the known miRNA processing system where a miRNA precursor would be cleaved in the cell nucleus (Papp et al. 2003), it is expected that both cytoplasmic replicating DNA and RNA viruses would be less exposed to the miRNA processing system. However, RNA viruses such as turnip mosaic virus (Chantal and Jean-Francois 2007) and hibiscus chlorotic ringspot virus (Gao et al. 2012) were found to enter the nucleus and therefore are prone to encode viral miRNAs. In fact, a recent report confirms the production of virus-encoded miRNAs from a plant RNA virus, sugarcane streak mosaic virus (Gao et al. 2012;

Viswanathan et al. 2014), and also geminiviruses such as African cassava mosaic virus and East African cassava mosaic virus—Uganda (Maghuly et al. 2014).

5 Geminiviruses and RNA Silencing

RNA silencing is an important mechanism of host defense against viruses. This phenomenon was revealed by studying that the synergism interaction between viruses in which symptom severity of a virus disease was increased by co-infection with an unrelated virus (Vance et al. 1995). In addition, recovery from the viral disease can occur in plants infected with both DNA viruses such as *Cauliflower mosaic virus* (CaMV), which replicates in the nucleus, and RNA viruses such as *Tobacco rattle virus*, which replicates in the cytoplasm (Al Kaff et al. 1998; Lecellier and Voinnet 2004; Vaucheret and Fagard 2001). Furthermore, mutants of some essential genes, such as *RdRP6* and *AGO1* which have key role in transgene-induced silencing, were found to enhance the susceptibility of Arabidopsis plants to virus infection (Ding et al. 2004).

Virus infection is sufficient to induce a PTGS-like response in the absence of sequence homology between viruses and host nuclear genes (Lecellier and Voinnet 2004). It seems that the dsRNA, produced as a replication intermediate for RNA viruses, or some ssRNA viruses with high secondary structure elicit this protection response. In addition, following virus infection, nuclear transgenes homologous to viral sequences became methylated, suggesting that viral RNAs present in the cytoplasm entered the nucleus and triggered DNA methylation. Similarly, RdDM has been also observed in the silenced tissue infected with cytoplasmically replicating RNA viruses (Hamilton et al. 2002; Jones et al. 1998). Therefore, both types of gene silencing pathways, PTGS and TGS, have been reported for plant viruses.

Geminiviruses can both induce and be a target for gene silencing (Akbergenov et al. 2006; Vanitharani et al. 2005). Production of siRNAs and PTGS has been reported from both monopartite (Lucioli et al. 2003) and bipartite geminiviruses (Chellappan et al. 2004). There was a significant correlation between recovery phenotype, siRNA production, and the level of viral DNA and mRNA in plants infected with African cassava mosaic virus (ACMV) (Chellappan et al. 2004). It is interesting that geminiviruses do not have a dsRNA replicative form in their disease cycle, yet they can induce PTGS in infected plants. The *dsRNA in geminiviruses can be produced* in three possible ways: bidirectional transcription (Hanley-Bowdoin et al. 1999; Townsend et al. 1985) in which viruses produce polycistronic mRNA with opposite polarity from the conserved region that overlaps at the 3′-end and results in the formation of a dsRNA. Supporting this possibility, the majority of siRNAs in ACMV-[CM]-infected plants are derived from the N-terminus of the AC2, which overlap the C-terminus of AC1 (Chellappan et al. 2004). Alternatively, the host RdRP can possibly use the C1 transcript as a template to produce dsRNA without using an exogenous primer (Dalmay et al. 2000; Tang et al. 2003), or it can extend the overlap between two 3′-ends of the mRNAs. However, DNA viruses do not code for an RdRP and it has shown that mutation of host RdRP2 and RdRP6 is also dispensable for the biogenesis of siRNAs from DNA viruses (Blevins et al. 2006). Finally, the possible hairpin structures in the geminivirus transcripts can be diced to produce siRNA (Chellappan et al. 2004).

5.1 Geminivirus and Methylation

In fact, geminiviruses are targeted by both *TGS and PTGS* mechanisms of RNA silencing, which controls the expression of viral genes (Akbergenov et al. 2006; Rodriguez-Negrete et al. 2008). Viral *siRNAs* (21–22 nt in size) were related to the coding regions (MP, Rep, TrAP, and REn genes), whereas longer size siRNAs (24 nt in size) were associated with the intergenic regions (Akbergenov et al. 2006; Rodriguez-Negrete et al. 2008). A counter correlation was observed between the methylation status of the intergenic region and the accumulation of viral DNA and symptom severity in plants infected with pepper golden mosaic virus. *The methylation* density was significantly higher in the intergenic region than that the one observed on the CP region (Rodriguez-Negrete et al. 2008). Similarly, methylation of the promoter region of *Vigna mungo* yellow mosaic virus, a bipartite geminivirus, through application of dsRNA was reported to interfere with the replication of the target virus (Pooggin et al. 2003). In another study, in vitro DNA methylation was shown to inhibit the accumulation of tomato golden mosaic virus in transfected protoplasts. Notably, this viral DNA methylation was not propagated in progeny viral DNA (Brough et al. 1992). However, a prolonged hypermethylation of a virus-derived transgene from tomato leaf curl virus was reported following the virus infection (Seemanpillai et al. 2003).

Plants use chromatin methylation as a defense against DNA viruses. Supporting this hypothesis, in vitro methylated viral DNA replicated at a lower level at about 20-fold lower than that in non-methylated viruses in plant cells (Brough et al. 1992). Mutation of genes with role in DNA methylation such as genes encoding cytosine or histone H3 lysine 9 (H3K9) methyltransferases (kyp2/suvh4), components of RNA-directed methylation pathway, or adenosine kinase (ADK) was shown to make plants more susceptible to geminivirus infection (Raja et al. 2008). Other components of RNA-directed methylation pathway such as AGO4 were also found to be required for geminiviral DNA methylation (Raja et al. 2008). Likewise, recovery in the new tissue of infected plants has been explained by viral genome methylation (Sunter et al. 2001; Wang et al. 2003) and *ago4* mutant plants could not recover from beet curly top virus (BCTV) or pepper golden mosaic virus infection (Raja et al. 2008). Interaction of DRB 3 (dsRNA binding protein 3) with DCL3 and AGO4 further supports the RNA-directed DNA methylation (RdDM) in geminivirus infection (Raja et al. 2014). These evidences suggest that methylation of viral genomic components is a basic defense pathway against geminiviruses in plants (Bisaro 2006) and geminiviruses can be useful models for genome methylation in plants (Raja et al. 2008).

5.2 Geminivirus and Small Noncoding RNAs

Small RNAs (sRNAs) are the key functional molecules of RNA silencing pathway that regulate gene expression in a sequence-dependent manner. They play a major role in fundamental cellular processes such as defense against viral invasion (Ding and Voinnet 2007) and also in the temporal and spatial regulation of gene expression (Carrington and Victor 2003), transposon mobility control, histone methylation (Matzke and Birchler 2005), and maintenance of chromatin status (Baulcombe 2004). Two main classes of sRNAs are siRNAs and miRNAs (Bologna and Voinnet 2014). In plants, sRNAs regulate the expression of both plant- and viral-derived genes.

Geminiviruses are targeted by all four DCL enzymes to produce three major sizes (21, 22, and 24 nt) of viral siRNAs from both coding and noncoding regions for DNA viruses. DCL4 and DCL3 were found to be more important for production of siRNAs in the plants infected with DNA viruses (Blevins et al. 2006). These viral siRNAs are methylated at the 5′-end and phosphorylated at the 3′-end (Akbergenov et al. 2006). Commonly, 21–22 nt siRNAs target viral coding regions, whereas 24 nt siRNAs were shown to target IR of the viral genome (Akbergenov et al. 2006). The most abundant siRNAs detected in geminivirus infected plants are 24 nt siRNAs. This suggests that *TGS* mode of gene silencing is the major antiviral mechanism (Akbergenov et al. 2006). On the other hand, production of viral 21 and 22 nt siRNAs by the action of DCL4 and DCL2 supports the slicing of the target virus transcript in the cytoplasm (Akbergenov et al. 2006).

In addition to the primary siRNAs, the *secondary siRNAs* are produced by the amplification of the silencing signals through the activity of host-derived RNA polymerases IV and V, RNA-dependent RNA polymerase II (RDR2), and the enzyme activity of DCL3. These siRNAs act in the *TGS* pathway (Garcia-Ruiz et al. 2010; Wang et al. 2011). Mutation of RDR6 also resulted in a small increase in viral DNA accumulation, suggesting that secondary siRNAs might play a vital role in the defense against viral infection (Aregger et al. 2012; Wang et al. 2011). Inactivation of RDR2 by viral suppressors of RNA silencing was shown to affect the methyl cycle in geminivirus infection (Hanley-Bowdoin et al. 2013), which further supports the role of secondary siRNAs in TGS pathway.

The other type of sRNAs is *MicroRNAs* (miRNAs), which are evolutionarily conserved endogenous noncoding small RNAs that control gene expression in eukaryotes. Profiling of miRNA in tomato plants infected with tomato leaf curl New Delhi virus showed downregulation of conserved miRNAs such as miR319 and miR172 (Naqvi et al. 2010). Similarly, most developmental miRNAs were upregulated in *N. benthamiana* plants infected with diverse geminiviruses such as tomato yellow leaf curl virus (TYLCV), Cotton leaf curl Multan virus (CLCuMV)/ cotton leaf curl Multan betasatellite (CLCuMB), and cabbage leaf curl virus (Amin et al. 2011b). Furthermore, in tomato plants infected with beet curly top Iran virus, miRNA target genes such as *MYB33* and *AP2* were differentially regulated in the susceptible and moderately resistant cultivars (Ghanbari et al. 2016), and analysis of host-derived miRNAs by microarray and qPCR in tomato (Naqvi et al. 2010) and soybean (Ramesh et al. 2017) also showed a significant change in the accumulation of conserved miRNAs after virus infection. Likewise, the *predicted miRNAs* and their target genes that have a role in the development of symptoms and resistance mechanism were found to be differentially regulated in resistant and susceptible genotypes of tomato plants infected with tomato leaf curl virus (Tousi et al. 2017). These findings suggest that regulation of host miRNAs may explain the disease symptom induction by geminiviruses.

In animals, endogenous miRNAs were found to bind to and regulate numbers of viral genomic RNA (Jopling et al. 2005) and consequently either reduce or increase the replication level of target viruses. In plants, computational approaches also provide evidences that plant viruses including geminiviruses could be targeted by host miRNAs (Amirnia et al. 2016; Feng and Chen 2013; Naqvi et al. 2011; Tousi and Eini 2016). For example, several tomato-derived miRNA strands were predicted that bind to the genome of tomato leaf curl New Delhi virus (ToLCNDV) (Perez-Quintero et al. 2010), beet curly top Iran virus (Amirnia et al. 2016), and ToLCV (Tousi and Eini 2016). Recently, it was experimentally proved that soybean-derived miRNAs target and direct cleavage of Mungbean yellow mosaic India virus (MYMIV) mRNA encoding movement protein (BC1) (Ramesh et al. 2017), which confirms the role of plant miRNAs in virus resistance. These studies reveal that miRNAs have a major role in plant–geminivirus interaction.

5.3 Artificial miRNA and Geminiviruses

Artificial miRNA (amiRNA) has been widely used for targeting and downregulating of endogenous genes and viruses in various plants. This efficient tool is based on using host-derived endogenous precursor miRNA in which the original 21 nt long miRNA sequence was replaced with a region complementary to the target genes or viral genome (Ramesh et al. 2014b; Schwab et al. 2006). AmiRNA strategy is highly accurate and able to degrade target genes without affecting expression of other genes. This strategy is heritable, environmentally safe, and highly stable in vivo (Li et al. 2013; Tiwari et al. 2014). AmiRNA strategy for antiviral resistance has been used successfully in various plant species including *N. benthamiana*, Arabidopsis, rice, wheat, tomato, maize, and grapevine, reviewed in (Liu et al. 2017).

This strategy has been also used for plant resistance against other pathogens such as bacteria, and fungi, reviewed in (Liu et al. 2017). Therefore, amiRNA strategy could be widely applied for in increasing plant resistance against various pathogens. However, due to the high sequence divergence of plant viruses designing a broad-spectrum amiRNAs is a challenge. In addition, the durability of amiRNAs is another obstacle, when amiRNA targets the non-conserved regions of plant viruses or when a strong pressure from plant viruses occurs in the field. Indeed, testing the stability of amiRNA-mediated resistance against turnip mosaic virus (TuMV) showed that

TuMV escapes this RNA silencing by rapidly accumulating mutations in the target regions (Lin et al. 2009). To overcome these obstacles, the efficiency of amiRNA strategy has been improved by targeting highly conserved RNA motifs in the RNA genome of viruses (Lafforgue et al. 2013) or using a polycistronic amiRNA system to mediate simultaneous resistance to plant viruses (Fahim et al. 2012; Kis 2016; Sun et al. 2016). For example, Arabidopsis expressing the recombinant miRNA precursors containing complementary sequences to turnip yellow mosaic virus (TYMV) and TuMV showed specific resistance to both viruses (Ai et al. 2011; Niu et al. 2006). In wheat, integrating five amiRNAs within one polycistronic amiRNA precursor, miR395, was developed to control wheat streak mosaic virus (WSMV) (Fahim et al. 2012). Finally, the efficiency of miRNAs on target viral RNAs depends on their nature, the accessibility of target sequences, and also the structures of the target mRNAs (Duan et al. 2008). Plant small RNA Maker Site (P-SAMS) is a tool that has been developed recently to design highly efficient amiRNA constructs (Fahlgren and Carrington 2010). Therefore, further studies on amiRNA engineering can make this strategy more practical and applicable.

The first report for using amiRNA technology to control geminiviruses was on wheat dwarf virus, a Mastrevirus (Kis 2016). A polycistronic amiRNA precursor construct (VirusBuster171) was built to express three amiRNAs simultaneously targeting conserved regions from wheat dwarf virus (Kis 2016) to make resistant plants.

5.4 Plant Virus Silencing Suppressors

Plant viruses have evolved various strategies to counteract host RNA silencing to overcome this antiviral defense system. These strategies include: (1) Evasion of RNA silencing. Some viruses localize and replicate in subcellular sites that are not exposed to the RNA silencing machinery. Examples are replication of brome mosaic virus, a *Bromovirus,* which occurs in membrane-bound vesicles, keeping viral RNAs away from host ribonucleases (Schwartz et al. 2002). (2) Protection of the viral genome from silencing. The secondary structure of viroids protects them from RNA silencing. Although their genomes are substrates for DCLs, viroid sequences are inaccessible to the RISC (Wang et al. 2004). In addition, protection of some viral genomes might also result from their association with proteins. For example, encapsidation protects viral genomes from silencing (Angell and Baulcombe 1997). (3) Overwhelming of silencing. It is believed that some viruses may replicate and spread at such high rates than the defensive capacity of the RNA silencing acts. (4) Most plant viruses encode a suppressor of RNA silencing (VSR) as an adaptive response to plant defense (Roth et al. 2004). VSRs interfere with host RNA silencing through multiple modes (Burgyán 2008; Csorba and Burgyán 2016; Ding and Voinnet 2007; Zhao et al. 2016). Most of these proteins are viral pathogenicity determinants (Voinnet 2005) and produce abnormal phenotypes, which resemble those of *ago1*, *hyl1*, or *dcl1* mutants (Dunoyer et al. 2004; Zhang et al. 2006b). VSRs act in viral symptom production by facilitating virus accumulation and modifying host miRNA-mediated regulation (Burgyán 2008; Silhavy and Burgyan 2004). VSRs are multifunction genes that also act in viral replication, encapsidation, or movement (Jiang et al. 2012; Zhao et al. 2016).

To identify plant viral suppressors of silencing, different *methods* have been used (Li and Ding 2006; Moissiard and Voinnet 2004; Roth et al. 2004). These methods are Agrobacterium-mediated transient suppression assay (Patch test), heterologous complementation of a suppressor protein, reversal of a silenced transgenic reporter gene, and stable expression assay (grafting). The last method is the most free of complications due to unrelated effects of pathogens.

VSRs encoded by plant viruses are *structurally diverse*. They interfere with different steps of RNA silencing including initiation, maintenance, and the systemic steps of RNA silencing. Therefore, they act by different mechanisms. Notably, incompatible results from different studies have made it difficult to find clear and firm conclusions about many suppressors (Ding et al. 2004).

Various *strategies* have been reported for plant viral suppressors (Csorba and Burgyán 2016; Ding and Voinnet 2007) as follows: blocking the initiation of host antiviral response. In this strategy, VSRs inhibit the DCLs activities, sequester the vsiRNAs, block systemic silencing, or interfere with the AGO loading. For example, physical interactions with AGO1 have been reported for 2b from CMV (Zhang et al. 2006b) and P0 protein from beet western yellows virus (Bortolamiol et al. 2007); direct interference with silencing machinery has been reported for the tombusviral P19, P21 of beet yellows virus, and HC-Pro of tobacco etch virus protein which bind to siRNAs (Lakatos et al. 2006); furthermore, inhibition of RISC assembly by capturing single-stranded siRNA was shown for ACMV AC4 protein (Chellappan et al. 2005a); interaction with host factors and regulation of these factors to modulate the silencing system. For example, in HC-Pro (Voinnet 2005) or AC2 from ACMV and mungbean yellow mosaic virus (Trinks et al. 2005) activation of some negative regulators of silencing has been reported. Some VSRs regulate R genes by direct interaction or indirectly through miRNA regulation. For example, 2b protein from CMV was found to suppress salicylic acid-mediated defense response (Ji and Ding 2001), while the HC-Pro from potato virus Y (PVY) was found to induce defense responses (Shams-Bakhsh et al. 2007); finally, some VSRs inhibit TGS. For example, L2 protein from curtoviruses interacts with adenosine kinase, which maintains the cellular methylation level (Wang et al. 2005).

Suppression of gene silencing has been reported for DNA viruses such as begomoviruses and curtoviruses (Bisaro 2006). For example, AC4 protein coded by ACMV-CM and Sri Lankan cassava mosaic virus (SLCMV) acts as a suppressor of PTGS, and complementation effect of AC4 and AC2 genes in both viruses has been linked to their synergism interaction (Vanitharani et al. 2004a). Plants infected with multiple geminivirus produce more severe symptoms due to *synergic interaction*, which can be explained by the action of VSRs (Vanitharani et al. 2004a).

5.5 Geminiviral Encoded Suppressor of Gene Silencing

5.5.1 Rep

Rep (replication-associated protein, also named C1, AL1, or AC1) is an essential gene for replication of geminiviruses. Rep is a multifunctional protein with site-specific nicking and ligation, DNA binding, helicase, and ATPase activities that enable it to initiate, elongate, and terminate rolling circle replication (Elmer et al. 1988; Hanley-Bowdoin et al. 1999). Rep protein from several geminiviruses was shown to interact with retinoblastoma-related proteins (RBRs), key regulators of the plant cell cycle (Ach et al. 1997; Kong et al. 2000) (Qi Xie and Gutierrez 1996). It was found that Rep protein interferes with the plant DNA methylation machinery and suppresses TGS through repression of the plant DNA methyltransferases, METHYLTRANSFERASE 1 (MET1) and HROMOMETHYLASE 3 (CMT3) (Rodríguez-Negrete et al. 2013). Bisulfite sequencing analyses also revealed that the expression of Rep from geminiviruses causes a significant reduction in the levels of DNA methylation at CG sites (Rodríguez-Negrete et al. 2013). A list of geminiviral suppressors is provided in Table 1.

5.5.2 C2 Protein

C2 Protein of monopartite geminiviruses is a homolog of AC2 or AL2 of bipartite geminiviruses. This transcriptional activator protein is required for the expression of late viral genes (Sunter and Bisaro 1992) and also suppresses PTGS by a mechanism that depends on its ability to activate transcription (Trinks et al., 2005). AC2 protein from ACMV and TYLCV was demonstrated to reverse the established RNA silencing in plants (van Wezel et al. 2002; Voinnet et al. 1999). Localization to the nucleus and presence of DNA binding domains in AC2 proteins facilitate their suppression activity. AC2 proteins do not bind any sRNA (miRNAs and siRNAs); therefore, their mechanism of function differs from other VSRs such as P19 from tomato bushy stunt virus. Silencing suppression by AC2 was found to correlate with the transactivation of host transcript(s) in plants infected with MYMIV (Rahman et al. 2012). Interestingly, WEL1 (Werner exonuclease-like 1), one of the upregulated genes by AC2, was shown to suppress RNA silencing in *N. benthamiana* plants (Trinks et al. 2005), whereas further studies showed that silencing suppression could be achieved by truncated AL2 that lacked the activation domain (Wang et al. 2003). Therefore, the activation domain has an indirect effect on the suppression activity of C2 protein from geminiviruses. Interaction of AC2 from tomato golden mosaic virus (TGMV) with a calmodulin-like protein (rgs-CaM), endogenous regulator of gene silencing, shows that this protein may sequester rgsCaM in the nucleus to prevent targeting of AL2 for degradation (Yong Chung et al. 2014). A similar study showed that AC2 of MYMIV interacts with RDR6 and AGO1 to suppress siRNA biogenesis and retract the RISC activity, respectively (Zhang et al. 2011).

There are other evidences for the effect of C2 protein on the effector steps of TGS. AL2 and C2 proteins encoded by TGMV and BCTV were shown to interact with and inactivate host serine/threonine kinase-related kinase (SnRK1) and adenosine kinase (ADK), which have a role in the host metabolism and methyl cycle maintenance,

Table 1 Suppressor genes from geminiviruses and their mode of action

Gene	Virus	Mode of action	Reference
Rep (C1)	Tomato yellow leaf curl Sardinia virus (TYLCSV) Tomato yellow leaf curl virus, mild strain (TYLCV) Tomato golden mosaic virus (TGMV) Cotton leaf curl Rajasthan Alfasatellite	Repression of the plant DNA methyltransferases, MET1 and CMT3 Suppressor of PTGS	Rodríguez-Negrete et al. (2013) Nawaz-ul-Rehman et al. (2010)
AC2 (AL2)	African cassava mosaic virus (ACMV)	Suppressor of PTGS, both local and systemic silencing	Voinnet et al. (1999)
	Tomato yellow leaf curl China virus (TYLCCV)	Suppressor of PTGS. The putative zinc-finger motif is essential.	van Wezel et al. (2002)
	Mungbean yellow mosaic India virus	Transactivation of hosttranscript(s), interacts with RDR6 and AGO1 to suppress siRNA biogenesis	Trinks et al. (2005) Kumar et al. (2015)
	Tomato golden mosaic virus (TGMV)	Inactivate host serine/threonine related kinase and adenosine kinase interaction with a calmodulin-like protein (rgs-CaM)	Buchmann et al. (2009), Raja et al. (2008); Wang et al. (2005), Yong Chung et al. (2014)
C2	Beet curly top virus (BCTV) Beet severe curly top virus (BSCTV)	Inhibits TGS via inhibition of adenosine kinase interact with S-adenosyl methionine decarboxylase 1 (SAMDC1) to suppress DNA methylation	Zhang et al. (2011)
AC4	African cassava mosaic virus Cameroon strain (ACMV-C)	Arrest the RISC activity, interacts with miRNAs	Chellappan et al. (2005a)
C4	Cotton leaf curl Multan virus (CLCuMV) Tomato leaf curl virus (ToLCV) Cotton leaf curl Kokhran virus (CLCuKoV)	Interacts with both short and long RNAs Interaction with a shaggy-like kinase Preventing the spread of systemic silencing	Amin et al. (2011a) Dogra et al. (2009) Saeed et al. (2015)
AC5	Mungben yellow mosaic Indian virus (MYMIV)	Interfere with dsRNA production; Reverse TGS probably by inhibiting the expression of a CHH cytosine methyltransferase	Li et al. (2015)
V2	Tomato yellow leaf curl virus (TYLCV)	SGS3 interaction (competition by dsRNA);	Glick et al. (2008), Fukunaga and Doudna

(continued)

Table 1 (continued)

Gene	Virus	Mode of action	Reference
	Tomato yellow leaf curl China virus (TYLCCV) Beet curly top virus (BCTV)	Direct interaction and inhibition of histone deacetylase 6 Acts downstream of the DCLs in RNA silencing pathway PTGS suppressor, possibly by impairing the RDR6	(2009), Wang et al. (2018), Zhang et al. (2012), Zrachya et al. (2007), Luna et al. (2017)
βC1	Tomato yellow leaf curl China betasatellite (TYLCCB) Cotton leaf curl Multan betasatellite (CLCuMuB)	SAHH inhibition; interacts with calmodulin like protein (Nb-rgs CAM); alter host methylation-mediated virus defence pathway by inhibiting S-adenosyl homocysteine hydrolase Interaction and regulation of AGO1; Binds long dsRNAs;	Yang et al. (2011), Fangfang et al. (2014), Yang et al. (2011), Eini (2017), Saeed et al. (2015), Amin et al. (2011a)

respectively (Buchmann et al. 2009; Raja et al. 2008; Wang et al. 2005). In line with this study, inhibition of ADK by adenosine analogs and/or silencing of ADK was also shown to mimic the effect of geminiviruses on ADK function (Wang et al. 2003). In addition, C2 protein encoded by BSCTV was found to interact with S-adenosyl methionine decarboxylase 1 (SAMDC1) and attenuates the degradation of SAMDC1 by arresting its ubiquitylation to suppress DNA methylation-mediated gene silencing in Arabidopsis (Zhang et al. 2011). Three cytosine residues in the putative zinc finger motif of C2 protein from Tomato yellow leaf curl virus-China were found to be essential for its anti-RNA silencing function and pathogenesis (van Wezel et al. 2002). AL2 from TGMV and L2 from BCTV were also found to inhibit TGS via ADK inhibition. However, at one exceptional locus, ADK inhibition was insufficient and transcriptional activation domain of AL2 was required to reverse the TGS. Finally, a genome-wide reduction in cytosine methylation was observed in transgenic plants expressing AL2 and L2 proteins (Buchmann et al. 2009). Therefore, C2 protein and its homologs such as AL2/L2 mainly interfere with host methylation mediated suppression of both host and viral gene expression.

5.5.3 AC4/C4

AC4/C4 is a multifunctional protein which have a role in virus movement (Teng et al. 2010), virus pathogenicity (Mills-Lujan and Deom 2010; Park et al. 2010), and suppression of gene silencing (Dogra et al. 2009; Fondong et al. 2007; Peretz et al. 2011; Vanitharani et al. 2004b). Unveiling the mechanism of suppression of gene silencing pathway by *C4 gene* has shown that C4 arrests the programmed RISC activity by targeting the guide RNA in this complex. In fact, C4 protein acts on the downstream step of the small RNA synthesis through interaction with single-stranded miRNAs and siRNAs but not with duplex siRNAs (Chellappan et al. 2005b). This interaction affects the miRNA-mediated gene regulation and leads in developmental defects by alternating in the stress signaling pathways linked to the hormonal regulations (Bazzini et al. 2007; Chellappan et al. 2005b). Notably, AC4 protein from ACMV has been shown to bind directly to certain miRNAs, thereby making miRNA-RISC non-functional. Similarly, overexpressing of AC4 in transgenic plants reduced the accumulation of miRNAs in Arabidopsis (Chellappan et al. 2005a) and plants expressing AC4 from ToLCNDV (Naqvi et al. 2010). This may suggest that C4 protein interacts and destabilizes host miRNAs. In addition, interaction of C4 protein from CLCuMV with both long and short RNAs with a preferential binding of dsRNA shows that this multifunctional protein acts in both upstream and downstream of siRNA synthesis by sequestering both long dsRNA from DCL cleavage and siRNA from RISC incorporation (Amin et al. 2011a). Additionally, C4 protein was shown to act as an ancillary player in suppression of TSG by Rep protein via downregulation of MET1 (Rodríguez-Negrete et al. 2013). This may reflect the diversity in the amino acid sequences of C4 proteins from various groups of geminiviruses.

5.5.4 AC5

The *AC5* protein encoded by most bipartite begomoviruses has a role in geminiviral DNA replication (Li et al. 2015) (Raghavan et al. 2004). Mutation of AC5 in two bipartite begomoviruses, tomato chlorotic mottle virus and watermelon chlorotic stunt virus, revealed that this protein is not essential for the virus infection cycle (Fontenelle et al. 2007; Kheyr-Pour et al. 2000). However, AC5 from an isolate of tomato leaf deformation virus was reported to play a role in the viral infectivity and symptom development (Melgarejo et al. 2013). Similarly, the AC5 from MYMIV was found to play a critical role in the virus infection and suppression of host gene silencing system (Li et al. 2015). Further investigation showed that this protein interferes with dsRNA production as it suppresses sense RNA-induced gene silencing but not RNA silencing triggered by dsRNA (Li et al. 2015). AC5 from MYMIV was also shown to effectively suppress ssRNA-induced PTGS and to reverse TGS of a GFP transgene, probably by inhibiting the expression of a CHH cytosine methyltransferase in *N. benthamiana* (Li et al. 2015).

5.5.5 AV2/V2

The *AV2/V2* ORF is present in members of different geminivirus genera, but not in the New World bipartite begomoviruses. This gene codes for the movement protein in mastreviruses and appears to be a symptom determinant (Fondong 2013) and a suppressor of PTGS in BCTV (Luna et al. 2017). In begomoviruses, the function of this gene varies depending on the group of viruses. In some monopartite viruses such as TYLCV and TYLCCV, *V2 protein* was reported to suppress RNA silencing. This VSR acts downstream of the DCLs in the RNA silencing pathway to affect the amplification of the antiviral silencing signals (Zhang et al. 2012; Zrachya et al. 2007). Suppression of TGS by V2 protein from TYLCV was determined in TGS-based GFP silenced plants (Glick et al. 2008) and recently was proved to limit histone deacetylase enzymatic activity of a histone deacetylase 6 in a direct interaction (Wang et al. 2018). Similarly, transgenic plants expressing V2 gene showed a significant reduction in the methylation of host genomic regions, confirming the role of V2 protein in the suppression of TGS (Zhang et al. 2012). Furthermore, V2 protein was found to prevent the spread of silencing signals. V2 protein inhibits the activity of suppressor of gene silencing3 (SGS3), the cofactor of RDR6, and therefore suppresses the amplification of silencing signals (Glick et al. 2008). Alternatively, V2 protein may compete with SGS3 for the dsRNA with 5′ overhang ends that may be an RDR6/SGS3 intermediate/substrate during vsiRNA amplification (Fukunaga and Doudna 2009). In plants, RDR6 and SGS3 are required to convert ssRNAs to dsRNAs in the initial step of RNAi-based antiviral response and to produce both exogenous and endogenous short-interfering RNAs (Allen et al. 2005; Chapman et al. 2004; Fukunaga and Doudna 2009).

5.5.6 βC1 Protein

The *βC1 protein* encoded by betasatellites is a pathogenicity determinant (Zhang et al. 2015) and also suppresses gene silencing (Amin et al. 2011a; Eini et al. 2012; Gopal et al. 2007; Kon et al. 2007; Saeed et al. 2015). This protein was found to bind ssDNA or dsDNA in a nonspecific manner and localizes in the nucleus (Cui et al. 2005). βC1 protein was shown to bind long dsRNAs, suggesting its effect on the activity of DCL and also on sequestration of siRNAs (Amin et al. 2011a). The βC1 encoded by DNA satellite associated with TYLCCV interacts with calmodulin-like protein (Nb-rgs CAM) causing its upregulation in *N. benthamiana* plants. This interaction is required to repress host RDR6 expression and ultimately impedes the production of secondary siRNAs (Fangfang et al. 2014). βC1encoded by CLCuMB suppresses systemic gene silencing and reduces the accumulation of viral siRNAs (Eini et al. 2012). This protein has been found to physically interact, in yeast-two hybrid system, with AGO1, an important component of host RNAi pathway, and regulate the expression of DLC1 and AGO1 genes in transgenic Arabidopsis plants (Eini 2017). In addition, expression of βC1 in plants downregulates the expression of host miRNAs such as miR165/166 which have a role in plant developmental, which can explain developmental abnormalities in plants infected with the betasatellite complex (Amin et al. 2011a; Yang et al. 2008).

Alternatively, the begomovirus betasatellite-encoded βC1 has been found to alter host methylation-mediated virus defense pathway by inhibiting S-adenosyl homocysteine hydrolase, a methyl cycle enzyme that is also required for TGS (Yang et al. 2011). Reduction in S-adenosyl homocysteine hydrolase activities indirectly blocks the methyl cycle, and thereby interfered with the epigenetic modification and methylation of the viral genome (Yang et al. 2011). Therefore, βC1 is a suppressor of both PTGS and TGS to facilitate helper virus replication and to enhance symptom production in infected plants. A schematic figure shows the main steps of gene

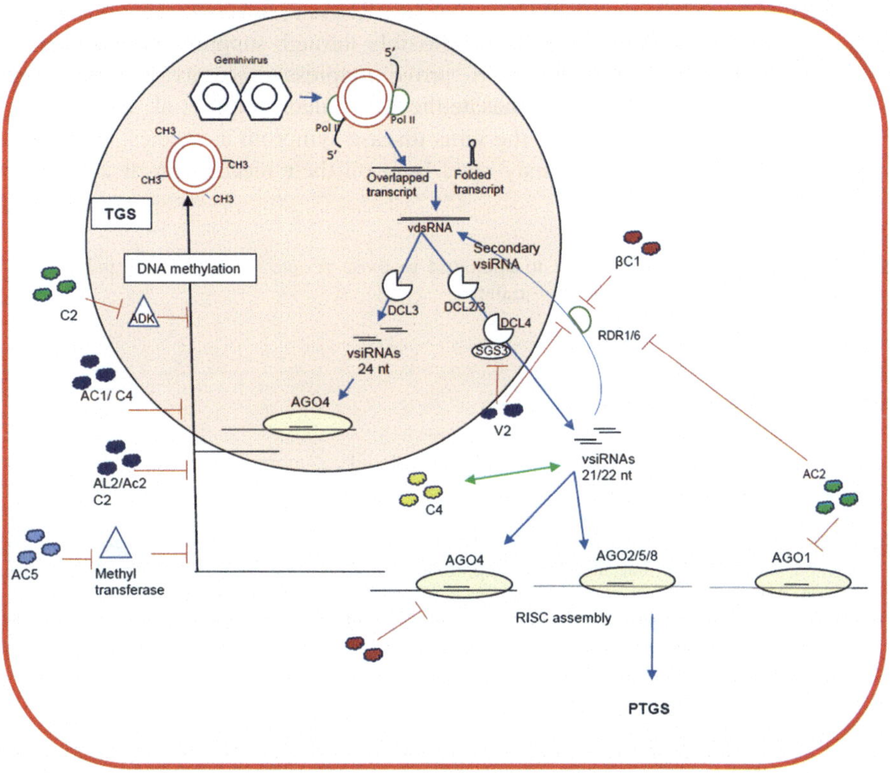

Fig. 1 Schematic for the plant antiviral silencing and suppressor of gene silencing from geminiviruses. The main components of gene silencing pathway and the steps in which a geminiviral suppressor act are shown

silencing and the steps in which the geminiviral suppressors interact with this antiviral pathway (Fig. 1).

Alphasatellites are circular ssDNAs with approximate size of 1400 nt associated with begomovirus/betasatellite complexes originating from the "Old World." They require the helper begomovirus for insect transmission and spread in plants (Briddon et al. 2004). An initially study showed that ageratum yellow vein alphasatellite had little effect on the accumulation of the Ageratum yellow vein virus in *N. benthamiana* plants (Saunders and Stanley 1999) and alphasatellites associated with okra leaf curl disease and tobacco curly shoot virus attenuated disease symptoms caused by begomovirus/betasatellite complex and reduced the accumulation of betasatellite (Kumar et al. 2015). Whereas more severe symptoms and a higher viral DNA accumulation was reported in wheat plants co-infected with wheat dwarf India virus and two alphasatellites, cotton leaf curl Multan alphasatellite and Guar leaf curl alphasatellite (Kumar et al. 2014). This thought to be occurs through suppression of RNA silencing-mediated host defense, since in the presence of alphasatellite the production of WDIV-derived siRNAs in infected wheat plants was reduced (Nawaz-ul-Rehman et al. 2010) possibly through suppression of PTGS by this satellite. However, the attempt to prove suppressor activity for alpha-Rep protein cotton leaf curl Multan alphasatellite was failed (Amin et al. 2011a). It is possible that alphasatellites affect the virus disease symptom in species- or isolate-specific manner. Further study may shed light on their mechanism of action for alpha-Rep proteins.

Acknowledgements We would like to apologize to those people whose relevant publications could not be cited because of space constraints.

References

Ach RA, Durfee T, Miller A, Taranto P, Hanley-Bowdoin L, Zambryski PC, Gruissem W (1997) RRB1 and RRB2 encode maize retinoblastoma-related proteins that interact with a plant D-type cyclin and geminivirus replication protein. Mol Cell Biol 17:5077–5086

Ai T, Zhang L, Gao Z, Zhu CX, Guo X (2011) Highly efficient virus resistance mediated by artificial microRNAs that target the suppressor of PVX and PVY in plants. Plant Biol 13:304–316

Akbergenov R, Si-Ammour A, Blevins T, Amin I, Kutter C, Vanderschuren H, Zhang P, Gruissem W, Meins F Jr, Hohn T, Pooggin MM (2006) Molecular characterization of geminivirus-derived small RNAs in different plant species. Nucleic Acids Res 34:462–471

Al Kaff NS, Covey SN, Kreike MM, Page AM, Pinder R, Dale PJ (1998) Transcriptional and posttranscriptional plant gene silencing in response to a pathogen. Science 279:2113–2115

Allen E, Xie Z, Gustafson AM, Carrington JC (2005) microRNA-directed phasing during trans-acting siRNA biogenesis in plants. Cell 121:207–221

Amin I, Hussain K, Akbergenov R, Yadav JS, Qazi J, Mansoor S, Hohn T, Fauquet CM, Briddon RW (2011a) Suppressors of RNA silencing encoded by the components of the cotton leaf curl begomovirus-betasatellite complex. Mol Plant-Microbe Interact 24:973–983

Amin I, Patil B, Briddon R, Mansoor S, Fauquet C (2011b) A common set of developmental miRNAs are upregulated in Nicotiana benthamiana by diverse begomoviruses. Virol J 8:143

Amirnia F, Eini O, Koolivand D (2016) In silico analysis of microRNA binding to the genome of Beet curly top Iran virus in tomato. Arch Phytopathol Plant Protect 49:434–444

Angell SM, Baulcombe DC (1997) Consistent gene silencing in transgenic plants expressing a replicating potato virus X RNA. EMBO J 16:3675–3684

Aregger M, Borah BK, Seguin J, Rajeswaran R, Gubaeva EG, Zvereva AS, Windels D, Vazquez F, Blevins T, Farinelli L, Pooggin MM (2012) Primary and secondary siRNAs in geminivirus-induced gene silencing. PLoS Pathog 8:e1002941

Arguello-Astorga GR, Guevara-Gonzalez RG, Herrera-Estrella LR, Rivera-Bustamante RF (1994) Geminivirus replication origins have a group-specific organization of iterative elements: a model for replication. Virology 203:90–100

Aufsatz W, Mette MF, Winden JVD, Matzke AJM, Matzke M (2002) RNA-directed DNA methylation in arabidopsis. Proc Natl Acad Sci USA 99:16499–16506

Bartel DP (2004) MicroRNAs: genomics, biogenesis, mechanism, and function. Cell 116:281–297

Baulcombe D (2004) RNA silencing in plants. Nature 431:356–363

Baulcombe DC (2007) Molecular biology: amplified silencing. Science 315:199–200

Bazzini AA, Hopp HE, Beachy RN, Asurmendi S (2007) Infection and coaccumulation of tobacco mosaic virus proteins alter microRNA levels, correlating with symptom and plant development. PNAS 104:12157–12162

Bender J (2004) Chromatin-based silencing mechanisms. Curr Opin Plant Biol 7:521–526

Bisaro DM (2006) Silencing suppression by geminivirus proteins. Virology 344:158–168

Blevins T, Rajeswaran R, Shivaprasad PV, Beknazariants D, Si-Ammour A, Park H-S, Vazquez F, Robertson D, Meins F, Hohn T, Pooggin MM (2006) Four plant dicers mediate viral small RNA biogenesis and DNA virus induced silencing. Nuclic Acids Res 34:6233–6246

Bologna NG, Voinnet O (2014) The diversity, biogenesis, and activities of endogenous silencing small RNAs in arabidopsis. Annu Rev Plant Biol 65:473–503

Bortolamiol D, Pazhouhandeh M, Marrocco K, Genschik P, Ziegler-Graff V (2007) The polerovirus F box protein P0 targets ARGONAUTE1 to suppress RNA silencing. Curr Biol 17:1615–1621

Bouché N, Lauressergues D, Gasciolli V, H V. (2006) An antagonistic function for Arabidopsis DCL2 in development and a new function for DCL4 in generating viral siRNAs. EMBO J 25:3347–3356

Boutet S, Vazquez F, Liu J, Beclin C, Fagard M, Gratias A, Morel J-B, Crete P, Chen X, Vaucheret H (2003) Arabidopsis HEN1: a genetic link between endogenous miRNA controlling development and siRNA controlling transgene silencing and virus resistance. Curr Biol 13:843–848

Briddon RW, Mansoor S, Bedford ID, Pinner MS, Saunders K, Stanley J, Zafar Y, Malik KA, Markham PG (2001) Identification of DNA components required for induction of cotton leaf curl disease. Virology 285:234–243

Briddon RW, Bull SE, Amin I, Idris AM, Mansoor S, Bedford ID, Dhawan P, Rishi N, Siwatch SS, Abdel-Salam AM, Brown JK, Zafar Y, Markham PG (2003) Diversity of DNA beta, a satellite molecule associated with some monopartite begomoviruses. Virology 312:106–121

Briddon RW, Bull SE, Amin I, Mansoor S, Bedford ID, Rishi N, Siwatch SS, Zafar Y, Abdel-Salam AM, Markham PG (2004) Diversity of DNA 1: a satellite-like molecule associated with monopartite begomovirus-DNA beta complexes. Virology 324:462–474

Brodersen P, Voinnet O (2006) The diversity of RNA silencing pathways in plants. Trends Genet 22:268–280

Brough CL, Gardiner WE, Inamdar NM, Zhang X-Y, Ehrlich M, Bisaro DM (1992) DNA methylation inhibits propagation of tomato golden mosaic virus DNA in transfected protoplasts. Plant Mol Biol 18:703–712

Brown JK, Fauquet CM, Briddon RW, Zerbini M, Moriones E, Navas CJ (2012) Geminiviridae. In: King AMQ et al (eds) Virus taxonomy: ninth report of the international committee on taxonomy of viruses. Elsevier, London, pp 351–373

Buchmann RC, Asad S, Wolf JN, Mohannath G, Bisaro DM (2009) Geminivirus AL2 and L2 proteins suppress transcriptional gene silencing and cause genome-wide reductions in cytosine methylation. J Virol 83:5005–5013

Burgyán J (2008) Role of silencing suppressor proteins. In: Foster GD et al (eds) Plant virology protocols: from viral sequence to protein function. Humana Press, Totowa, NJ, pp 69–79

Carbonell A, Carrington JC (2015) Antiviral roles of plant argonautes. Curr Opin Plant Biol 27:111–117

Carrington JC, Victor A (2003) Role of microRNAs in plant and animal development. Science 301:336–338

Chantal B, Jean-Francois L (2007) The poly(A) binding protein is internalized in virus-induced vesicles or redistributed to the nucleolus during turnip mosaic virus infection. J Virol 81:10905–10913

Chapman EJ, Prokhnevsky AI, Gopinath K, Dolja VV, Carrington JC (2004) Viral RNA silencing suppressors inhibit the microRNA pathway at an intermediate step. Genes Dev 18:1179–1186

Chellappan P, Vanitharani R, Fauquet CM (2004) Short interfering RNA accumulation correlates with host recovery in DNA virus-infected hosts, and gene silencing targets specific viral sequences. J Virol 78:7465–7477

Chellappan P, Vanitharani R, Fauquet CM (2005a) MicroRNA-binding viral protein interferes with Arabidopsis development. PNAS 102:10381–10386

Chellappan P, Vanitharani R, Ogbe F, Fauquet CM (2005b) Effect of temperature on geminivirus-induced RNA silencing in plants. Plant Physiol 138:1828–1841

Cogoni C, Giuseppe M (1997) Isolation of quelling-defective (qde) mutants impaired in posttranscriptional transgene-induced gene silencing in Neurosporacrassa. PNAS 94:10233–10238

Csorba T, Burgyán J (2016) Antiviral silencing and suppression of gene silencing in plants. In: Wang A, Zhou X (eds) Current research topics in plant virology. Springer, Cham, pp 1–33
Csorba T, Bovi A, Dalmay T, Burgyan J (2007) The p122 subunit of tobacco mosaic virus replicase is a potent silencing suppressor and compromises both small interfering RNA- and microRNA-mediated pathways. J Virol 81:11768–11780
Cui X, Li G, Wang D, Hu D, Zhou X (2005) A begomovirus DNA β encoded protein binds DNA, functions as a suppressor of RNA silencing, and targets the cell nucleus. J Virol 79 (16):10764–10775
Dalmay T, Hamilton A, Rudd S, Angell S, Baulcombe DC (2000) An RNA-dependent RNA polymerase gene in arabidopsis is required for posttranscriptional gene silencing mediated by a transgene but not by a virus. Cell 101:543–553
Deleris A, Gallego-Bartolome J, Bao J, Kasschau KD, Carrington JC, Voinnet O (2006) Hierarchical action and inhibition of plant dicer-like proteins in antiviral defense. Science 313:68–71
Ding S-W, Voinnet O (2007) Antiviral immunity directed by small RNAs. Cell 130:413–426
Ding SW, Li HW, Lu R, Li F, Li WX (2004) RNA silencing: a conserved antiviral immunity of plants and animals. Virus Res 102:109–115
Dogra S, Eini O, Rezaian M, Randles J (2009) A novel shaggy-like kinase interacts with the tomato leaf curl virus pathogenicity determinant C4 protein. Plant Mol Biol 71:25–38
Duan Y-P, Powell CA, Purcifull DE, Broglio P, Hiebert E (1997) Phenotypic variation in transgenic tobacco expressing mutated geminivirus movement/pathogenicity (BC1) proteins. Mol Plant-Microbe Interact 10:1065–1074
Duan C-G, Wang C-H, Fang R-X, Guo H-S (2008) Artificial microRNAs highly accessible to targets confer efficient virus resistance in plants. J Virol 82:11084–11095
Dunoyer P, Voinnet O (2005) The complex interplay between plant viruses and host RNA-silencing pathways. Curr Opin Plant Biol 8:415–423
Dunoyer P, Lecellier CH, Parizotto EA, Himber C, Voinnet O (2004) Probing the microRNA and small interfering RNA pathways with virus-encoded suppressors of RNA silencing. Plant Cell 16:1235–1250
Eini O (2017) A betasatellite-encoded protein regulates key components of gene silencing system in plants. Mol Biol 51:579–585
Eini O, Dogra SC, Dry IB, Randles JW (2012) Silencing suppressor activity of a begomovirus DNA β encoded protein and its effect on heterologous helper virus replication. Virus Res 167:97–101
Elbashir SM, Lendeckel W, Tuschl T (2001) RNA interference is mediated by 21- and 22-nucleotide RNAs. Genes Dev 15:188–200
Elmer JS, Brand L, Sunter G, Gardiner WE, Bisaro DM, Rogers SG (1988) Genetic analysis of the tomato golden mosaic virus. II. The product of the AL1 coding sequence is required for replication. Nucleic Acids Res 16:7043–7060
Fagard M, Boutet S, Morel J-B, Bellini C, Vaucheret H (2000) AGO1, QDE-2, and RDE-1 are related proteins required for post-transcriptional gene silencing in plants, quelling in fungi, and RNA interference in animals. PNAS 97:11650–11654
Fahim M, Millar AA, Wood CC, Larkin PJ (2012) Resistance to wheat streak mosaic virus generated by expression of an artificial polycistronic microRNA in wheat. Plant Biotechnol J 10(2):150–163
Fahlgren N, Carrington J (2010) miRNA target prediction in plants. In: Meyers BC, Green PJ (eds) Plant microRNAs. Humana Press, New York, pp 51–57
Fangfang L, Changjun H, Zhenghe L, Xueping Z (2014) Suppression of RNA silencing by a plant DNA virus satellite requires a host calmodulin-like protein to repress RDR6 expression. PLoS Pathog 10:e1003921
Feng J, Chen J (2013) In silico analysis the complementarity of tomato microRNA/microRNA* sequences with cucumber mosaic virus (CMV) genomic RNAs. J Nanosci Nanotechnol 13:4421–4426
Finnegan EJ, Matzke MA (2003) The small RNA world. J Cell Sci 116:4689–4693
Fire A, Xu S, Montgomery MK, Kostas SA, Driver SE, Mello CC (1998) Potent and specific genetic interference by double-stranded RNA in *Caenorhabditis elegans*. Nature 391:806–811
Fondong VN (2013) Geminivirus protein structure and function. Mol Plant Pathol 14:635–649

Fondong V, Reddy R, Lu C, Hankoua B, Felton C, Czymmek K, Achenjang F (2007) The consensus N-myristoylation motif of a geminivirus AC4 protein is required for membrane binding and pathogenicity. Mol Plant-Microbe Interact 20:380–391

Fontenelle MR, Luz DF, Gomes APS, Florentino LH, Zerbini FM, Fontes EPB (2007) Functional analysis of the naturally recombinant DNA-A of the bipartite begomovirus tomato chlorotic mottle virus. Virus Res 126:262–267

Fukudome A, Fukuhara T (2017) Plant dicer-like proteins: double-stranded RNA-cleaving enzymes for small RNA biogenesis. J Plant Res 130:33–44

Fukunaga R, Doudna JA (2009) dsRNA with 5′ overhangs contributes to endogenous and antiviral RNA silencing pathways in plants. EMBO J 28:545–555

Fusaro AF, Matthew L, Smith NA, Curtin SJ, Dedic-Hagan J, Ellacott GA, Watson J, Wang M-B, Brosnan C, Carroll BJ, Waterhouse PM (2006) RNA interference-inducing hairpin RNAs in plants act through the viral defence pathway. EMBO Rep 7:1168–1175

Gao R, Liu P, Wong S-M (2012) Identification of a plant viral RNA genome in the nucleus. PLoS One 7:e48736

Garcia-Ruiz H, Takeda A, Chapman EJ, Sullivan CM, Fahlgren N, Brempelis KJ, Carrington JC (2010) Arabidopsis RNA-dependent RNA polymerases and dicer-like proteins in antiviral defense and small interfering RNA biogenesis during turnip mosaic virus infection. Plant Cell 22:481

Gehring M, Henikoff S (2007) DNA methylation dynamics in plant genomes. Biochim Biophys Acta 1769:276–286

Ghanbari M, Eini O, Ebrahimi S (2016) Differntial expression of MYB33 and AP2 genes and response of Ty resistant plants beet curly top Iran vitus infection in tomato. J Plant Pathol 98:555–562

Glick E, Zrachya A, Levy Y, Mett A, Gidoni D, Belausov E, Citovsky V, Gafni Y (2008) Interaction with host SGS3 is required for suppression of RNA silencing by tomato yellow leaf curl virus V2 protein. PNAS 105:157–161

Gopal P, Pravin Kumar P, Sinilal B, Jose J, Kasin Yadunandam A, Usha R (2007) Differential roles of C4 and βC1 in mediating suppression of post-transcriptional gene silencing: evidence for transactivation by the C2 of Bhendi yellow vein mosaic virus, a monopartite begomovirus. Virus Res 123(1):9–18

Gottwein E, Cullen BR (2008) Viral and cellular microRNAs as determinants of viral pathogenesis and immunity. Cell Host Microbe 3:375–387

Gu M, Liu W, Meng Q, Zhang W, Chen A, Sun S, Xu G (2014) Identification of microRNAs in six solanaceous plants and their potential link with phosphate and mycorrhizal signalings. J Integr Plant Biol 56:1164–1178

Hamilton A, Voinnet O, Chappell L, Baulcombe D (2002) Two classes of short interfering RNA in RNA silencing. EMBO J 21:4671–4679

Hanley-Bowdoin L, Settlage SB, Orozco BM, Nagar S, Robertson D (1999) Geminiviruses: models for plant DNA replication, transcription, and cell cycle regulation. Crit Rev Plant Sci 18:71–106

Hanley-Bowdoin L, Bejarano ER, Robertson D, Mansoor S (2013) Geminiviruses: masters at redirecting and reprogramming plant processes. Nat Rev Microbiol 11(11):777

Hardcastle TJ, Lewsey MG (2016) Mobile small RNAs and their role in regulating cytosine methylation of DNA. RNA Biol 13:1060–1067

Huang J, Yang M, Lu L, Zhang X (2016) Diverse functions of small RNAs in different plant–pathogen communications. Front Microbiol 7:1552

Hull R (2002) Mattews' plant virology, 4th edn. Academic Press, London

Hutvagner G, Simard MJ (2008) Argonaute proteins: key players in RNA silencing. Nat Rev Mol Cell Biol 9:22–32

Ji L-H, Ding S-W (2001) The suppressor of transgene RNA silencing encoded by cucumber mosaic virus interferes with salicylic acid-mediated virus resistance. Mol Plant-Microbe Interact 14:715–724

Jiang L, Wei C, Li Y (2012) Viral suppression of RNA silencing. Sci China Life Sci 55:109–118

Jones AL, Thomas CL, Maule AJ (1998) De novo methylation and co-suppression induced by a cytoplasmically replicating plant RNA virus. EMBO J 17:6385–6393

Jopling CL, Yi M, Lancaster AM, Lemon SM, Sarnow P (2005) Modulation of hepatitis C virus RNA abundance by a liver-specific microRNA. Science 309:1577–1581

Kalantidis K, Schumacher HT, Alexiadis T, Helm JM (2008) RNA silencing movement in plants. Biol Cell 100:13–26

Kheyr-Pour A, Bananej K, Dafalla GA, Caciagli P, Noris E, Ahoonmanesh A, Lecoq H, Gronenborn B (2000) Watermelon chlorotic stunt virus from the Sudan and Iran: sequence comparisons and identification of a whitefly-transmission determinant. Phytopathology 90:629–635

Kis A (2016) Polycistronic artificial miRNA mediated resistance to wheat dwarf virus in barley is highly efficient at low temperature. Mol Plant Pathol 17:427–437

Kon T, Sharma P, Ikegami M (2007) Suppressor of RNA silencing encoded by the monopartite tomato leaf curl Java begomovirus. Arch Virol 152:1273–1282

Kong LJ, Orozco BM, Roe JL, Nagar S, Ou S, Feiler HS, Durfee T, Miller AB, Gruissem W, Robertson D, Hanley-Bowdoin L (2000) A geminivirus replication protein interacts with the retinoblastoma protein through a novel domain to determine symptoms and tissue specificity of infection in plants. EMBO J 19:3485–3495

Kumar J, Kumar J, Singh SP, Tuli R (2014) Association of satellites with a mastrevirus in natural infection: complexity of wheat dwarf India virus disease. J Virol 88:7093–7104

Kumar V, Mishra SK, Rahman J, Taneja J, Sundaresan G, Mishra NS, Mukherjee SK (2015) Mungbean yellow mosaic Indian virus encoded AC2 protein suppresses RNA silencing by inhibiting Arabidopsis RDR6 and AGO1 activities. Virology 486:158–172

Lafforgue G, Martínez F, Niu Q-W, Chua N-H, Daròs J-A, Elena SF (2013) Improving the effectiveness of artificial microRNA (amiR)-mediated resistance against turnip mosaic virus by combining two amiRs or by targeting highly conserved viral genomic regions. J Virol 87:8254–8256

Lakatos L, Csorba T, Pantaleo V, Chapman EJ, Carrington JC, Liu Y-P, Dolja VV, Calvino LF, López-Moya JJ, Burgyán J (2006) Small RNA binding is a common strategy to suppress RNA silencing by several viral suppressors. EMBO J 25:2768–2780

Lecellier C, Voinnet O (2004) RNA silencing: no mercy for viruses? Immunol Rev 198:285

Li F, Ding S-W (2006) Virus counterdefense: diverse strategies for evading the RNA-silencing immunity. Annu Rev Microbiol 60:503–531

Li J-F, Chung HS, Niu Y, Bush J, McCormack M, Sheen J (2013) Comprehensive protein-based artificial microRNA screens for effective gene silencing in plants. Plant Cell 25:1507

Li F, Xu X, Huang C, Gu Z, Cao L, Hu T, Ding M, Li Z, Zhou X (2015) The AC5 protein encoded by Mungbean yellow mosaic India virus is a pathogenicity determinant that suppresses RNA silencing-based antiviral defenses. New Phytol 208:555–569

Lin S-S, Wu H-W, Elena SF, Chen K-C, Niu Q-W, Yeh S-D, Chen C-C, Chua N-H (2009) Molecular evolution of a viral non-coding sequence under the selective pressure of amiRNA-mediated silencing. PLoS Pathog 5:e1000312

Liu S-R, Zhou J-J, Hu C-G, Wei C-L, Zhang J-Z (2017) MicroRNA-mediated gene silencing in plant defense and viral counter-defense. Front Microbiol 8:1801

Lu Y-D, Gan Q-H, Chi X-Y, Qin S (2008) Roles of microRNA in plant defense and virus offense interaction. Plant Cell Rep 27:1571–1579

Lucioli A, Noris E, Brunetti A, Tavazza R, Ruzza V, Castillo AG, Bejarano ER, Accotto GP, Tavazza M (2003) Tomato yellow leaf curl Sardinia virus rep-derived resistance to homologous and heterologous geminiviruses occurs by different mechanisms and is vercome if virus-mediated transgene silencing is activated. J Virol 77:6785–6798

Luna AP, Rodríguez-Negrete EA, Morilla G, Wang L, Lozano-Durán R, Castillo AG, Bejarano ER (2017) V2 from a curtovirus is a suppressor of post-transcriptional gene silencing. J Gen Virol 98:2607–2614

Maghuly F, Ramkat RC, Laimer M (2014) Virus versus host plant microRNAs: who determines the outcome of the interaction? PLoS One 9:e98263

Matzke MA, Birchler JA (2005) RNAi-mediated pathways in the nucleus. Nat Rev Genet 6:24–35

Melgarejo TA, Kon T, Rojas MR, Paz-Carrasco L, Zerbini FM, Gilbertson RL (2013) Characterization of a new world monopartite begomovirus causing leaf curl disease of Tomato in Ecuador and Peru reveals a new direction in geminivirus evolution. J Virol 87:5397–5413

Mills-Lujan K, Deom C (2010) Geminivirus C4 protein alters Arabidopsis development. Protoplasma 239(1–4):35–110

Mlotshwa S, Voinnet O, Mette MF, Matzke M, Vaucheret H, Ding SW, Pruss G, Vance VB (2002) RNA silencing and the mobile silencing signal. Plant Cell 14:s289–s301

Moissiard G, Voinnet O (2004) Viral suppression of RNA silencing in plants. Mol Plant Pathol 5:71–82

Morel J-B, Godon C, Mourrain P, Beclin C, Boutet S, Feuerbach F, Proux F, Vaucheret H (2002) Fertile hypomorphic ARGONAUTE (ago1) mutants impaired in post-transcriptional gene silencing and virus resistance. Plant Cell 14:629–639

Nair V, Zavolan M (2006) Virus-encoded microRNAs: novel regulators of gene expression. Trends Microbiol 14:169–175

Napoli C, Lemieux C, Jorgensen R (1990) Introduction of a chimeric chalcone synthase gene into petunia results in reversible co-suppression of homologous genes in trans. Plant Cell 2:279–289

Naqvi A, Haq Q, Mukherjee S (2010) MicroRNA profiling of tomato leaf curl New Delhi virus (ToLCNDV) infected tomato leaves indicates that deregulation of mir159/319 and mir172 might be linked with leaf curl disease. Virol J 7:281–297

Naqvi AR, Choudhury NR, Mukherjee SK, Haq QMR (2011) In silico analysis reveals that several tomato microRNA/microRNA* sequences exhibit propensity to bind to tomato leaf curl virus (ToLCV) associated genomes and most of their encoded open reading frames (ORFs). Plant Physiol Biochem 49:13–17

Nawaz-ul-Rehman MS, Nahid N, Mansoor S, Briddon RW, Fauquet CM (2010) Post-transcriptional gene silencing suppressor activity of two non-pathogenic alphasatellites associated with a begomovirus. Virology 405:300–308

Niu Q-W, Lin S-S, Reyes JL, Chen K-C, Wu H-W, Yeh S-D, Chua N-H (2006) Expression of artificial microRNAs in transgenic *Arabidopsis thaliana* confers virus resistance. Nat Biotechnol 24:1420

Papp I, Mette MF, Aufsatz W, Daxinger L, Schauer SE, Ray A, Winden JVD, Matzke M, Matzke AJM (2003) Evidence for nuclear processing of plant micro RNA and short interfering RNA precursors. Plant Physiol 132:1382–1390

Paprotka T, Metzler V, Jeske H (2010) The first DNA 1-like α satellites in association with New World begomoviruses in natural infections. Virology 404:148–157

Park J, Hwang H-S, Buckley K, Park J-B, Auh C-K, Kim D-G, Lee S, Davis K (2010) C4 protein of beet severe curly top virus is a pathomorphogenetic factor in Arabidopsis. Plant Cell Rep 29:1377–1389

Peretz Y, Eybishtz A, Sela I (2011) Silencing of ORFs C2 and C4 of tomato yellow leaf curl virus engenders resistant or tolerant plants. The Open Virology Journal 5:141–147

Perez-Quintero A, Neme R, Zapata A, Lopez C (2010) Plant microRNAs and their role in defense against viruses: a bioinformatics approach. BMC Plant Biol 10:138–150

Pfeffer S, Zavolan M, Grasser FA, Chien M, Russo JJ, Ju J, John B, Enright AJ, Marks D, Sander C, Tuschl T (2004) Identification of virus-encoded microRNAs. Science 304:734–736

Pickford AS, Cogoni C (2003) RNA-mediated gene silencing. Cell Mol Life Sci 60:871–882

Pooggin M, Shivaprasad PV, Veluthambi K, Hohn T (2003) RNAi targeting of DNA virus in plants. Nat Biotech 21:131–132

Poornima Priyadarshini CG, Ambika MV, Tippeswamy R, Savithri HS (2011) Functional characterization of coat protein and V2 involved in cell to cell movement of cotton leaf curl kokhran virus-dabawali. PLoS One 6:e26929

Pumplin N, Voinnet O (2013) RNA silencing suppression by plant pathogens: defence, counter-defence and counter-counter-defence. Nat Rev Micro 11:745–760

Qi Xie AR-B, Gutierrez GJHAC (1996) Plant cells contain a novel member of the retinoblastoma family of growth regulatory proteins. EMBO J 15:4900–4908

Raghavan V, Malik PS, Choudhury NR, Mukherjee SK (2004) The DNA-A component of a plant geminivirus (Indian mung bean yellow mosaic virus) replicates in budding yeast cells. J Virol 78:2405–2413

Rahman J, Karjee S, Mukherjee S (2012) MYMIV-AC2, a geminiviral RNAi suppressor protein, has potential to increase the transgene expression. Appl Biochem Biotechnol 167:758–775

Raja P, Sanville BC, Buchmann RC, Bisaro DM (2008) Viral genome methylation as an epigenetic defense against geminiviruses. J Virol 82:8997–9007

Raja P, Jackel JN, Li S, Heard IM, Bisaro DM (2014) Arabidopsis double-stranded RNA binding protein DRB3 participates in methylation-mediated defense against geminiviruses. J Virol 88:2611–2622

Ramesh S, Ratnaparkhe M, Kumawat G, Gupta G, Husain S (2014a) Plant miRNAome and antiviral resistance: a retrospective view and prospective challenges. Virus Genes 48:1–14

Ramesh SV, Ratnaparkhe MB, Kumawat G, Gupta GK, Husain SM (2014b) Plant miRNAome and antiviral resistance: a retrospective view and prospective challenges. Virus Genes 48(1):1–14

Ramesh SV, Chouhan BS, Kumar G, Praveen S, Chand S (2017) Expression dynamics of *Glycine max* (L.) Merrill microRNAs (miRNAs) and their targets during mungbean yellow mosaic India virus (MYMIV) infection. Physiol Mol Plant Pathol 100:13–22

Reinhart BJ, Weinstein EG, Rhoades MW, Bartel B, Bartel DP (2002) MicroRNAs in plants. Genes Dev 16:1616–1626

Roberts APE, Lewis AP, Jopling CL (2011) The role of microRNAs in viral infection. Prog Mol Biol Transl Sci 102:101–139

Rodriguez-Negrete EA, Carrillo-Tripp J, Rivera-Bustamante RF (2008) RNA silencing against geminivirus: complementary action of PTGS and TGS in host-recovery. J Virol 83 (3):1332–1340

Rodríguez-Negrete E, Lozano-Durán R, Piedra-Aguilera A, Cruzado L, Bejarano ER, Castillo AG (2013) Geminivirus rep protein interferes with the plant DNA methylation machinery and suppresses transcriptional gene silencing. New Phytol 199:464–475

Roth BM, Gail JP, Vicki BV (2004) Plant viral suppressors of RNA silencing. Virus Res 102:97–108

Saeed M, Behjatnia SAA, Shahid M, Yusuf Z, Shahida H, Rezaian MA (2005) A single complementary-sense transcript of a geminiviral DNA beta satellite is determinant of pathogenicity. Mol Plant-Microbe Interact 18:7–14

Saeed M, Briddon RW, Dalakouras A, Krczal G, Wassenegger M (2015) Functional analysis of cotton leaf curl kokhran virus/cotton leaf curl Multan betasatellite RNA silencing suppressors. Biology 4:697–714

Sanderfoot AA, Lazarowitz SG (1995) Cooperation in viral movement: the geminivirus BL1 movement protein interacts with BR1 and redirects it from the nucleus to the cell periphery. Plant Cell 7:1185–1194

Saunders K, Stanley J (1999) A nanovirus-like DNA component associated with yellow vein disease of *Ageratum conyzoides*: evidence for interfamilial recombination between plant DNA viruses. Virology 264:142–152

Saunders K, Bedford ID, Briddon RW, Markham PG, Wong SM, Stanley J (2000) A unique virus complex causes Ageratum yellow vein disease. PNAS 97:6890–6895

Scaria V, Hariharan M, Maiti S, Pillai B, Brahmachari S (2006) Host-virus interaction: a new role for microRNAs. Retrovirology 3:68

Schwab R, Ossowski S, Riester M, Warthmann N, Weigel D (2006) Highly specific gene silencing by artificial microRNAs in Arabidopsis. Plant Cell 18:1121–1133

Schwartz M, Chen J, Janda M, Sullivan M, den Boon J, Ahlquist P (2002) A positive-strand RNA virus replication complex parallels form and function of retrovirus capsids. Mol Cell 9:505–514

Seemanpillai M, Dry I, Randles J, Rezaian A (2003) Transcriptional silencing of geminiviral promoter-driven transgenes following homologous virus infection. Mol Plant-Microbe Interact 16:429–438

Shams-Bakhsh M, Canto T, Palukaitis P (2007) Enhanced resistance and neutralization of defense responses by suppressors of RNA silencing. Virus Res 130:103–109

Shimura H, Pantaleo V, Ishihara T, Myojo N, Inaba J-I, Sueda K, Burgyán J, Masuta C (2011) A viral satellite RNA induces yellow symptoms on tobacco by targeting a gene involved in chlorophyll biosynthesis using the RNA silencing machinery. PLoS Pathog 7:e1002021

Sijen T, Fleenor J, Simmer F, Thijssen KL, Parrish S, Timmons L, Plasterk RHA, Fire A (2001) On the role of RNA amplification in dsRNA-triggered gene silencing. Cell 107:465–476

Silhavy D, Burgyan J (2004) Effects and side-effects of viral RNA silencing suppressors on short RNAs. Trends Plant Sci 9:76–83

Smith NA, Eamens AL, Wang M-B (2011) Viral small interfering RNAs target host genes to mediate disease symptoms in plants. PLoS Pathog 7:e1002022

Sun L, Lin C, Du J, Song Y, Jiang M, Liu H, Zhou S, Wen F, Zhu C (2016) Dimeric artificial microRNAs mediate high resistance to RSV and RBSDV in transgenic rice plants. Plant Cell Tissue Organ Cult 126(1):127–139

Sunter G, Bisaro DM (1992) Transactivation of geminivirus AR1 and BR1 gene expression by the viral AL2 gene product occurs at the level of transcription. Plant Cell 4:1321–1331

Sunter G, Sunter JL, Bisaro DM (2001) Plants expressing tomato golden mosaic virus AL2 or beet curly top virus L2 transgenes show enhanced susceptibility to infection by DNA and RNA viruses. Virology 285:59–70

Tang G, Reinhart BJ, Bartel DP, Zamore PD (2003) A biochemical framework for RNA silencing in plants. Genes Dev 17:49–63

Teng KCH, Lai J, Zhang Z, Fang Y, Xia R, Zhou X, Guo H, Xie Q (2010) Involvement of C4 protein of beet severe curly top virus (family Geminiviridae) in virus movement. PLoS One 24:11280

Tiwari M, Sharma D, Trivedi PK (2014) Artificial microRNA mediated gene silencing in plants: progress and perspectives. Plant Mol Biol 86:1–18

Tousi N, Eini O (2016) Various tomato microRNAs could target a mild and a severe strain of tomato leaf curl virus. Genet Eng Biosaf J 5:15–22

Tousi N, Eini O, Ahmadvand R, Carra A, Miozzi L, Noris E, Accotto GP (2017) In silico prediction of miRNAs targeting ToLCV and their regulation in susceptible and resistant tomato plants. Australas Plant Pathol 46:379–386

Townsend R, Stanley J, Curson SJ, Short MN (1985) Major polyadenylated transcripts of cassava latent virus and location of the gene encoding coat protein. EMBO J 4:33–37

Trinks D, Rajeswaran R, Shivaprasad PV, Akbergenov R, Oakeley EJ, Veluthambi K, Hohn T, Pooggin MM (2005) Suppression of RNA silencing by a geminivirus nuclear protein, AC2, correlates with transactivation of host genes. J Virol 79:2517–2527

Vaistij FE, Jones L, Baulcombe DC (2002) Spreading of RNA targeting and DNA methylation in RNA silencing requires transcription of the target gene and a putative RNA-dependent RNA polymerase. Plant Cell 14:857–867

van Wezel R, Dong X, Liu H, Tien P, Stanley J, Hong Y (2002) Mutation of three cysteine residues in tomato yellow leaf curl virus-China C2 protein causes dysfunction in pathogenesis and posttranscriptional gene-silencing suppression. Mol Plant-Microbe Interact 15(3):203–208

Vance V, Berger P, Carrington J, Hunt A, Shi X (1995) 5′ proximal potyviral sequences mediate potato virus X/potyviral synergistic disease in transgenic tobacco. Virology 206:538–590

Vanitharani R, Chellappan P, Pita J, Fauquet C (2004a) Differential roles of AC2 and AC4 of cassava geminiviruses in mediating synergism and suppression of post-transcriptional gene silencing. J Virol 78:9487

Vanitharani R, Chellappan P, Pita JS, Fauquet CM (2004b) Differential roles of AC2 and AC4 of cassava geminiviruses in mediating synergism and suppression of posttranscriptional gene silencing. J Virol 78:9487–9498

Vanitharani R, Chellappan P, Fauquet CM (2005) Geminiviruses and RNA silencing. Trends Plant Sci 10:144–151

Varsani A, Roumagnac P, Fuchs M, Navas-Castillo J, Moriones E, Idris A, Briddon RW, Rivera-Bustamante R, Murilo Zerbini F, Martin DP (2017) Capulavirus and grablovirus: two new genera in the family geminiviridae. Arch Virol 162:1819–1831

Vaucheret H, Fagard M (2001) Transcriptional gene silencing in plants: targets, inducers and regulators. Trends Genet 17:29–35

Viswanathan C, Anburaj J, Prabu G (2014) Identification and validation of sugarcane streak mosaic virus-encoded microRNAs and their targets in sugarcane. Plant Cell Rep 33:265–276

Voinnet O (2005) Induction and suppression of RNA silencing: insights from viral infections. Nat Rev Genet 6:206–220

Voinnet O (2009) Origin, biogenesis, and activity of plant microRNAs. Cell 136:669–687

Voinnet O, Vain P, Angell S, Baulcombe DC (1998) Systemic spread of sequence-specific transgene RNA degradation in plants is initiated by localized introduction of ectopic promoterless DNA. Cell 95:177–187

Voinnet O, Pinto YM, Baulcombe DC (1999) Suppression of gene silencing: a general strategy used by diverse DNA and RNA viruses of plants. Plant J 96:14147–14152

Wang H, Hao L, Shung C-Y, Sunter G, Bisaro DM (2003) Adenosine kinase is inactivated by geminivirus AL2 and L2 proteins. Plant Cell 15:3020–3032

Wang M-B, Bian X-Y, Wu L-M, Liu L-X, Smith NA, Isenegger D, Wu R-M, Masuta C, Vance VB, Watson JM, Rezaian A, Dennis ES, Waterhouse PM (2004) On the role of RNA silencing in the pathogenicity and evolution of viroids and viral satellites. PNAS 101:3275–3280

Wang H, Buckley KJ, Yang X, Buchmann RC, Bisaro DM (2005) Adenosine kinase inhibition and suppression of RNA silencing by geminivirus AL2 and L2 proteins. J Virol 79:7410–7418

Wang X-B, Jovel J, Udomporn P, Wang Y, Wu Q, Li W-X, Gasciolli V, Vaucheret H, Ding S-W (2011) The 21-nucleotide, but not 22-nucleotide, viral secondary small interfering RNAs direct potent antiviral defense by two cooperative argonautes in *Arabidopsis thaliana*. Plant Cell 23:1625

Wang M-B, Masuta C, Smith NA, Shimura H (2012) RNA silencing and plant viral diseases. Mol Plant-Microbe Interact 25:1275–1285

Wang B, Yang X, Wang Y, Xie Y, Zhou X (2018) Tomato yellow leaf curl virus V2 interacts with host HDA6 to suppress methylation-mediated transcriptional gene silencing in plants. J Virol 92: e00036-18

Wassenegger M, Heimes S, Riedel L, Sanger HL (1994) RNA-directed de novo methylation of genomic sequences in plants. Cell 76:567–576

Xie Z, Allen E, Fahlgren N, Calamar A, Givan SA, Carrington JC (2005a) Expression of arabidopsis MIRNA genes. Plant Physiol 138:2145–2154

Xie Z, Allen E, Wilken A, Carrington JC (2005b) DICER-LIKE 4 functions in trans-acting small interfering RNA biogenesis and vegetative phase change in *Arabidopsis thaliana*. PNAS 102:12984–12989

Yang J-Y, Iwasaki M, Machida C, Machida Y, Zhou X, Chua N-H (2008) βC1, the pathogenicity factor of TYLCCNV, interacts with AS1 to alter leaf development and suppress selective jasmonic acid responses. Genes Dev 22:2564–2577

Yang X, Yan X, Raja P, Sizhun L, Wolf JN, Shen Q, Bisaro DM, Zhou X (2011) Suppression of methylation-mediated transcriptional gene silencing by βC1-SAHH protein interaction during geminivirus-betasatellite infection. PLoS Path 7:1–13

Yong Chung H, Lacatus G, Sunter G (2014) Geminivirus AL2 protein induces expression of, and interacts with, a calmodulin-like gene, an endogenous regulator of gene silencing. Virology 460–461:108–118

Zhang B, Pan X, Anderson TA (2006a) Identification of 188 conserved maize microRNAs and their targets. FEBS Lett 580:3753–3762

Zhang X, Yuan Y-R, Pei Y, Lin S-S, Tuschl T, Patel DJ, Chua N-H (2006b) Cucumber mosaic virus-encoded 2b suppressor inhibits Arabidopsis Argonaute1 cleavage activity to counter plant defense. Genes Dev 20:3255–3268

Zhang Z, Chen H, Huang X, Xia R, Zhao Q, Lai J, Teng K, Li Y, Liang L, Du Q, Zhou X, Guo H, Xie Q (2011) BSCTV C2 attenuates the degradation of SAMDC1 to suppress DNA methylation-mediated gene silencing in Arabidopsis. Plant Cell 23:273–288

Zhang J, Dong J, Xu Y, Wu J (2012) V2 protein encoded by tomato yellow leaf curl China virus is an RNA silencing suppressor. Virus Res 163:51–58

Zhang J, Dang M, Huang Q, Qian Y (2015) Determinants of disease phenotype differences caused by closely-related isolates of begomovirus betasatellites inoculated with the same species of helper virus. Viruses 7:4945–4959

Zhao J-H, Hua C-L, Fang Y-Y, Guo H-S (2016) The dual edge of RNA silencing suppressors in the virus–host interactions. Curr Opin Virol 17:39–44

Zilberman D, Cao X, Jacobsen SE (2003) ARGONAUTE4 control of locus-specific siRNA accumulation and DNA and histone methylation. Science 299:716–719

Zrachya A, Glick E, Levy Y, Arazi T, Citovsky V, Gafni Y (2007) Suppressor of RNA silencing encoded by tomato yellow leaf curl virus-Israel. Virology 358:159–165

Geminivirus Resistance Strategies

Abhinav Kumar and Jawaid A. Khan

Abstract

Geminiviruses are a major threat to world agriculture, and breeding resistant crops against these viruses is one of the major challenges faced by both plant pathologists and biotechnologists. In the past, most of these strategies follow the conceptual development ranging from coat protein-mediated restricted viral propagation to the expression of mutant or truncated viral proteins that interfere with virus infection, or RNA molecule-mediated gene silencing approach transcription of viral RNA sequences that silence the expression of virus genes. However much of the progress has been made so far in this direction observes limited success in field, but still research is running and new approaches such as CRISPR/Cas9 have found space in laboratories. To date, no comparative data has been published or available that examines the merit of different approaches which have been used against this class of viruses. There is a common belief among the geminivirologists across the globe about the recombination and mutation capacity as the main reason for the appearance of new species and breaking resistance. This chapter deals with different strategies which have been used to curb geminivirus spread.

A. Kumar (✉)
Research Projects Lab, Department of Biotechnology, IILM College of Engineering and Technology, Greater Noida, UP, India
e-mail: abhinav.kumar@iilmcet.ac.in

J. A. Khan
Plant Virology Lab, Department of Biosciences, Jamia Millia Islamia, New Delhi, India
e-mail: jkhan1@jmi.ac.in

R. V. Kumar (ed.), *Geminiviruses*, https://doi.org/10.1007/978-3-030-18248-9_11

1 Introduction

Plant viruses are important pathogens causing enormous losses to agricultural crops, thus affecting the economy of a country. Any plant which is grown by humans for food, fiber, or any other use is virtually infected by one or more viruses. Although the direct effects of plant viruses on human beings are not substantial as compared to animal viruses, the damage caused by the former is considerable when it comes to food supply.

Geminiviruses, grouped into family *Geminiviridae*, are widely distributed plant viruses infecting a wide range of plants from monocots such as maize to dicots such as cassava and tomato (Hanley-Bowdoin et al. 1999). They all share two distinctive features: (1) the geminate morphology of the virion particle and (2) the nature of their genetic material that consists of one or two single-stranded DNA (ssDNA) molecules (2.5–3.0 kb in length), depending on the genera. Differences in the genetic organization of their genomes as well as their host range and insect vectors serve as criteria to recognize nine different genera.

The single-stranded (ss) circular DNA genome of geminiviruses is packed into twin-shaped virions (Zhang et al. 2001). Virions are twinned, geminate, icosahedral, non-enveloped, 38-nm long, and 22 nm in diameter and contain 22 capsomeres per nucleocapsid. Genomic DNA is either mono- or bipartite, circular ssDNA. There are coding regions in both the virion sense (+) and complementary (−) sense strands. They have a single coat protein (CP) of 28–34 kDa and a replication (Rep) protein of 41 kDa, which initiates rolling circle replication. There is a potential stem-loop structure in the intergenic region that includes a conserved nonanucleotide sequence (TAATATTAC), where single-stranded DNA synthesis is initiated.

1.1 Replication of Geminivirus

Geminivirus DNA replication follows a rolling circle strategy (Saunders et al. 1992), which resembles that of prokaryotic ssDNA replicons (Fig. 1). The initial stage encompasses the conversion of the ssDNA genome into a dsDNA intermediate product (Saunders et al. 1992). It requires a priming step at the so-called (−)- strand or c-strand origin. In the second stage, ds DNA intermediate acts as a template for genome amplification through a RCR mechanism, a mechanism common among bacterial systems, where it is used frequently for viral and plasmid DNA replication (de la Campa et al. 1990). The initial priming event depends on the interaction of the viral initiator protein (rep protein) with cis-acting signals of the genetically defined origin. A 9-nt sequence (TAATATT↓AC) is present invariably in all the geminivirus sequences to date. This sequence contributes to form a stem-loop structure to which Rep can get access and carry out the initiation reaction, a single-stranded, site-specific, endo-nucleolytic cleavage that provides a free 3′-OH primer terminus for further elongation during the RCR stage (Khan and Dijkstra 2006).

A mastrevirus, transmitted by leafhopper, generally infects monocots and possesses monopartite genome (type species: *Maize streak virus*). A curtovirus is

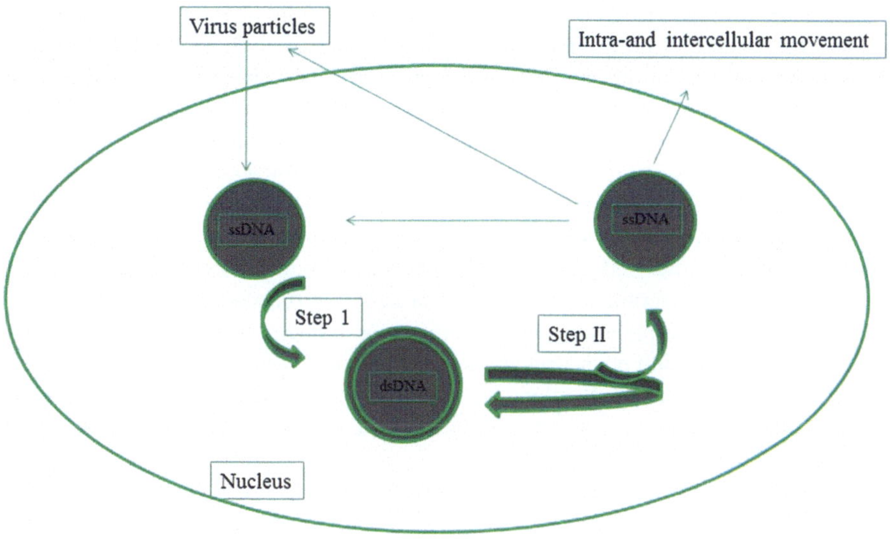

Fig. 1 Replication strategy of plant DNA virus. Simplified scheme of DNA replication cycle of geminiviruses

leafhopper-transmitted, infects dicots, and has monopartite genome (type species: *Beet curly top virus*). Begomoviruses are whitefly-transmitted, infect dicots, and have either monopartite or bipartite genome (type species: *Bean golden mosaic virus-Puerto Rico*). The members of genus *Topocuvirus* are transmitted by treehoppers, infect dicots, and have monopartite genome (type species: *Tomato pseudo curly top virus*). Mastreviruses mostly infect monocots in the family *Poaceae*; two species, namely *Bean yellow dwarf virus* (BeYDY, Liu et al. 1997) reported from South Africa and *Tobacco yellow dwarf virus* (Morris et al. 1991) from Australia, are also known to infect dicots. *Chickpea chlorotic dwarf virus* from India is transmitted by leafhopper and is considered to be a tentative mastrevirus species (Horn et al. 1993). Different leafhopper species transmit mastreviruses in a persistent circulative manner and the genus includes about a dozen species. MSV causes maize streak disease, one of the oldest known plant viral diseases reported about a century ago (Fuller 1901). It is one of the three economically most important plant virus diseases in Africa (Shepherd et al. 2009). Other mastreviruses occur in Asia, Australia, or Pacific islands and *Wheat dwarf virus* is prevalent in Europe.

Curtoviruses are transmitted by leafhoppers and possess a monopartite genome; Begomoviruses are spread by whiteflies and most of them have a bipartite genome referred to as DNA-A and DNA-B. Sequence comparison between the genera has led to the suggestion that Curtoviruses have evolved from a recombination of ancient mastreviruses and begomoviruses (Chen et al. 2010).

Type species of genus Topocuvirus is *Tomato pseudo curly top virus*. Its genome encodes six proteins and members of these genera are known to spread by treehoppers. Like most of the geminiviruses, these genera also infect dicotyledonous plant.

The name of genus Begomovirus is derived from the first two letters in the name of the type species, *Bean golden mosaic virus* (Padidam et al. 1996). DNA-A and DNA-B components of bipartite Begomoviruses are each 2.6 Kb in size and share a noncoding intergenic or common region (CR) of 180–200 nucleotides (nt) that is typically identical for cognate components of the same virus species. The CR contains modular cis-acting elements of the origin of replication (ori) (Fontes et al. 1994), while five ORFs capable of encoding proteins >10 kilodaltons (Kd) in size are conserved among the DNA-A component of Begomoviruses. The coat protein, CP, is encoded by the ORF (AV1) and is the most highly conserved gene among Begomoviruses. The replication-associated protein (*Rep*) encoded by the AC1 ORF initiates viral DNA replication, and specificity of replication is mediated through interactions of REP with cis-acting elements of the ori (Jupin et al. 1994; Laufs et al. 1995; Bisaro 2006). DNA B encodes two polypeptides, BV1 and BC1; both are essential for systemic movement and have been shown to influence host range (Lazarowitz et al. 1992).

Genus Becurtovirus contains two recognized species, namely *Beet curly top Iran virus* and *Spinach curly top Arizona virus*. Members of both species unusually have the "TAATATTAC" nonanucleotide instead of "TAAGATTCC" nonanucleotide. Becurtoviruses infect dicotyledonous plants. Members of the genus Eragrovirus have been reported to infect monocotyledonous *Eragrostis curvula* (weeping love grass) in the Kwa-Zulu Natal region of South Africa.

Turnip curly top virus (TCTV) is currently the only species within the genus Turncurtovirus. All the known TCTV isolates reported so far possess "TAATATTAC" nonanucleotide sequence motif. It infects dicotyledonous plants.

Genus Capulavirus contains four species, transmitted by an aphid, and genus Grablovirus has only one species. Not much information is available so far.

2 Economic Losses Due to Geminivirus Disease

Though geminivirus may be the subject of interest in molecular virology because of their relatively small genome, and their ability to manipulate and reprogram host cellular processes to their advantage (Rojas et al. 2005), they cause significant economic losses to several crops across the globe. Some notable examples are losses in cotton in Asian countries such as India and Pakistan (Briddon and Markham 2001; Khan and Ahmad 2005). In Pakistan, the cotton leaf curl disease causes loss of 5 billion dollars, and it seriously challenged national economy. In India, every year cotton leaf curl disease caused severe losses, ranging between 10 and 50%, and in 2015, a serious epidemic break caused yield loss up to 100% in several cotton growing areas in western parts of India. Cassava and maize in Africa are prone to geminiviral infection (Thresh and Cooter 2005; Shepherd et al. 2009). The loss observed due to this is several million dollars every year. Legumes in India are infected with begomovirus that cause annual yield losses estimated at $300 million (Varma and Malathi 2003). An annual loss of $140 million is estimated due to

Tomato leaf curl viruses (ToLCVs) in Florida, USA, and continues to be a major constraint to tomato production worldwide.

3 Geminivirus Management Strategies

3.1 Conventional Approaches

Conventional management of geminivirus diseases is based on the applications of different insecticides to control insect vectors (Singh 1990; Lapidot and Friedmann 2002). The insect control required massive use of insecticides and chemicals every year resulting in the development of resistance in the vector. Furthermore, chemicals and insecticides cause environmental pollution, health hazards, and phytotoxicity besides their high cost. Physical barriers such as fine mesh screens have been used in the Mediterranean basin to protect crops (Cohen and Antignus 1994). UV-absorbing plastic screens have shown to inhibit penetration of whiteflies into greenhouses (Antignus et al. 2001). However, use of physical barriers is not the best solution owing to high cost and creates problems of shading, overheating, and poor ventilation (Lapidot and Friedmann 2002).

Cultural practices such as crop-free periods, altering dates, crop rotation, weed and crop residue disposal, high planting densities, floating row cover, mulches, trap crop, or living barriers performed well against several plant viruses in general. This control is based on the removal of infected plants, production, and using of virus-free planting stock. Use of virus-free seed is advisable, as seed-borne viruses are transmitted in the embryo of the seed, but some viruses, such as *Tobacco mosaic virus* (TMV) in tomato, are transmitted through contamination of the seed coat. In this case, seed transmission may be controlled by sterilizing the seed coat in hydrochloric acid or sodium hypochlorite. The suggested methods of cultural practices are, however, not much successful against geminiviruses. Alternatively, it was suggested that cultural practices when combined with insecticides are more effective (Hilje et al. 2001).

3.1.1 Control Through Resistant Cultivars

Development of resistant plant varieties against virus or its vector through breeding techniques is yet another attractive approach for controlling viral diseases. Virus-resistant crops increase profitability for the breeders, as this approach requires no extra input toward production of virus-free planting materials or control of virus vectors (Valkonen 1998).

There seems to be disagreement between plant breeders and plant pathologists. On one end, breeders are interested in improving the performance of a plant variety under field conditions, but on the other end, plant pathologists focus more on the fate of the plant virus.

Another challenge to the breeders are emergence of new strains of geminiviruses due to frequent recombination and changes in cultivation habits (Padidam et al. 1999). Co-infection by two or more viruses to a crop plant is another issue. The different challenges posed by viruses have necessitated the development of plants, which confers multi-virus resistance. Some of the approaches addressed by several authors involve in assembling of different resistance genes, which will ultimately give multi-virus resistance (Lapidot and Friedmann 2002). However, this idea poses another problem to the breeders in distinguishing two different traits/genes to be combined and a continuous system to check the resistance against newly emerged virus.

3.1.2 Cross-Protection

It has been long observed that plants infected by mild strain can be protected against infection by more severe strain of the related virus, a biological term called cross-protection. The cross-protection test has been previously regarded as an important means for identifying the same strain or a distinct species of plant virus. This method is also quoted as plant vaccination by some authors (Nicaise 2014). This phenomenon was first discovered by McKinney in 1926; since then several reports of cross-protection have been identified (McKinney 1926; Crowdy and Posnette 1947; Fletcher 1978; Wang 1991; Wen et al. 1991; Hugues and Ollennu 1994; Nakazono-Nagaoka et al. 2009; Kurth et al. 2012).

In detail, this strategy is dependent on the prowess of the primary virus, whose infection is weak (either symptomless with low viral load or mild symptom with low virus titer). This triggers virus-induced gene silencing (VIGS), which targets both mild infection virus and challenge viruses (Nishiguchi and Kobayashi 2011). Primary virus acts as vaccines and is further classified as the attenuated and the engineering viruses. An attenuated virus corresponds to a weak isolate that triggers cross-protection against virulent isolates of the same virus or closed related viruses (Ziebell and Carr 2010; Nishiguchi and Kobayashi 2011). However, cross-protection strategy also bears some disadvantages. Infection by the mild strain might cause significant yield loss sometimes. Another fear is that a severe strain might evolve from the mild strain leading to more serious disease incidence.

Conventional methods are effective but also protracted and expensive. To overcome these challenges and limitations, nonconventional methods of genetic engineering are practiced.

3.2 Nonconventional Methods

3.2.1 Pathogen-Derived Resistance

Concept of pathogen-derived resistance or parasite-derived resistance was developed by Sanford and Johnston in 1985. Subsequently, Powell-Abel et al. (1986) applied this approach and opened new horizons for plant pathologists following the development of virus-resistant crops. Since then, several attempts were made to develop transgenic plants using virus-derived genes or genome fragments and led to the development of virus-resistant plants for commercial application (Beachy 1993; Wilson 1993; Baulcombe 1994, Lomonossoff 1995).

Coat Protein

CP gene was the first and one of the most widely used genes to confer pathogen-derived resistance (PDR) against plant viruses (Prins 2003). Virus resistance has been achieved by transforming the plants with viral *CP* gene. Remarkable success has been achieved in transformed tobacco showing resistance to (TMV) (Powell-Abel et al. 1986) and transgenic papaya plants resistant to Papaya ringspot virus (Gonsalves 1998).

The CP gene of tobacco mosaic virus TMV was manipulated for the first time demonstrating virus-derived resistance in transgenic plants (Powell-Abel et al. 1986). In these experiments, transgenic tobacco plants expressing high levels of TMV CP were more resistant to TMV virions than to TMV RNA inocula. Based on the observation, they suggested that CP-mediated protection against TMV was through the inhibition of virion disassembly in the initially infected cells (Register and Beachy 1988). CP-mediated resistance has been successfully applied to numerous crop species (Beachy 1997) (see Table 1).

Table 1 List of approaches used against different geminiviruses

Approach/target gene	Target virus	References
Coat protein-mediated resistance	*Tobacco mosaic virus* *Tomato yellow leaf curl virus* *Papaya ringspot virus* *Papaya leaf curl virus*	Powell-Abel et al. (1986) Kunik et al. (1994) Gonsalves (1998) Sinha et al. (2017)
Movement protein	*Tobacco rattle virus, tobacco ringspot nepovirus, alfalfa mosaic alfamovirus, cucumber mosaic virus*	Cooper et al. (1995)
Replication associated protein (*Rep*)	*African cassava mosaic virus* *Tomato yellow leaf curl Sardinia virus* *Cotton leaf curl virus* *Cassava-infecting geminivirus* *Tomato yellow leaf curl virus*	Hong and Stanley, (1996) Noris et al. (1996) Brunetti et al. (1997) Asad et al. (2003) Chellapan et al. (2004) Yang et al. (2004)
siRNA mediated	*Tomato leaf curl China virus* *Cotton leaf curl virus*	Cui et al. (2005) Khatoon et al. (2016)
miRNA mediated	*Cotton leaf curl virus* *Cotton leaf curl virus*	Akmal et al. (2017) Gazal and Jawaid (2018)
amiRNA mediated	*Tomato leaf curl virus*	Vu et al. (2013)
Antisense mediated	*Cotton leaf curl virus* *Papaya leaf curl virus*	Asad et al. (2003) Sinha et al. (2017)
CRISPR/Cas9	*Tomato yellow leaf curl virus* *Cotton leaf curl Kokhran virus*	Ali et al. (2015, 2016)

Movement Protein

Cell-to-cell movement of plant viruses in host plant is associated with the movement protein encoded by viruses. MP interacts with the plasmodesmata and thus modifies it to facilitate cell-to-cell movement of virus. Transgenic tobacco plants that expressed a gene encoding a defective TMV movement protein showed resistance to *Tobacco rattle virus, Tobacco ringspot nepovirus, Alfalfa mosaic alfamovirus,* and *Cucumber mosaic virus* (Cooper et al. 1995). Resistance shown by transgenic expression of a dysfunctional TMV MP is probably due to competition for plasmodesmatal binding sites between the mutant MP and the wild-type MP of the inoculated virus (Lapidot et al. 1993). An interesting and potentially useful attribute of MP-mediated protection is the broad spectrum efficacy of the resistance mechanism (Table 1).

Replication-Associated Protein

Rep-mediated approach is the second most widely used method to control plant viruses (Lomonossoff 1995; Wintermantel et al. 1997). The engineering of geminivirus resistance using *Rep* gene has been achieved in model host species against Begomoviruses. Expression of full-length or a truncated N-terminal portion of *Rep* gene of *African cassava mosaic virus* (ACMV) inhibits replication of ACMV in *Nicotiana tabacum* protoplasts. A modest degree of ACMV resistance was achieved by the expression of full-length *Rep* gene in experimental plant tobacco. None of transgenic tobacco plants was resistant to distantly related viruses TGMV and Beet curly top virus (sharing Ca. 60% Rep amino acid sequence identity with ACMV), suggesting that resistance was probably ACMV specific or to its closely related viruses (Hong and Stanley 1996).

Rep gene has been successfully manipulated to engineer resistance against *Tomato yellow leaf curl Sardinia virus* (TYLCSV) in *N. benthamiana* (Noris et al. 1996) and tomato (Brunetti et al. 1997) against cotton leaf curl disease in experimental plant tobacco (Asad et al. 2003) (Table 1).

3.3 RNA Silencing

In recent years, resistance-mediated by RNA silencing seems to be one of the most promising approaches. RNA silencing is an ancient mechanism involved in different fundamental processes, such as gene regulation, de novo histone and DNA methylation, establishment of heterochromatin, defense against viruses, and control of transposon mobility (Baulcombe 2004; Voinnet 2005). RNA silencing involves suppression of gene expression by sequence-specific degradation of mRNA in diverse eukaryotes. The RNA silencing phenomena was first discovered and termed post-transcription gene silencing (PTGS) in plants (Napoli et al. 1990), quelling in fungi (Cogoni and Macino 1997), and RNA interference (RNAi) in animals (Cogoni et al. 1996; Fire et al. 1998).

The key molecules which are involved in the RNA silencing pathways are ribonuclease Dicer (RNA-dependent RNA polymerase, RDR) and Argonaute (AGO). The RNA silencing machinery in plants is more evolved than in fungal and animal systems. Its pathway follows a dsRNA trigger: a processor called Dicer or a Dicer-like (DCL) protein generating small RNAs (siRNAs or miRNAs) of 21–24 nt in length in an effector complex called RISC (RNA-induced silencing complex) in which the AGO protein plays a key role. The siRNA-guided AGO actually cleaves target RNA, which is recognized by RDR. The *Arabidopsis* genome encodes four DCL enzymes, 6 RDRs, and 10 AGO proteins.

There are three different pathways in the gene silencing mechanism: (1) cytoplasmic short interfering (siRNA) silencing, (2) silencing of endogenous mRNAs by microRNAs (miRNAs), and (3) DNA methylation and suppression by transcription (Vanitharani et al. 2005). The siRNA silencing is actually post-transcriptional gene silencing (PTGS) (Bisaro) in 2006 resulting in the production of 21–25 nucleotide siRNA from inducing dsRNA leading to the degradation of mRNA (Hamilton and Baulcombe 1999). Practically, linking the sense and antisense sequences by an intron, which is eventually spliced, resulted in efficient silencing in plants (Smith et al. 2000; Wesley et al. 2001). The mechanism is now better understood and widely used to engineer plant against virus infection (Tenllado et al. 2003). Several vectors have been developed for the efficient expression of such hairpin dsRNA in plants (Wesley et al. 2001; Khatoon et al. 2016; Table 1). The dsRNA region is processed into small interfering RNAs (siRNAs), which guide silencing complexes to target regions on RNA or DNA (Fig. 2).

Geminiviruses have an ability to suppress the induced RNA silencing. About 35 RNA silencing suppressor proteins have been identified in recent years from several plant and animal viruses. There are mainly three distinct phases reported in the RNA silencing process: initiation, maintenance, and systemic signaling (Llave et al. 2002). These suppressor proteins do not share homology at either sequence or viral functional levels; it is assumed that these suppressor proteins might target similar or different steps of the RNA silencing pathway.

A study on cassava-infecting geminiviruses demonstrated that *AC4* region has the capacity to suppress the induced post-transcriptional gene silencing (Vanitharani et al. 2004).

It is a well-known fact that geminiviruses have no dsRNA stage in their replication cycle, but they do induce the production of virus-specific siRNA and have been shown to trigger PTGS in infected plants, as demonstrated by Chellappan et al. (2004). Another observation is that an increased accumulation of cassava-infecting geminivirus-derived siRNAs in infected cassava is associated with a corresponding decrease in disease symptom severity (Chellappan et al. 2004), providing a clue for RNAi as an adaptive defense against geminiviral infection in plants. According to the reports of Vanitharani et al. (2005), both silencing mechanism PTGS and TGS are applicable to geminiviruses, whereas for RNA viruses only TGS is applicable.

Betasatellite molecule is associated with a number of monopartite begomoviruses, and it induces symptom. The betasatellite associated with *Tomato yellow leaf curl China virus*-Y10 has shown to behave as a silencing suppressor in *N. benthamiana* 16c plants (Cui et al. 2005). It was shown that TYLCCV along with betasatellite DNA could prevent silencing in newly emerging leaves of infected plants.

Fig. 2 siRNA-mediated cleavage of target RNA

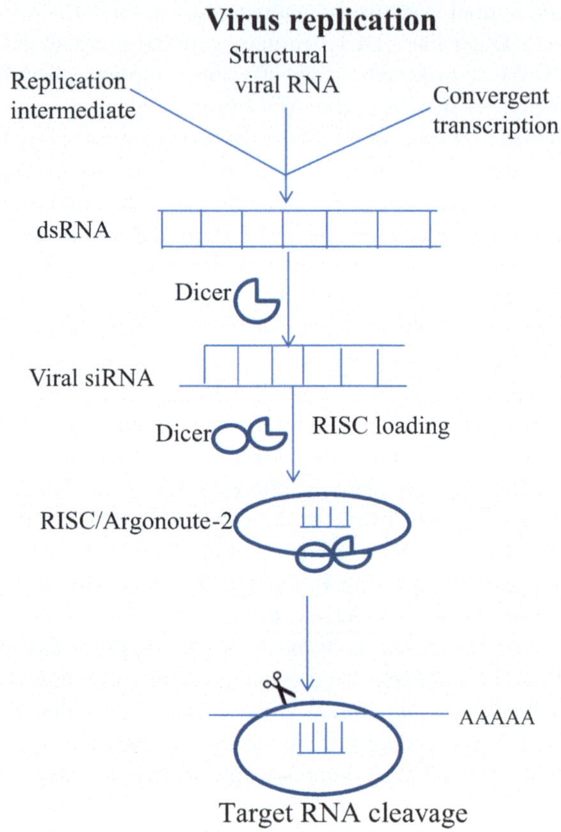

Intergenic region (IR) is also a potential candidate which could play a crucial role for the generation of siRNA-mediated resistance. IR contains origin of replication and divergent promoter (Zulma et al. 2002). Methylation of this region can hamper the virus's ability to thrive in the plant host. The region of homology ranges from 80 to 100 bp in length and includes some common motifs found in Begomoviruses, such as nonanucleotide sequence (TAATATTAC). This finding is successfully applied in the generation of transgenic cotton, which showed complete protection against CLCuD (Khatoon et al. 2016). In an attempt to generate siRNA-mediated *African cassava mosaic virus* (ACMV, genus *Begomovirus*) resistant transgenic plants, a 360 nt fragment corresponding to IR of ACMV DNA-A was cloned in sense and antisense orientation, interrupted with a synthetic plant intron. *N. benthamiana* plants were stably transformed with an intron-spliced dsRNA construct cognate to bidirectional promoter of ACMV DNA. The transgenic lines expressed multiple siRNAs species upon ACMV inoculation. It was demonstrated that the mRNA transcribed from ACMV genome was degraded by 21–22 nt siRNA and the begomoviral genomic DNA appeared to be methylated by 24–25 nt siRNA. It was demonstrated that silencing was associated with hypermethylation of promoter sequence and did not occur with heterologous begomovirus infection (Dogar 2006).

PTGS occurs as a natural defense mechanism against virus infection. Virus or its derivative or replication intermediate acts as a pathogenic agent by the host. This triggers a response responsible for the progressive slowdown in virus accumulation (Ratcliff et al. 1997, 1999). However, in this situation, viruses counteract the host response by encoding suppressors of PTGS (Voinnet et al. 1999; Hamilton et al. 2002). PTGS in plants can be triggered due to the presence of an inverted repeat in the transcribed region of a transgene (Jones et al. 1999). Tobacco plants transformed with constructs that produce RNAs capable of duplex formation induced virus immunity or gene silencing when targeted against virus or endogenous genes (Smith et al. 2000; Waterhouse et al. 1998). There are strong evidences in support of dsRNA as an inducer of PTGS in both the plant and animal kingdoms.

3.4 MicroRNA-Based Resistance

MicroRNA is a small noncoding sequences, thought to be useful in gene regulation during development and in the stress conditions in eukaryotes. They are about 18–25 nucleotides in length and play a major role in the negative regulation at post-transcriptional level for the normal activity of the organisms under stress. These miRNAs sharing the homology with the target mRNAs in plant are capable of causing the RNA-induced silencing lead to mRNA cleavage (Fig. 3).

3.4.1 Origin and Evolutionary Role of miRNA

The animal miRNA is believed to be originated about 420 million years ago from the metazoans as common ancestors (Pasquinelli et al. 2003). The hypothesis regarding origin of plant miRNAs is not clear. Many experimental and computational predictions of miRNA and its targets led to identification of a large number of miRNA families that are evolutionarily conserved across all major lineages of plants including bryophytes, lycopods, ferns, and seed plants (Axtell and Bartel 2005; Zhang et al. 2005, 2006; Jones-Rhoades et al. 2006), and even many are reported to be conserved between monocots and dicots (Sunkar and Jagadeeswaran 2008). There are reports of 21 miRNA families (predominantly 156, 159, 160, 162, 164, 166–169, 171, 172, 319, 390, 393–399, and 408) which are identified. These miRNAs are highly conserved between all three sequenced plant genomes: *Arabidopsis*, *Oryza sativa*, and *Populus trichocarpa* (Axtell and Bowman 2008). Several plant miRNAs are believed to be universal among land plants; they are less conserved than animal miRNA (Axtell and Bowman 2008). Recent advanced techniques of DNA sequencing and research into miRNA gene complements of individual species have listed several "non-conserved" miRNAs (nearly 48 in Arabidopsis), which outnumbered the "conserved" miRNAs. On the other hand, several (16 out of 48) non-conserved Arabidopsis miRNA genes are believed to exhibit significant sequence similarity outside the miRNA binding sites within their putative target genes. These features signify that non-conserved miRNAs originated

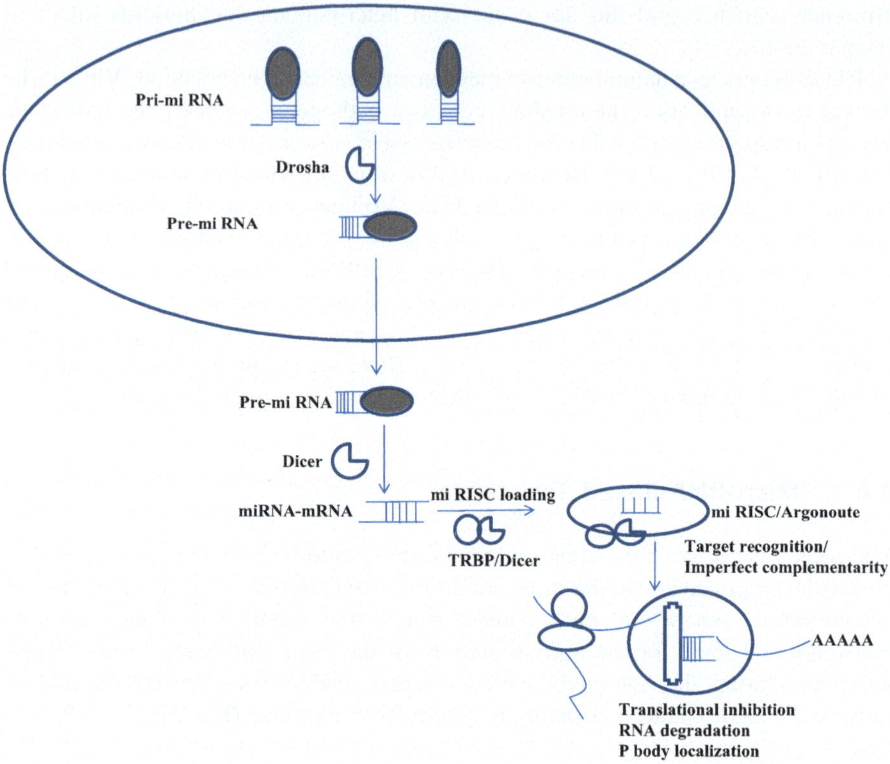

Fig. 3 miRNA-mediated cleavage of target RNA

recently with high birth and death rates (Rajagopalan et al. 2006; Fahlgren et al. 2007). Various roles of microRNAs are described in the recent past, but plant defense mechanism to viruses is one of the several anticipated roles, which is yet to be explored in full potential. There has been a successful application of this strategy in recent times against cotton leaf curl virus (CLCuV) (Akmal et al. 2017; Shweta et al. 2018)

3.4.2 Biogenesis and Mechanism of miRNAs

MicroRNAs are small non-protein coding RNAs consisting of 21–24 nucleotides present in intergenic regions of genome. The extensive complementarily between the target mRNA and the miRNA leads to target mRNA cleavage and gene silencing in plants. Biogenesis of plant miRNA occurs in multiple steps to form mature miRNAs from miRNA genes. Initially, the miRNA genes are transcribed by their own promoters, resulting in primary transcripts (pri-miRNAs) (Tang et al. 2003). Further, these pri-miRNAs fold up into unique stem-loop structures. This structure is further identified and cleaved by the Dicer-like (DCL) enzyme of the RNase III family (Tang et al. 2003). In the plant nucleus, DCL-1, in association with HYL1 protein, processes the pri-miRNAs. The mature miRNA then unwinds into single-strand miRNA by helicase (Bartel 2004) and assembled into the RISCs complex, which contains Arganoute (AGO1) protein to carryout silencing reactions. The plant miRNA will have the complementarity with the target mRNA and lead to cleavage of it (Llave et al. 2002; Bartel 2004; Dugas and Bartel 2004). The high base pairing requirement of plant miRNAs results in limited number of targets compared to animal miRNA. The mRNA cleavage is considered to be predominant mechanism used by plant miRNAs, but reports also prove the presence of translational inhibition by plant miRNA and other siRNAs (Chen et al. 2010; Aukerman and Sakai 2003). The method of overexpression of miRNAs has proved to a promising approach against several pathogens including geminivirus also (Baldrich and Segundo 2016). Recently, overexpression of *Gossypium hirsutum* miRNAs (miR398/miR2950) in *G. hirsutum* has been demonstrated to suppress symptoms of cotton leaf curl disease (CLCuD) caused by *Cotton leaf curl Multan virus* (genus *Begomovirus*, family *Geminiviridae*) in association with circular, single-stranded DNA molecule satellite molecule (Akmal et al. 2017).

3.5 Artificial microRNA

The amiRNA acts as a specific, powerful, and robust tool that can be applied to study metabolic pathways, gene functions, and for improving favorable traits. The AmiRNAs have also been used as a powerful tool to produce antiviral transgenic plants. The transgenic *Arabidopsis* expressing amiRNAs targeting the viral mRNA sequences encoding gene silencing suppressor P69 of *turnip yellow mosaic virus* (TYMV) and HC-Pro of *turnip mosaic virus* (TuMV) are specifically resistant to TYMV and TuMV (Niu et al. 2006). There have also been reports of developing resistant tobacco and *Arabidopsis* by expressing amiRNAs. These amiRNAs were also used for generating resistance against *Watermelon Silver Mottle Virus* in tobacco (Kung et al. 2012). The idea of amiRNA was also successfully used against monopartite and bipartite Tomato leaf curl virus (Vu et al. 2013).

In principle, artificial microRNA (amiRNA) technology is based on designing miRNA or engineering miRNA artificially by mimicking the intact secondary structure of endogenous miRNA precursors (Ossowski et al. 2008; Sablok et al. 2011). It was demonstrated that altering several nucleotides within sense and antisense strands of miRNA has no effect on its biogenesis and maturation, as long as secondary structure of its precursor remains unaltered. It was also demonstrated that amiRNAs when expressed under constitutive or tissue-specific promoters can downregulate a number of endogenous genes without affecting the expression of other unrelated genes (Alvarez et al. 2006).

3.6 Antisense RNA Approach

Antisense RNA approach, based on the manipulation of potential gene sequence of targeted virus, would prove to be a significant approach against the virus. It ends up into degradation of mRNA in a processing body inside the cytoplasm of the cell.

Antisense technology involves the cloning of a gene in reverse orientation with respect to the promoter such that the coding strand acts as a template and the sequence of mRNA is the same as the opposite strand or the coding "sense" strand. The gene cloned in reverse orientation when transcribed gives rise to mRNA having the sequence complementary to the sense mRNA. The RNA-RNA binding of the sense-antisense RNA strand leads to inhibition of sense mRNA expression.

Naturally occurring antisense RNAs are known to regulate gene expression in plants too. Antisense RNA arises when transcription of a gene proceeds in the strand opposite to template in the absence of a strong transcription termination site in the short intergenic region. On the basis of certain basic regulatory mechanisms, they are classified into three classes:

Class I–antisense RNAs are directly complementary to the coding region of target gene, resulting in direct inhibition of translation or mRNA destabilization.

Class II–RNAs include those that bind to the non-coding regions of target RNA resulting in indirect effects produced by, e.g., alternative secondary structure formations that sequesters the ribosome binding site.

Class III–antisense RNA regulates transcription of target mRNA by a mechanism similar to transcription attenuation.

The antisense inhibition may also take place at translational phase as antisense transcript would compete with the ribosomes to bind 5′ end of the sense RNA, hence inhibiting the translation. Antisense RNA technology is a proven strategy that has been widely used for crop improvement. The most commercial example is the development of Flavr Slavr tomatoes in which tomato plants were transformed with antisense Polygalacturonase gene (gene responsible for cell wall degradation and fruit softening) (Kumria et al. 1998 and references therein). The resultant transgenic plants showed longer shelf life. Other examples are antisense inhibition of chitinase gene expression which resulted in enhancement of fungal disease susceptibility in *Arabidopsis* plants and alteration of lignin composition by inhibiting lignin biosynthetic enzymes in tobacco (Kumria et al. 1998 and references therein). Its application is not limited in enhancing quality of food crops, but it has also been used to confer resistance to viral plant infections. Transgenic potato plants expressing antisense RNA to potato leaf roll luteovirus coat protein were resistant to the infection (Kumria et al. 1998 and references therein). Antisense RNA construct against cotton leaf curl virus has been successfully developed by Asad et al. (2003). Antisense construct targeting coat protein region of the begomovirus isolated from leaf curl disease affected papaya plant was also developed by Sinha et al. (2017).

3.7 Application of CRISPR/cas9 in the Generation of Geminivirus Resistance

CRISPR/Cas9 is a molecular immunity system, exclusively found in prokaryotic system (Fig. 4). This system actually acts against invading nucleic acids, following

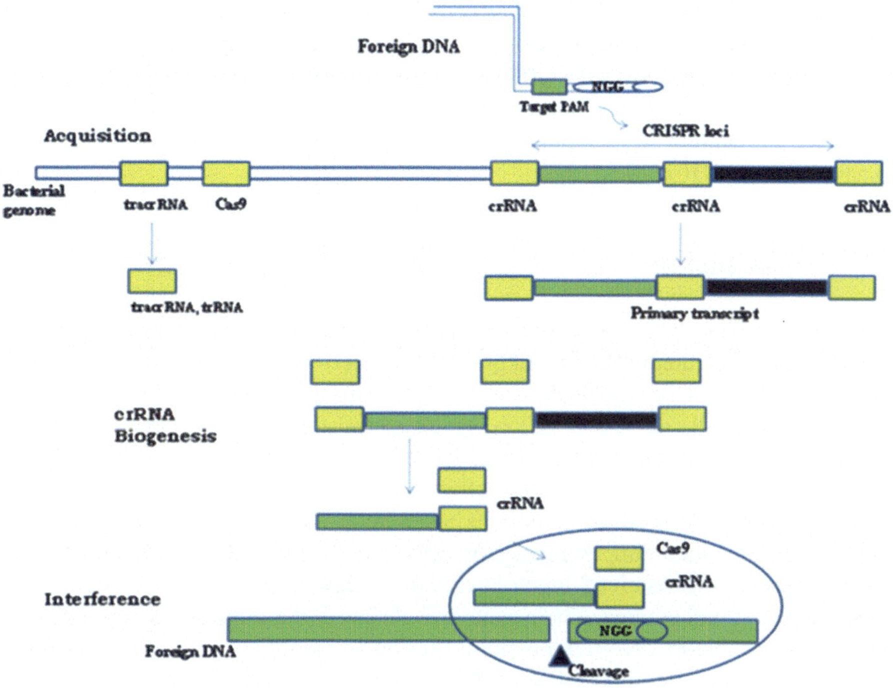

Fig. 4 A simplified CRISPR mechanism: In the acquisition phase (I phase), foreign DNA is incorporated into the bacterial genome at the CRISPR loci. CRISPR loci is then transcribed and processed into crRNA during crRNA biogenesis (II phase). During interference, Cas9 endonuclease form a complex with crRNA, separate tracrRNA cleaves foreign DNA containing a 20-nucleotide crRNA complementary sequence adjacent to the PAM sequence (III phase)

methods of horizontal gene transfer and phages, as suggested by Marraffini and Sontheimer (2008). This molecular memory concept further follows the process in which bacteria and archaea acquire short pieces or even spacers from these invading nucleic acids and incorporate them within their genome (Bolotin et al. 2005). In case of subsequent infection(s), these short pieces are transcribed as part of CRISPR array. Further, after transcription and maturation, the CRISPR RNA (CrRNA) can help guide the Cas9 endonuclease to scan the invading DNA and cleave the target sequence (Nunez et al. 2016).

Some recent studies demonstrated the efficiency of this system against geminiviruses. In one of the studies (Ali et al. 2015), *N. benthamiana* plants expressing CRISPR/Cas9 exhibited resistance against Tomato yellow leaf curl virus, Beet curly top virus, Merremia mosaic virus. Other studies (Ji et al. 2015) demonstrated interfere in virus activities in *N. benthamiana* against Bean yellow dwarf virus (BeYDV) and BCTV, respectively. Also, later it was suggested that catalytically inactive Cas9 can be used to mediate virus interference. This action eliminates concerns of off-target activities in the plant genome. In model plant *N. benthamiana* targeting conserved nonanucleotide sequence of *Cotton leaf curl*

Kokhran virus (CLCuKoV), broad spectrum resistance was achieved (Ali et al. 2016). However, all the studies used *N. benthamiana*, which is a model plant system. According to Woo et al. (2015), CRISPR/Cas9 system can be used to engineer "non-transgenic" virus-resistant varieties. Major advantage to use this system is that progenies of genome-edited plant carrying desired edits can be selected easily (Kanchiswamy 2016).

4 Conclusion

There is a fine and balanced battle occurring between plants and pathogens which consistently attack them. The role of genetic engineering should be in favor of plant. According to Harrison and Robinson (2002), resistant genes against a particular virus should ideally have the following characteristics: (1) they should provide protection against at least the entire range of virus variants, strains, and species that cause the disease; (2) they should provide robust protection that will require the virus to accumulate multiple mutations to overcome the resistance; and (3) they should confine incoming viruses to cells into which they are inoculated. Geminiviruses have received much attention, as this group of virus is one of the most important and studied. Their main prevalence in the tropics and subtropics is due to climatic factors favoring the multiplication and ability of vectors for transmitting them in a more composed way. The incidence and severity of the disease is increasing every year due to the emergence of new geminiviruses through recombination or pseudo-recombination among strains and/or species in various crops.

Conventional control strategies are generally effective, but they are time-consuming, and sound knowledge of agronomic practices, chemicals, and their effects on environment is must. A serious limitation of breeding program is the availability of resistance traits. It is often difficult to transfer the new character (s) from one species or variety to another while maintaining the agronomical qualities of the target cultivar. There are several other methods such as field sanitation, eradication of infected plants serving as primary source of virus inoculum, removal of weeds and alternate host plants from fields, plantation practices, spraying of insecticides, and use of virus-free planting material. The genetic engineering approach has many advantages, as it can give significant protection against viruses. Application of pathogen-derived resistance in various disciplines of biological science has raised a number of questions on its possible impact on the environment and human health. In the light of available knowledge, there is less or no environmental hazard, it seems. Moreover, PDR, if applied responsibly, is a powerful and safe means for combating plant pathogens, which cannot be controlled otherwise.

The gene silencing approach targeted against virus genes is nowadays a major approach against most of the plant viruses. It occurs either through repression of transcription (TGS) or through mRNA degradation (PTGS). PTGS results from a marked decrease in transcription and hypermethylation of the gene occurs. In TGS, mRNA synthesis is greatly reduced or absent. In addition to genetic engineering, genome engineering has recently emerged as a potential tool to improve various eukaryotic species, which also include variety of plant species. This concept follows introduction of trait of interest through the site-specific modification of the genome (Sovova et al. 2016). Genome engineering refers to the use of site-specific nucleases (SSNs), which can be designed to bind and cleave a specific nucleic acid sequence by introducing double-stranded breaks (Stella and Montoya 2016). CRISPR/Cas9 is one of the classes of SSNs. However, potential of this technique is yet to be tested in open field. The coming years will provide and witness more details on these technologies and development of marketable crops.

Acknowledgements AK is funded by Science and Engineering Research Board (SERB), Govt of India, and is thankful to Mr. Ajay Pratap Singh, Sr Director, IILM-CET Greater Noida for facilities and keen interest.

References

Akmal M, Baig MS, Khan JA (2017) Suppression of cotton leaf curl disease symptoms in *Gossypium hirsutum* through over expression of host-encoded miRNAs. J Biotechnol 10 (263):21–29

Ali Z, Abulfaraj A, Idris A, Ali S, Tashkandi M, Mahfouz MM (2015) CRISPR/Cas9-mediated viral interference in plants. Genome Biol 16:238

Ali Z, Ali S, Tashkandi M, Zaidi SS, Mahfouz MM (2016) CRISPR/Cas9-mediated immunity to geminiviruses: differential interference and evasion. Sci Rep 6:26912. https://doi.org/10.1038/srep30223

Alvarez JP, Pekker I, Goldshmidt A, Blum E, Amsellem Z, Eshed Y (2006) Endogenous and synthetic microRNAs stimulate simultaneous, efficient, and localized regulation of multiple targets in diverse species. Plant Cell 18:1134–1151

Antignus Y, Nestel D, Cohen S, Lapidot M (2001) Ultraviolet-deficient greenhouse environment affects whitefly attraction and flight-behavior. Environ Entomol 30:394–399

Asad S, Haris WA, Bashir A, Zafar Y, Malik KA, Malik NN, Lichtenstein CP (2003) Transgenic tobacco expressing geminiviral RNAs are resistant to the serious viral pathogen causing cotton leaf curl disease. Arch Virol 148:2341–2352

Aukerman MJ, Sakai H (2003) Regulation of flowering time and floral organ identity by a MicroRNA and its APETALA2-like target genes. Plant Cell 15:2730 2741

Axtell MJ, Bartel DP (2005) Antiquity of microRNAs and their targets in land plants. Plant Cell 17:1658–1673

Axtell MJ, Bowman JL (2008) Evolution of plant microRNAs and their targets. Trends Plant Sci 13:343–349

Baldrich P, Segundo BS (2016) MicroRNAs in rice innate immunity. Rice (NY) 9(6). https://doi.org/10.1186/s12284-016-0078-5

Bartel DP (2004) MicroRNAs: genomics, biogenesis, mechanism, and function. Cell 116:281–297

Baulcombe DC (1994) Novel strategies for engineering virus resistance in plants. Curr Opin Biotechnol 5:117–124

Baulcombe DC (2004) RNA silencing in plants. Nature 431:356–363

Beachy RN (1993) Transgenic resistance to plant viruses. Arch Virol 4:327–416

Beachy RN (1997) Mechanisms and application of pathogen-derived resistance in transgenic plants. Curr Opin Biotechnol 8:215–220

Bisaro DM (2006) Silencing suppression by geminivirus proteins. Virology 344:158–168

Bolotin A, Quinquis B, Sorokin A, Ehrlich SD (2005) Clustered regularly interspaced short palindrome repeats (CRISPRs) have spacers of extra chromosomal origin. Microbiol 151:2551–2561

Briddon RW, Markham PG (2001) Cotton leaf curl virus disease. Virus Res 71:151–159

Brunetti A, Tavazza M, Noris E, Tavazza R, Caciagli P, Ancora G, Crespi S, Accotto GP (1997) High expression of truncated viral rep protein confers resistance to tomato yellow leaf curl virus in transgenic tomato plants. Mol Plant-Microbe Interact 10:571–579

Chellappan P, Masona MV, Vanitharani R, Taylor NJ, Fauquet CM (2004) Broad spectrum resistance to ssDNA viruses associated with transgene-induced gene silencing in cassava. Plant Mol Biol 56:601–611

Chen LF, Brannigan K, Clark R, Gilbertson RL (2010) Characterization of curtoviruses associated with curly top disease of tomato in California and monitoring for these viruses in beet leafhoppers. Plant Dis 94:99–108

Cogoni C, Macino G (1997) Isolation of quelling-defective (qde) mutants impaired in posttranscriptional transgene-induced gene silencing in Neurospora crassa. Proc Natl Acad Sci USA 94:10233–10238

Cogoni C, Irelan JT, Schumacher M, Schmidhauser TJ, Selker EU, Macino G (1996) Transgene silencing of the al-1 gene invegetative cells of Neurospora is mediated by a cytoplasmiceffector and does not depend on DNA-DNA interactions or DNA methylation. EMBO J 15:3153–3163

Cohen S, Antignus Y (1994) Tomato yellow leaf curl virus, a whitefly-borne geminivirus of tomatoes. Adv Dis Vec Res 10:259–288

Cooper B, Lapidot M, Heick JA, Dodds JA, Beachy RN (1995) A defective movement protein of TMV in transgenic plants confers resistance to multiple viruses whereas the functional analog increases susceptibility. Virology 206:307–313

Crowdy SH, Posnette AF (1947) Virus diseases of cacao in West Africa II. Cross immunity experiments with virus 1A, 1B and 1C. Ann Appl Biol 34:403–411. https://doi.org/10.1111/j.1744-7348.1947.tb06373.x

Cui X, Li G, Wang D, Hu D, Zhou X (2005) A begomovirus DNAbencoded protein binds DNA, functions as a suppressor of RNA silencing, and targets the cell nucleus. J Virol 79:10764–10775

de Campa l, del Solar AG, Espinosa M (1990) Initiation of replication of plasmid pLS1: the initiator protein RepB acts on two distant DNA regions. J Mol Biol 213:247–262

Dogar AM (2006) RNAi dependent epigenetic marks on geminivirus promoter. Virol J 3:5

Dugas DV, Bartel B (2004) MicroRNA regulation of gene expression in plants. Curr Opin Plant Biol 7:512–520

Fahlgren N, Howell MD, Kasschau KD, Chapman EJ, Sullivan CM, Cumbie JS, Givan SA, Law TF, Grant SR, Dangl JL et al (2007) High-throughput sequencing of Arabidopsis microRNAs: Evidence for frequent birth and death of MIRNA genes. PLoS One 2:e219

Fire A, Xu S, Montgomery MK, Kostas SA, Driver SE, Mello CC (1998) Potent and specific genetic interference by double stranded RNA in *Caenorhabditis elegans*. Nature 391:806–811

Fletcher JT (1978) The use of a virulent virus strain to protect plants against the effects of virulent strains. Ann Appl Biol 89:110–114. https://doi.org/10.1111/j.1744-7348.1978.tb02581.x

Fontes EPB, Gladfelter HJ, Schaffer RL, Petty ITD, Hanley-Bowdoin L (1994) Geminivirus replication origins have a modular organization. Plant Cell 6:405–416

Fuller C (1901) First Rep. Gov Entomol Natal 1899–1901:17–19

Gazal W, Jawaid AK (2018) Overexpression of ghr-miR166b generates resistance against Bemisia tabaci infestation in Gossypium hirsutum plants. Planta 247:1175–1189

Gonsalves D (1998) Control of papaya ringspot virus in papaya: a case study. Annu Rev Phytopathol 36:415–437

Hamilton AJ, Baulcombe DC (1999) A species of small antisense RNA in posttranscriptional gene silencing in plants. Science 286:950–952

Hamilton AJ, Voinnet O, Chappell L, Baulcombe DC (2002) Two classes of short interfering RNA in RNA silencing. EMBO J 21:4671–4679

Hanley-Bowdoin L, Settlage SB, Orozco BM, Nagar S, Robertson D (1999) Geminiviruses: models for plant DNA replication, transcription, and cell cycle regulation. Crit Rev Plant Sci 18:71–106

Harrison BD, Robinson DJ (2002) Green shoots of geminivirology. Physiol Mol Plant Pathol 60:215–218

Hilje L, Costa HS, Stansly PA (2001) Cultural practices for managing *Bemisia tabaci* and associated viral diseases. Crop Prot 20:801–812

Hong Y, Stanley J (1996) Virus resistance in *Nicotiana benthamiana* conferred by African cassava mosaic virus replication associated (ACI) transgene. Mol Plant Micro Interact 9:219–225

Horn NM, Reddy SV, Roberts IM, Reddy DVR (1993) Chickpea chlorotic dwarf virus, a new leafhopper-transmitted geminivirus of chickpea in India. Ann Appl Biol 122:467–479

Hugues JA, Ollennu LAA (1994) Mild strain protection of cocoa in Ghana against cocoa swoollen shoot virus—a review. Plant Pathol 43:442–457. https://doi.org/10.1111/j.1365-3059.1994.tb01578.x

Ji X, Zhang H, Zhang Y, Wang Y, Gao C (2015) Establishing a CRISPR-Cas-like immune system conferring DNA virus resistance in plants. Nat Plants 1:15144

Jones L, Hamilton AJ, Voinnet O, Thomas CL, Maule AJ, Baulcombe DC (1999) RNA-DNA interactions and DNA methylation in post-transcriptional gene silencing. Plant Cell 11:2291–2301

Jones-Rhoades MW, Bartel DP, Bartel B (2006) MicroRNAs and their regulatory roles in plants. Annu Rev Plant Biol 57:19–53

Jupin I, De Kouchkovsky F, Jouanneau F, Gronenborn B (1994) Movement of tomato yellow leaf curl geminivirus (TYLCV): involvement of the protein encoded by ORF C4. Virology 204:82–90

Kanchiswamy CN (2016) DNA-free genome editing methods for targeted crop improvement. Plant Cell Rep 35:1469–1474

Khan JA, Ahmad J (2005) Diagnosis, monitoring and transmission characteristics of *Cotton leaf curl virus*. Curr Sci 88:1803–1809

Khan JA, Dijkstra J (2006) Plant viruses as molecular pathogens. The Haworth Press, New York

Khatoon S, Kumar A, Sarin NB, Khan JA (2016) RNAi-mediated resistance against cotton leaf curl disease in elite Indian cotton *Gossypium hirsutum* cultivar. Virus Genes 52(4):530–537

Kumria R, Verma R, Rajam MV (1998) Potential applications of antisense RNA technology in plants. Curr Sci 74:35–41

Kung YJ, Lin SS, Huang YL, Chen TC, Harish SS, Chua NH et al (2012) Multiple artificial microRNAs targeting conserved motifs of the replicase gene confer robust transgenic resistance to negative-sense single-stranded RNA plant virus. Mol Plant Pathol 13(3):303–317

Kunik T, Salomon R, Zamir D, Navot N, Zeidan M, Michelson I et al (1994) Transgenic tomato plants expressing the tomato yellow leaf curl virus capsid protein are resistant to the virus. Nat Biotechnol 12(5):500–504

Kurth EG, Peremyslov VV, Prokhnevsky AI, Kasschau KD, Miller M, Carrington JC et al (2012) Virus-derived gene expression and RNA interference vector for grapevine. J Virol 86:6002–6009. https://doi.org/10.1128/JVI.00436-12

Lapidot M, Friedmann M (2002) Breeding for resistance to whitefly-transmitted geminiviruses. Ann Appl Biol 140:109–127

Lapidot M, Gafny R, Ding E, Wolf S, Lucas WJ, Beachy RN (1993) A dysfunctional movement protein of tobacco mosaic virus that partially modifies the plasmodesmata and limits virus spread in transgenic plants. Plant J 4:959–970

Laufs J, Traut W, Heyraud F, Matzeit V, Rogers SG, Schell J, Gronenborn B (1995) In vivo cleavage and joining at the viral origin of replication by the replication initiator protein of tomato yellow leaf curl virus. Proc Natl Acad Sci USA 92:3879–3883

Lazarowitz SG, Wu LC, Rogers SG, Elmer JS (1992) Sequence-specific interaction with the viral AL1 protein identifies a geminivirus DNA replication origin. Plant Cell 4:799–809

Liu L, van Tonder T, Pietersen G, Davies JW, Stanley J (1997) Molecular characterisation of a subgroup I geminivirus from a legume in South Africa. J Gen Virol 78:2113–2117

Llave C, Xie Z, Kasschau KD, Carrington JC (2002) Cleavage of scarecrow-like mRNA targets directed by a class of Arabidopsis miRNA. Science 297:2053–2056

Lomonossoff GP (1995) Pathogen-derived resistance to plant viruses. Annu Rev Phytopathol 33:323–343

Marraffini LA, Sontheimer EJ (2008) CRISPR interference limits horizontal gene transfer in staphylococci by targeting DNA. Science 322:1843–1845

McKinney HH (1926) Virus mixtures that may not be detected in young tobacco plants. Phytopathology 16:883

Morris B, Richardson KA, Eddy P, Zhan X, Haley A, Gardner R (1991) Mutagenesis of the AC3 open reading frame of African cassava mosaic virus DNA a reduces DNA B replication and ameliorates disease symptoms. J Gen Virol 72:1205–1213

Nakazono-Nagaoka E, Takahashi T, Shimizu T, Kosaka Y, Natsuaki T, Omura T et al (2009) Cross-protection against bean yellow mosaic virus (BYMV) and clover yellow vein virus by attenuated BYMV isolate M11. Phytopathology 99:251–257. https://doi.org/10.1094/PHYTO-99-3-0251

Napoli C, Lemieux C, Jorgensen R (1990) Introduction of a chimeric chalcone synthase gene into Petunia results in reversible co-suppression of homologous genes in trans. Plant Cell 2:279–289

Nicaise V (2014) Crop immunity against viruses: outcomes and future challenges. Front Plant Sci 5:660. https://doi.org/10.3389/fpls.2014.00660

Nishiguchi M, Kobayashi K (2011) Attenuated plant viruses: preventing virus diseases and understanding the molecular mechanism. J Gen Plant Pathol 77:221–229

Niu Q-W, Lin S-S, Reyes JL, Chen K-C, Wu H-W, Yeh S-D, Chua N-H (2006) Expression of artificial microRNAs in transgenic Arabidopsis thaliana confers virus resistance. Nat Biotechnol 24:1420–1428

Noris E, Accotto GP, Tavazza R, Brunetti A, Crespi S, Tavazza M (1996) Resistance to tomato yellow leaf curl geminivirus in *Nicotiana benthamiana* plants transformed with a truncated viral C1 gene. Virology 224:130–138

Nunez JK, Harrington LB, Doudna JA (2016) Chemical and biophysical modulation of Cas9 for tunable genome engineering. ACS Chem Biol 11:681–688

Ossowski S, Schwab R, Weigel D (2008) Gene silencing in plants using artificial microRNAs and other small RNAs. Plant J 53:674–690

Padidam M, Beachy RN, Fauquet CM (1996) The role of AV2 ("precoat") and coat protein in viral replication and movement in tomato leaf curl geminivirus. Virology 224:390–404

Padidam M, Sawyer S, Fauquet CM (1999) Possible emergence of new geminiviruses by frequent recombination. Virology 265:218–225

Pasquinelli AE, McCoy A, Jiménez E, Saló E, Ruvkun G, Martindale MQ et al (2003) Expression of the 22 nucleotide let-7 heterochronic RNA throughout the Metazoa: a role in life history evolution. Evol Develop 5:372–378

Powell-Abel P, Nelson RS, De B, Hoffman N, Rogers SG, Fraley RT, Beachy RN (1986) Delay of disease development in transgenic plants that express the tobacco mosaic virus coat protein gene. Science 232:738–743

Prins M (2003) Broad virus resistance in transgenic plants. Trends Biotechnol 21:373–375

Rajagopalan R, Vaucheret H, Trejo J, Bartel DP (2006) A diverse and evolutionarily fluid set of microRNAs in Arabidopsis thaliana. Genes Dev 20:3407–3425

Ratcliff F, Harrison BD, Baulcombe DC (1997) A similarity between viral defense and gene silencing in plants. Science 276(5318):1558–1560

Ratcliff FG, MacFarlane SA, Baulcombe DC (1999) Gene silencing without DNA: RNA-mediated cross-protection between viruses. Plant Cell 11(7):1207–1215

Register JC 3rd, Beachy RN (1988) Resistance to TMV in transgenic plants results from interference with an early event in infection. Virology 166:524–532

Rojas MR, Hagen C, Lucas WJ, Gilbertson RL (2005) Exploiting chinks in the plant's armor. Evolution and emergence of geminiviruses. Annu Rev Phytopathol 43:361–394

Sablok G, Pérez-Quintero ÁL, Hassan M, Tatarinova TV, López C (2011) Artificial microRNAs (amiRNAs) engineering–on how microRNA-based silencing methods have affected current plant silencing research. Biochem Biophy Res Comm 406(3):315–319

Saunders K, Lucy A, Stanley J (1992) RNA-primed complementary-sense DNA synthesis of the geminivirus African cassava mosaic virus. Nucleic Acids Res 20:6311–6315

Shepherd DN, Martin DP, Van der Walt E, Dent K, Vasrani A, Rybicki EP (2009) Maize streak virus: an old and complex emerging pathogen. Mol Plant Pathol 11:1–12

Shweta, Akhter Y, Khan JA (2018) Genome wide identification of cotton (*Gossypium hirsutum*)-encoded microRNA targets against cotton leaf curl Burewala virus. Gene 638:60–65

Singh I (1990) Papaya. IBH Publishing, New Delhi

Sinha V, Sarin NB, Bhatnagar D (2017) The efficacy of antisense-based construct for inducing resistance against Croton yellow vein mosaic virus in Nicotiana tabacum. Virus Genes 53 (6):906–912

Smith NA, Singh SP, Wang MB, Stoutjesdijk P, Green A, Waterhouse PM (2000) Total silencing by intron-spliced hairpin RNAs. Nature 407:319–320

Sovova T, Kerins G, Demnerova K, Ovesna J (2016) Genome editing with engineer ednuclease sin economically important animals and plants: state of the art in the research pipeline. Curr Issues MolBiol 21:41–62

Stella S, Montoya G (2016) The genome editing revolution: a CRISPR- Cas TALE off-target story. BioEssays 38:S4–S13

Sunkar R, Jagadeeswaran G (2008) In silico identification of conserved microRNAs in large number of diverse plant species. BMC Plant Biol 8:37

Tang G, Reinhart BJ, Bartel DP, Zamore PD (2003) A biochemical framework for RNA silencing in plants. Genes Dev 17:49–63

Tenllado F, Martinez-Garcia B, Vargas M, Diaz-Ruiz JR (2003) Crude extracts of bacterially expressed dsRNA can be used to protect plants against virus infections. BMC Biotechnol 3:3

Thresh JM, Cooter RJ (2005) Strategies for controlling cassava mosaic virus disease in Africa. Plant Pathol 54:587–614

Valkonen J (1998) Virus disease control in plants using natural and engineered resistance and some consideration regarding biosafety. Currents 17:51–55

Vanitharani R, Chellappan P, Pita JS, Fauquet CM (2004) Differential roles of AC2 and AC4 of cassava geminiviruses in mediating synergism and suppression of posttranscriptional gene silencing. J Virol 78(17):9487–9498

Vanitharani R, Chellappan P, Fauquet CM (2005) Geminiviruses and RNA silencing. Trends Plant Sci 10:144–151

Varma A, Malathi VG (2003) Emerging geminivirus problems: a serious threat to crop production. Ann Appl Biol 142:145–164

Voinnet O (2005) Induction and suppression of RNA silencing: insights from viral infections. Nat Rev Genet 6:206–220. https://doi.org/10.1038/nrg1555

Voinnet O, Pinto YM, Baulcombe DC (1999) Suppression of gene silencing: a general strategy used by diverse DNA and RNA viruses of plants. Proc Natl Acad Sci USA 96:14147–14152

Vu TV, Choudhury NR, Mukherjee SK (2013) Transgenic tomato plants expressing artificial microRNAs for silencing the pre-coat and coat proteins of a begomovirus, Tomato leaf curl New Delhi virus, show tolerance to virus infection. Virus Res 172:35–45. https://doi.org/10.1016/j.virusres.2012.12.008

Wang HL (1991) Effectiveness of cross protection by a mild strain of zucchini yellow mosaic virus in cucumber, melon, and squash. Plant Dis 75:203

Waterhouse PM, Graham MW, Wang MB (1998) Virus resistance and gene silencing in plants can be induced by simultaneous expression of sense and antisense RNA. Proc Natl Acad Sci U S A 95:13959–13964

Wen F, Lister RM, Fattouh FA (1991) Cross-protection among strains of barley yellow dwarf virus. J Gen Virol 72:791–799

Wesley SV, Helliwell CA, Smith NA, Wang MB, Rouse DT, Liu Q, Gooding PS, Singh SP, Abbott D, Stoutjesdijk PA, Robinson SP, Gleave AP, Green AG, Waterhouse PM (2001) Construct design for efficient, effective and high-throughput gene silencing in plants. Plant J 27:581–590

Wilson TMA (1993) Strategies to protect crop plants against viruses: pathogen-derived resistance blossoms. Proc Natl Acad Sci USA 90:3134–3141

Wintermantel WM, Banerjee N, Oliver JC, Paolillo DJ, Zaitlin M (1997) Cucumber mosaic virus is restricted from entering minor veins in transgenic tobacco exhibiting replicase-mediated resistance. Virology 231:248–257

Woo JW, Kim J, Kwon SI, Corvalan C, Cho SW, Kim H et al (2015) DNA- free genome editing in plants with pre-assembled CRISPR- Cas9 ribonucleoproteins. Nat Biotechnol 33:1162–1164

Yang Y, Sherwood T, Patte C, Hiebert E, Polston J (2004) Use of Tomato yellow leaf curl virus (TYLCV) Rep gene sequences to engineer TYLCV resistance in tomato. Phytopathology 94 (5):490–496

Zhang SC, Wege C, Jeske H (2001) Movement proteins (BC1 and BV1) of Abutilon mosaic geminivirus are cotransported in and between cells of sink but not of source leaves as detected by green fluorescent tagging. Virology 290:249–260

Zhang BH, Pan XP, Wang QL, Cobb GP, Anderson TA (2005) Identification and characterization of new plant microRNAs using EST analysis. Cell Res 15:336–360

Zhang BH, Pan XP, Cannon CH, Cobb GP, Anderson TA (2006) Conservation and divergence of plant microRNA genes. Plant J 46:243–259

Ziebell H, Carr JP (2010) Cross-protection: a century of mystery. Adv Virus Res 76:211–264

Zulma IM, Argüello-Astorga GR, Bustamante RFR (2002) Geminivirus replication and gene expression. In: Khan JA, Dijkstra J (eds) Plant viruses as molecular pathogens. The Haworth Press, New York, p 537

Integrated Pest Management Approaches

S. U. Mohammed Riyaz and K. Kathiravan

Abstract

Integrated pest management (IPM) is an internationally recognized approach to pest and disease control. IPM embraces diversity, is knowledge intensive, and varies by crop, scale, and geographical location. All farmers practice IPM to some degree, including the cultural control techniques that underpin all good farming practices. In reality, most farming practice is neither IPM nor non-IPM, but can be defined at a point along the so-called IPM continuum from chemically intensive systems to bio-intensive systems. IPM was initially conceptualized to reduce dependence on pesticides and their effects on the environment. It has been built into virus control strategies from the beginning of plant virology because of the known in vivo insensitivity of viruses to chemical agents. Several methodologies are available for implementing IPM for *Bemisia tabaci* populations: chemical control with selective insecticides, biological control, crop plant resistance, and physical/mechanical methods. Insecticides, by their poisonous nature, are often harmful to natural enemies and therefore are disruptive to overall pest management. However, the more modern materials that are effective for *B. tabaci* control are relatively specific to the target pests and therefore less harmful to natural enemies and the environment; consequently, they are also more suitable for integrative combination with other methods. Conventional IPM technologies, such as intercropping, will yield mixed results with little, if any, beneficial impact on pest population in crops. This chapter reviews the known measures used for

S. U. M. Riyaz (✉)
Molecular Plant Virology Laboratory, Department of Biotechnology, University of Madras, Guindy Campus, Chennai, India

Centre for Ocean Research, Sathyabama Institute of Science and Technology, Chennai, India

K. Kathiravan
Molecular Plant Virology Laboratory, Department of Biotechnology, University of Madras, Guindy Campus, Chennai, India
e-mail: kathir68@unom.ac.in

R. V. Kumar (ed.), *Geminiviruses*, https://doi.org/10.1007/978-3-030-18248-9_12

reducing populations of *B. tabaci*, advocating the view that only a comprehensive approach incorporating IPM programs will offer effective and sustainable strategies for managing whiteflies.

1 Introduction

Integrated pest management (IPM) is the comprehensive and coordinated use of cultural, biological, and chemical tactics to reduce a pest population below an acceptable threshold. Cultural IPM practices include nonchemical tactics, host plant resistance, planting dates, cover crops, traps, scouting, crop rotation, and sanitation. Biological IPM practices include natural enemy conservation and enhancement, whereas chemical IPM practices include pesticide selection and spray timing.

IPM is a systems-based approach designed to reduce environmental, health, and economic risks. IPM is implemented as an ongoing series of science-based pest management evaluations, decisions, and interventions. IPM practitioners use knowledge of pest biology and environmental conditions and technology to prevent, avoid, monitor, and suppress pests. IPM practices may be basic or advanced. Basic IPM practices include scouting or sampling crops for pests and pest damage (visually or with devices), monitoring weather and other conditions, and acting when pests approach economically damaging levels. Advanced IPM practices include planting pest-resistant crop varieties, rotating crops, adjusting planting times, using reduced-risk pesticides, implementing mating disruption, planting companion crops, and incorporating beneficial insects.

Some IPM practices, such as organic soil amendments with poultry refuse, mustard oil-cake, neem oil-cake, cow dung, vermicompost, and *Trichoderma harzianum*, can significantly reduce plant parasitic nematodes and increase or induce the growth of various beneficial fungal- and bacterial-feeding nematodes. Neither the cropping pattern nor the crop production region (e.g., Jessore and Sirajganj) influences the effects of IPM and non-IPM systems.

2 Tools for IPM in Greenhouse Production

In greenhouse production, IPM tools may include trap crops, indicator plants, and banker plants. Trap crops are most often used for insect pest control, such as perimeter trap cropping in field vegetables and trap crops interspersed in greenhouse ornamentals, with characteristics that are more attractive to pests than are crops. Because of the inability to cure plant viral diseases and the need to protect the environment from toxic pesticides, alternative indirect strategies of disease control are required. In recent decades, virologists have developed non-pesticidal, cultural control practices aimed at reducing the damage caused by these viral diseases by interrupting their epidemiological cycle. One of the most important crop improvements has been the enhancement of tolerance to biotic stresses. The identification and use of resistance sources in plant breeding programs have resulted in substantial gains in crop productivity.

Despite the ongoing efforts, India's productivity for major crops is far below the global average, largely due to persisting problems of pests and diseases. Some IPM interventions have been used as novel strategies for overcoming the pests and diseases in India, including the following:

- Seed treatment with chemical pesticides to avoid sucking pest attacks
- Intercropping with legumes to augment natural enemy populations
- Trap cropping to reduce damage to the main crop from important pests
- Bird perches for alighting insectivorous birds to predate on insects
- Pheromone traps for monitoring or mass trapping of moths
- Scouting to monitor the status of pests and beneficial organisms at regular intervals
- Augmenting biocontrol agents, such as *Trichogramma*/*Chrysoperla*
- Spraying biopesticides, such as *Helicoverpa armigera* Nuclear Polyhedrosis Virus (Ha NPV) and neem seed kernel extract
- Topping the cotton plants at the time of high oviposition with *Helicoverpa*
- Periodic removal and destruction of dropped squares, dried flowers, premature bolls, and infested shoots
- Yellow sticky traps and light traps to control sucking pests such as whiteflies, jassids, and aphids
- Manipulation of wavelength-dependent behavior of insects to impede insects and restrict epidemics of insect-borne viruses

Using conventional practices and recent innovation techniques, scientists have designed some IPM strategies especially for tomato growers to promote and ensure the use of transplants to reduce herbicides and conserve water. In addition, the use of disease-resistant varieties can eliminate pesticide usage. Other IPM practices can reduce synthetic insecticide usage, including disease-free seeds, disease- and pest-resistant varieties, biological control (parasitic wasps), mating confusion (sex pheromones), biological pesticides, forecasting systems (TOM-CAST), risk assessment (GIS/GPS), the judicious use of synthetic pesticides, conservation tillage (which reduces fuel, dust, emission, water runoff, and soil erosion), 2–3 years of crop rotation to minimize diseases, cover cropping to improve soil texture, and habitat management, such as replanting ditches with native vegetation and preservation of wetlands.

3 Insect–Plant Communication: Visual Cues

The long evolutionary associations between insects and plants have led to mechanisms that enable insects to detect and select their preferred hosts for feeding and oviposition. Vision (color, shape, size) and olfaction (host odor) are the primary cues used by insects to orient to their plant hosts; sometimes, the two types of cues are complementary (Prokopy and Owens 1983; Dobson 1994; Terry 1997). Cues for detecting hosts may be general for polyphagous species or very specific for those that are monophagous. Once a potential host is contacted, then odor, tactile, and gustatory cues may predominate (Terry 1997).

The behavioral response of insects to colored surfaces or colored lights has been referred to as *color sensation* or *spectral sensitivity*. The first term describes a phenomenon that is governed by physical stimuli, sensorial receptors, and an integrative system. The second term refers to sensory cells or sensory organs. Visual cells may be sensitive to all wavelengths, but it is the integration of the sensorial inputs to the central nervous system that results in the specific phototactic response of a given insect species (Vaishampayan et al. 1975a).

Color and color contrasts are used by insects to distinguish between a host and the surrounding environment. From a biological perspective, there are three main parameters of color (Vaishampayan et al. 1975a; Terry 1997):

1. The hue or dominant wavelength remitted by the surface (λ_{max}).
2. The color saturation or purity of the hue. For example, adding white to yellow causes a significant increase in the blue-violet region.
3. Brightness (light intensity) refers to the overall reflection. Intensity affects the response when associated with a peak of a dominant wavelength.

4 Phototactic Action Spectrum for Whiteflies and Aphids

Mound (1962) suggested that the whitefly *Bemisia tabaci* (Gennadius) is attracted by two groups of wavelengths of transmitted light: the blue/ultraviolet and the yellow. He related the reaction to ultraviolet with the induction of migratory behavior, whereas yellow radiation induces vegetative behavior that may be a part of the host selection mechanism. It has also been found that *B. tabaci* has no detectable olfactory reactions. A close agreement was found between the phototactic action spectrum of the greenhouse whitefly (*Trialeurodes vaporariorum* (Westw.)) and the transmission spectrum of the leaf in the mid-visible wavelength (550 nm) (Macdowall 1972). Vaishampayan et al. (1975a) observed a strong positive response of *T. vaporariorum* to surfaces with maximum reflectance or transmittance in the yellow-green region (520–610 nm) and a moderately positive response to ultraviolet (360–380 nm). Light in the blue-violet region seemed to inhibit the response, and red (610–700 nm) may also be moderately inhibitory. Based on these findings, he suggested that the first steps in host selection, orientation, and landing of *T. vaporariorum* are mediated largely, if not exclusively, by a response to reflected yellow light (520–610 nm; Vaishampayan et al. 1975b).

Coombe (1982) reported that adults took off more readily and walked faster under light of 400 nm than under 500 nm. He confirmed Mound's hypothesis that the two types of radiation are complementary, thus eliciting a balance between migratory behavior induced by ultraviolet (UV) and a landing reaction controlled by sensitivity to yellow (Mound 1962). In flying aphids, it has been suggested that the primary function of color vision lies in distinguishing plants from sky. Moericke (1955) suggested that the sensitivity of aphids to color may be related to the host range for any given species.

5 Control of Insect Vectors by Altering Their Vision Behavior

Insects communicate with their environment and host plants by light signals that elicit photoreceptors in their compound eyes. The vision behavior of insects is linked to the sequence that begins with their orientation to the plant from a distance and ends with their establishment on plants for feeding and oviposition. By interfering with different links along this pathway, contact between the vector and the plant may be prevented and, therefore, virus spread will be decreased.

6 Attracting Insects Using Color

6.1 The Use of Colored Soil Mulches to Control Whitefly-Borne Viruses

The ability of mulches to attract or to repel insects can be very important in protecting plants from virus diseases. The attraction of whiteflies to yellow was utilized successfully to protect cucumber and tomato crops from infection with the whitefly-borne viruses cucumber vein yellowing virus and tomato yellow leaf curl virus (TYLCV), respectively. These viruses were controlled by soil mulches of saw dust, straw, or yellow polythene film (Nitzany et al. 1964; Cohen and Melamed-Madjar 1978; Cohen and Berlinger 1986; Cohen and Antignus 1994). Polyethylene sheets were the most effective of these materials in reducing the incidence of TYLCV (Cohen 1982). It was suggested that this protection mechanism is associated with the preferential attraction of *B. tabaci* to yellow, leading to subsequent death of the insect caused by the reflected heat (Cohen 1982). No protection from TYLCV occurred when a tomato field was surrounded by a strip of yellow sticky polyethylene erected vertically 70 cm above ground level (Cohen 1982). Csizinszky et al. (1995) reported that tomato plants grown on orange plastic mulch and exposed to whitefly-transmitted tomato mottle virus performed better in terms of delayed virus symptoms and yield of marketable fruit than those grown conventionally with white or black mulches.

Grass-feeding thrips show little preference for any wavelength, whereas all anthophilous thrips are attracted to colors that match those of flowers—that is, low-UV, white, blue, and yellow, whereas a few are attracted to green, red, and black. Matteson and Terry (1992) found that the degree of the color's attractiveness to the western flower thrips *Frankliniella occidentalis* (Pergande) corresponded to the brightness in the blue wavelength.

Because of economics, availability, limited capacity, and a lack of reliable information, synthetic pesticides have not been used extensively in small-scale cultivation worldwide. However, the limited published information available, together with analogous experience in other crops, suggests that the cost-benefit ratio of controlling pests and diseases using inorganic pesticides is favorable in some circumstances if highly standardized timing, dosing, and targeting are applied as a part of an IPM strategy. In particular, newer, more selective molecules, such as imidacloprod, which are applied as a spray or seed dressing, can be very effective at controlling sucking pests (and some disease vectors), with older molecules (principally pyrethroids) to control chewing and boring pests and low-cost, old molecules for fungal disease control. Non-target impacts on natural enemies and resistance management are important considerations in any successful regimen.

7 *Bemisia tabaci*: Whitefly

The whitefly *Bemisia tabaci* (Gennadius) (Hemiptera: Aleyrodidae) is a pest in many agricultural systems, including various vegetable, ornamental, and field crops (Byrne and Bellows 1991; Oliveira et al. 2001; Stansly and Naranjo 2010). It directly damages plants by feeding on phloem sap and excreting honeydew on the leaves and fruit. The sticky, sugary surface forms a substrate for the growth of black, sooty mold fungi that stains the crop and covers the leaves, thus preventing proper photosynthesis. The resulting stickiness and discoloration greatly reduce the value of agricultural crops such as ornamentals, vegetables, and cotton. In cotton, the honeydew may cause fiber stickiness that interferes with the spinning process in textile mills and greatly reduces the product's value (Hequet et al. 2007).

B. tabaci is a vector of several important families of plant viruses (Jones 2003; Hogenhout et al. 2008). In some crops (e.g., tomatoes and cassava), the resulting viral diseases are limiting-growth factors and may cause total crop loss. Most of the important virus diseases transmitted by *B. tabaci* belong to the geminivirus group (Family: *Geminiviridae*).

B. tabaci is known for its genetic diversity, which is expressed in a complex of biotypes (Brown et al. 1995a, b; Perring 2001; De Barro et al. 2005) or, as recently suggested, a complex of separate species (Xu et al. 2010; De Barro et al. 2011). The biotypes are largely differentiated based on biochemical or molecular polymorphisms. They differ in characteristics such as host plant range, the capacity to cause plant disorders, attraction of natural enemies, expression of resistance, and plant virus-transmission capabilities (Bedford et al. 1994; Brown et al. 1995a, b; Sánchez-Campos et al. 1999; Perring 2001; Horowitz et al. 2005). Reports have suggested that the floral composition of bacterial symbionts might be specific to certain biotypes (Gottlieb et al. 2006; Chiel et al. 2007) and might confer upon them resistance to insecticides (Kontsedalov et al. 2008). The most widespread biotype, B, was recognized in the late 1980s (Costa and Brown 1991; Costa et al. 1993) following extensive outbreaks of *B. tabaci* in the southwestern United States, and it has a worldwide distribution. An additional widespread biotype, Q, which probably originated in the Iberian Peninsula (Guirao et al. 1997), has since spread globally (Horowitz et al. 2003; Boykin et al. 2007; Chu et al. 2010).

Management of *B. tabaci* populations and, in particular, management of the viral plant diseases it transmits, is difficult. This is due to the pest's elevated population growth rates, rapid evolution of resistance to insecticides, and the relatively protected location of the individuals on the underside of the leaves. *B. tabaci* is highly polyphagous and is known to develop on more than 500 plant species, including a large number of fiber, vegetable, and ornamental crops (Mound and Halsey 1978; Oliveira et al. 2001). Another remarkable feature is its easy adaptation to changing environmental conditions, especially in subtropical and tropical agroecosystems and in greenhouse-grown crops, even in temperate climates (Brown 2007a, b; Castle et al. 2010). Brown (2007a, b) proposed that monoculture cropping together with year-round production practices are mostly responsible for the present whitefly and viral disease outbreaks. Because viral plant diseases transmitted by *B. tabaci* are not curable, the principal strategies for their management are based on prevention of transmission (Antignus 2007) and/or on utilization of host-plant resistance (Lapidot and Friedmann 2002). At present, the use of insecticides is the main approach employed to manage *B. tabaci* populations. This practice is greatly restricted, however, due to both environmental concerns and the widespread resistance that *B. tabaci* has developed to most of the insecticides in use (Palumbo et al. 2001; Horowitz et al. 2007; Castle et al. 2010). Consequently, increasing importance is being placed upon control by other methods (including cultural, mechanical, and biological) as a means of managing pest populations.

Worldwide outbreaks of *B. tabaci* whiteflies, especially biotype B, have facilitated the emergence of whitefly-transmitted geminiviruses (WTGs). These viruses cause economically important diseases of vegetable and fiber crops, especially in tropical and subtropical regions of the world. Because small populations of whiteflies can efficiently spread WTGs, management of these diseases is more challenging than for whiteflies alone. As the WTGs have emerged worldwide, key aspects of the biology of WTGs and *B. tabaci* have shaped the development of an IPM approach for these diseases. The generalized IPM package involves strategies for implementation before the growing season, such as the use of virus- and whitefly-free transplants, propagative stock, and resistant varieties. During the growing season, approaches may include whitefly population suppression, roguing virus-infected plants, floating row covers, and reflective mulches. After the growing season, strategies include region-wide sanitation, weed management, and implementation of a host-free period.

Because it is not possible to cure plants of WTG infections, efforts must be taken to keep plants from becoming infected or to manage the rate, timing, and severity of the infection to protect crop health. Growers emphasize whitefly management with insecticides to control WTGs; however, in most cases, this does not provide adequate protection. IPM approaches have been more successful in the management of WTGs (Jones 2003), using multiple strategies that target different levels of the plant–WTG–whitefly interaction. The first widely recognized concept of IPM stressed a combination of chemical, biological, and other control methods for insect pest management (Stern et al. 1959).

A number of very effective strategies can provide effective management of diseases caused by WTGs when combined into an IPM package. The specific strategies used for the IPM package in a given agroecosystem are dependent on knowledge of the crop plant, the cropping system, climatic conditions, and the biology of the virus and the vector.

A generalized scheme for the IPM of a whitefly-transmitted virus is divided into three parts: before the growing season, during the growing season, and after the growing season. Before the growing season, advance preparation in terms of the cultivar of the crop, the source of the planting material (seed, transplants, or propagative material), and field location are very important. The cultivar and seed selection of the proper cultivar are important for many reasons. However, in the case of IPM of WTGs, the key points are related to the availability of virus resistance or tolerance and certain horticultural aspects. Host plant resistance to whiteflies or WTGs provides an ideal pest management tool, with little or no environmental impact. Unfortunately, host plant resistance is not available for many whitefly-transmitted begomoviruses, and there are even fewer examples of resistance to the vector.

Coping with plant diseases in the field is relatively difficult because the causal organisms (bacteria, MLO, fungi, virus and nematodes) are very small and cannot be seen moving around like insects or rats. The most important first step in thinking about diseases is to realize that diseases must be managed and not controlled. What is the difference? Management means a complete set of activities that support each other. Management means that these activities are carefully planned and are implemented over several seasons, not controlled within a single season. Management includes control methods for prevention and control methods to slow down epidemics; diseases will never be completely eradicated; only populations reduced to very low levels. Management usually needs the cooperation of several farmers working together to reduce overall disease in an area. Management requires someone who can observe larger areas of disease incidence and levels of infection.

Many WTGs are important in developing countries where subsistence farmers are involved in vegetable production. In general, the earlier that a plant is infected with a virus, the more severe the disease symptoms and the greater the yield loss. Thus, it is critical to establish new plantings with virus-free and whitefly-free transplants or propagative stocks. The first step is to keep transplant propagation facilities free of whiteflies. Greenhouses should have induced positive airflow, double-door airlock entrances, and roofs covered with UV-absorbing films. All vents and other openings should be covered with whitefly-resistant, fine-mesh screening with 0.25×0.8 mm openings or less. Sanitation within and around the propagation facilities is also important. Potential whitefly or virus host plants must be eliminated in and around the facility, with discarded plant materials sealed in whitefly-proof containers or destroyed.

In addition, systemic neonicotinoid class insecticides (e.g., imidacloprid or thiamethoxam) applied as soil drenches, along with foliar insecticide sprays, can be used in greenhouse operations to suppress whitefly populations. Monitoring of whitefly adults with yellow sticky traps can be used to know when foliar insecticides need to be applied (Gillespe and Quiring 1987). One approach is to place one trap per 80 plants, or at least one per 6 m^2, at the beginning of the transplant production to be used as a monitoring and control measure.

8 Location and Time of Planting

New plantings should be established following a host-free period or during periods when virus and whitefly pressure are low. If there is a good information available on the seasonal patterns of whitefly populations and virus pressure, planting times can be modified to avoid periods of high pressure. If multiple staggered plantings are planned, barrier crops can be planted prior to the establishment of the plantings, which are established upwind of earlier plantings and in blocks such that minimal area of the field is exposed to wind. However, under heavy virus pressure, these approaches alone are unlikely to substantially reduce virus infection in the field.

Whitefly management during the growth season leads to the suppression of whitefly populations with insecticides. Especially in areas with histories of whitefly outbreaks, this is an important component of a successful IPM package for WTGs. Insecticides are most commonly applied as foliar sprays or injected into the soil, but may also be applied via chemigation through drip irrigation. Soil applications are typically systemic insecticides, mostly in the neonicotinoid chemical class. The prophylactic use of soil-applied systemic insecticides has been documented to slow, reduce, or delay virus transmission by whiteflies. However, the use of insecticides alone often does not deliver sufficient protection from WTGs to prevent economically important crop damage.

9 Roguing

A roguing strategy involves the physical removal of virus-infected plants over the course of the growing season. Roguing needs to be done soon after plots are established and is most helpful if the incidence of the virus is low (<5%). After roguing, it is also important that there is a minimal level of virus spread in the field, as well as a limited amount of introduction of virus form outside the field. If whitefly populations are high, plants should be treated with an insecticide to kill whitefly adults prior to roguing. If nymphs are present, rogued plants should be removed in plastic bags and disposed of well away from production fields.

Vegetables can be protected from whitefly damage by an exclusion method using protected culture in greenhouses and screen houses for virus infection by physical means (i.e., preventing the insects from contacting susceptible plants). In the most extreme case, the entire crop is grown in a greenhouse or screen house, and plants are protected from whiteflies for the entire production cycle. When these structures are kept free of whiteflies (e.g., through the use of glass, plastic, or screening; vents covered with screening; double doors with positive pressure), excellent management of whiteflies and WTGs can be achieved.

Another common method used for the protection of plants in the field with floating row covers is the covering of young plants, either those emerging from seeds or that have been transplanted, with protective netting. This netting is a spun-bonded polyester material (commercially available as Agribon or Agril) that is placed directly over the rows of emerging seedings or transplants. The covers are typically placed over the plants without any type of support, such that it is a floating row cover and moves with the growth of the plants. In other cases, semi-circular lengths of wire or piping are used to provide support and keep the netting from directly contacting the leaves. These materials allow passage of adequate amounts of light for normal plant growth, although there have been some reports that the microclimate formed under the row covers can favor the development of foliar diseases caused by bacteria and fungi. In general, the row covers are left on for 30 days or until pollination, such as in the case of cucurbits.

It is well established that the use of row covers can protect plants from whiteflies and reduce the spread of WTGs in crops such as cucurbits, pepper, and tomato (Natwick and Durazo 1985; Natwick and Laemmlen 1993; Orozco-Santos et al. 1995; Webb and Linda 1992). In cases of severe whitefly and virus pressure, this protection can make the difference in whether a marketable crop is produced. This approach has been shown to slow the spread of geminivirus and is being used in Guatemala to protect tomato and peppers form infection with various WTGs. Row covers have also been successfully used in Guatemala to protect melons from WTGs and the whitefly-transmitted crinivirus. However, the use of floating row covers is expensive; it should be used to protect plants during periods where whitefly and virus pressure are known to be high. Small-scale farmers can use row covers to protect seed beds or to produce small tunnels in which seedlings can be protected from whiteflies during this critical stage of growth (Hilje et al. 2001).

10 Barrier Crops and Mulches

A number of other cultural practices can be used to protect crops from whiteflies and thus slow the spread of WTGs. Physical barriers can be designed to prevent the movement of whiteflies into fields of susceptible crops. Barriers may be non-living, such as plastic (yellow plastic with sticky material to trap insects) or screening, or living, such as the planting of a tall plant species (non-hosts of the whitefly and WTGs) between fields of susceptible crops. The best barrier plants for WTGs are monocots such as corn, sorghum, and elephant grass. There is little evidence that barriers effectively reduce whitefly migration or virus spread because whiteflies can fly or be wind-carried over barriers and transmit WTGs for long periods of time, due to the persistent nature of transmission (Hilje et al. 2001). Thus, barriers are generally not an essential component of the IPM package for WTGs.

Mulches are designed to prevent insects from recognizing and landing on a crop that is susceptible to virus infection. Like barriers, mulches can be non-living (plastic or some other material) or living (plants grown among the susceptible crop). In terms of non-living mulches, the most effective materials are colored or UV-reflective plastic. These have been reported to have some success in reducing whitefly population densities as well as the incidence of WTGs (Antignus 2000). Living mulches involve planting low-growing ground cover-type plants, which are non-hosts for whiteflies and WTGs, among a susceptible crop. These living mulches reduce whitefly populations by causing the insects to leave the field due to the presence of the non-host plants (Hilje et al. 2001).

11 IPM Package for Tomato-Infecting Geminiviruses: Preplant Activities

11.1 Use of Virus-Free and Whitefly-Free Transplants

Tomato-infecting begomoviruses are not seed-transmitted, so transplants will not become infected via contaminated or infected seeds. However, whiteflies can transmit the virus to plants in the seedling stage; establishing fields with virus-infected transplants will lead to rapid spread of the virus within the field. Furthermore, infection of plants at such an early stage of growth will lead to the greatest economic losses. Therefore, an essential component of the IPM package is the use of virus-free and whitefly-free transplants. In the case of WTGs, this means keeping whiteflies physically separated from transplants.

11.2 Whitefly Monitoring and Management

Whitefly monitoring and management is costly and not good for the health of farmers or the environment. It is important to monitor whitefly populations to understand the population dynamics on a regional basis, especially to detect the build-up of populations early in the crop production cycle. This can be done by monitoring adult populations with yellow sticky cards or with the leaf turn method, in which adults and/or nymphs are directly counted on the undersurfaces of leaves. However, as mentioned earlier, with WTGs, the challenge is developing threshold populations that can trigger pesticide applications that will slow the spread of the virus.

During the growing season in row covers, the crop is being transplanted into the field in the presence of viruliferous whiteflies; plants can be physically protected with floating row covers. These materials are placed over the rows of plants, leaving the ground between rows uncovered. The covers can only be left over plants for approximately 30 days; if viruliferous whiteflies are still present, these plants will become infected. Cultivated tomato has been a good host for the evolution of new WTGs, and this has been facilitated by the worldwide dissemination of the polyphagous *B. tabaci* biotype B. In many cases, these viruses cause diseases of considerable economic importance, particularly in tropical and subtropical regions. Many components of the IPM package for these viruses do not require specific knowledge of the begomovirus(es) involved, but identification of the WTGs involved in a region may influence the selection of resistant varieties.

It is critical to start with virus-free and whitefly-free transplants and, if possible, resistant varieties. Ideally, transplants are planted during a period of low virus pressure, such as following a host-free period or away from established fields. If viruliferous whiteflies are present, then additional measures may be taken, such as floating row covers or management of whitefly populations with insecticides. Roguing infected plants early in the season may slow down spread of the virus, as may the use of reflective mulches. Following harvest, it is critical to uproot and destroy old plants through removal or tillage. In tropical and sub-tropical regions, the implementation of a 1- to 3-month tomato or whitefly host-free period can substantially reduce virus and whitefly pressure for the next crop. The free period provides an effective approach that is not based on pesticides, but this requires regional cooperation. By implementing this IPM package, there is a high probability that effective management of any tomato-infecting begomovirus can be accomplished.

11.3 Cassava Mosaic Disease

Cassava mosaic disease (CMD) is one of the most damaging diseases of cassava (Fargette et al. 2006; Thresh and Cooter 2005). It can cause substantial yield reductions and is very difficult to manage. CMD is characterized by a striking light to dark green mosaic of leaves, various degrees of leaf and stem distortion, and reduced numbers and weights of tubers. CMD is caused by a complex of whitefly-transmitted begomoviruses, including African cassava mosaic virus and East African cassava mosaic virus. Biological considerations for an effective IPM strategy for CMD must consider the perennial nature of the crop and the fact that is vegetatively propagated (Thresh and Cooter 2005). Also, the viruses that cause CMD have a relatively narrow host range, infecting only members of the plant family *Euphorbiaceae,* including cassava, castor bean, and certain wild hosts and weeds.

An IPM package for CMD includes preplanting activities, such as disease-free cuttings, resistant varieties, and cultural practices. A number of cultural practices can also be considered, although these may have only limited beneficial effects or be too difficult for farmers to utilize (Thresh and Cooter 2005). Elongated plots that are exposed to prevailing winds should be avoided, because this is where the highest infection rates tend to occur. Intercropping of cassava with other crops such as banana, sweet potato, and legumes can reduce virus spread through the reduction of whitefly populations. Finally, cassava should be grown under favorable conditions, as CMD spreads slower in fields with healthy plants.

During the growing season, the physical removal of virus-infected plants over the course of the growing season can be useful, particularly when disease incidences are relatively low (<5%). Thus, roguing will be most effective when used in combination with preplanting measures, such as planting disease-free cuttings. It is also an important method for the amplification of sources of disease-free cuttings; it needs to be done soon after plots are established. Fields should be monitored once or twice shortly after planting as cuttings show symptoms in newly emerging leaves.

With respect to whitefly management, although there is a correlation between the number of whiteflies and the rate of spread of CMD, managing the disease with insecticide sprays has not been effective, nor is it practical. This relates to the fact that cassava is grown on small plots by subsistence farmers who often lack an understanding of CMD, the training and equipment to apply pesticides, and the resources to purchase the appropriate insecticides. However, the suppression of whitefly populations, either via biological control or with natural or synthetic insecticides, may help to slow the spread of the CMD in certain situations.

After the growing season ,cassava is the main host of the viruses that cause CMD, It is critical to destroy cassava plants promptly after harvest, as well as any other known host plants. This should be done within and around fields (for reservoir hosts) following harvest and, if necessary, before establishing new plantings. Ideally, the planting and harvest times in defined regions or localities should be coordinated to avoid periods of high disease pressure (e.g. high populations of viruliferous whiteflies) and to possibly allow a period with minimal plantings of cassava to help cleanse the agroecosystem of the virus.

12 Summary

Different IPM strategies can be followed for different periods of implementation (Fig. 1). Before the growing season, IPM strategies include the use of virus-free propagative material, the use of resistant cultivars, modification of planting dates, and avoidance of the planting of new fields near old fields. During the growing season, IPM includes the roguing of plants showing mosaic symptoms, monitoring for whitefly populations using established means of sampling, application of insecticides only when necessary, and rotation of insecticides to minimize development of resistance (e.g., no more than two uses of any material per season). After the growing season, IPM requires the prompt removal of crops following harvest. Certain cultural practices can also help to reduce the incidence or spread of disease, as can a systematic roguing program. Extensive sanitation, in the form of prompt removal of other hosts of the virus, can reduce inoculum pressure for subsequent plantings.

An IPM program or its components can provide effective disease management, but IPM has not been widely implemented. This is related to a lack of understanding of the disease by farmers and a lack of extension programs to deliver the IPM package to farmers. It will take a major effort to develop regional coordination to

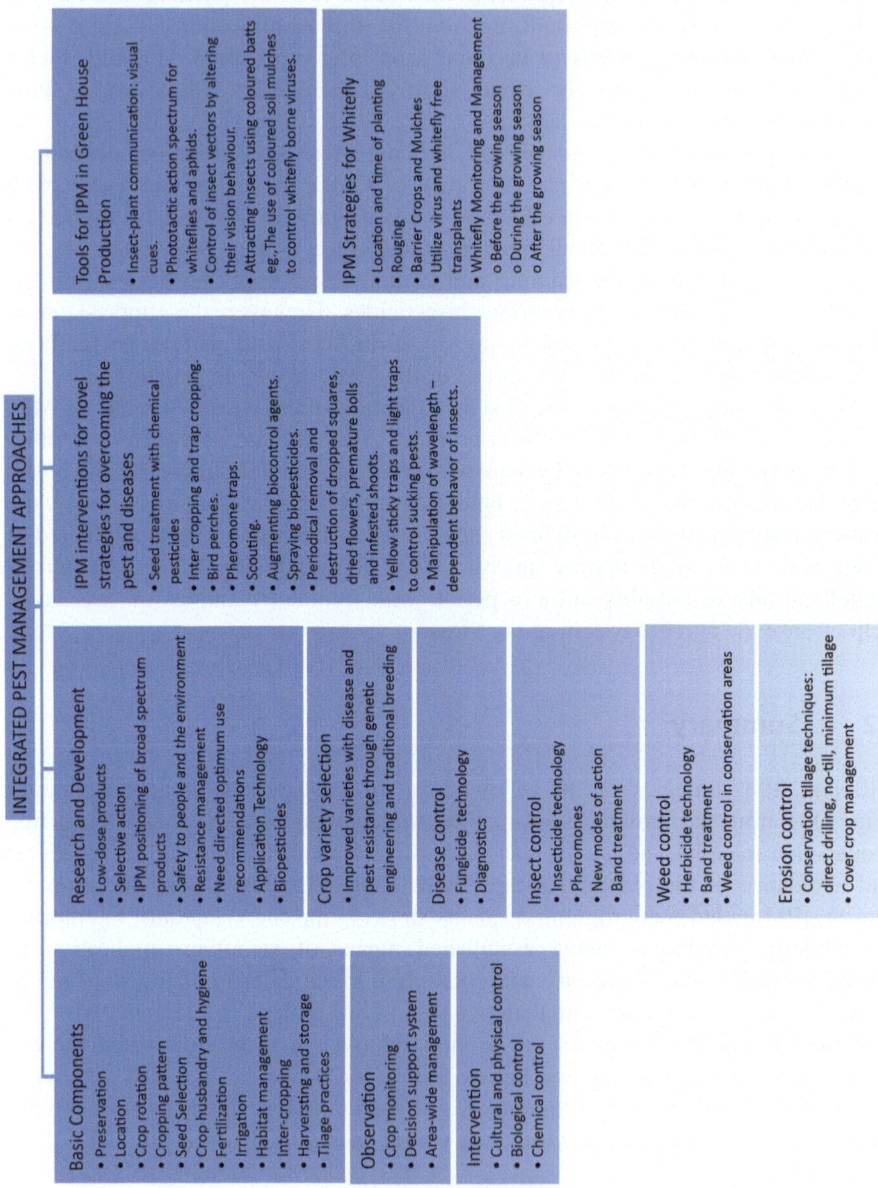

Fig. 1 List of integrated pest management (IPM) strategies available for implementation to curb plant viral diseases

implement relevant IPM packages. These packages may differ depending on the region or locality. IPM may require decentralized and participatory breeding efforts to generate resistant varieties that provide the desired horticultural properties preferred by local growers and consumers.

References

Antignus Y (2000) Manipulation of wavelength dependent behavior of insects: an IPM tool to impede epidemics and restrict spread of insect-borne viruses. Virus Res 71:213–220

Antignus Y (2007) The management of tomato yellow leaf curl virus in greenhouses and the open field, a strategy of manipulation. In: Czosnek H (ed) Tomato yellow leaf curl virus disease. Springer, Dordrecht, pp 263–278

Bedford ID, Briddon RW, Brown JK, Rosell RC, Markham PG (1994) Geminivirus transmission and biological characterisation of *Bemisia tabaci* Gennadius from different geographic regions. Ann Appl Biol 125:311–325

Boykin LM, Shatters RG Jr, Rosell RC, McKenzie CL, Bagnall RN, De Barro P, Frohlich DR (2007) Global relationships of *Bemisia tabaci* (Hemiptera: Aleyrodidae) revealed using Bayesian analysis of mitochondrial COI DNA sequences. Mol Phylogenet Evol 44:1306–1319

Brown JK (2007a) The *Bemisia tabaci* complex: genetic and phenotypic variability drives begomovirus spread and virus diversification. APSnetvFeature

Brown JK (2007b) The *Bemisia tabaci* complex: genetic and phenotypic variation and relevance to TYLCV-vector interactions. In: Czosnek H (ed) Tomato yellow leaf curl virus disease: management, molecular biology, breeding for resistance. Springer, Dordrecht, pp 25–56

Brown JK, Bird J, Banks G, Kiesler SM et al (1995a) First report of an epidemic in tomato caused by two whitefly-transmitted geminiviruses in Puerto Rico. Plant Dis 79:1250

Brown JK, Bird J, Banks G, Kiesler SM et al (1995b) First report of an epidemic in tomato caused by two whitefly-transmitted geminiviruses in Puerto Rico. Plant Dis 79:1250

Byrne DN, Bellows TS (1991) Whiteflies biology. Annu Rev Entomol 36:431–457

Castle SJ, Palumbo JC, Prabhaker N, Horowitz AR, Denholm I (2010) Ecological determinants of *Bemisia tabaci* resistance to insecticides. In: Stansly PA, Naranjo SE (eds) Bemisia: bionomics and management of a global pest. Springer, Dordrecht

Chiel E, Gottlieb Y, Zchori-Fein E, Mozes-Daube N, Katzir N, Inbar M, Ghanim M (2007) Biotype-dependent secondary symbiont communities in sympatric populations of *Bemisia tabaci*. Bull Entomol Res 97:407–413

Chu D, Wan FH, Zhang YJ, Brown JK (2010) Change in the biotype composition of *Bemisia tabaci* in Shandong Province of China from 2005 to 2008. Environ Entomol 39:1028–1036

Cohen S (1982) Control of whitefly vectors of viruses by color mulches. In: Harris KF, Maramorosch K (eds) Pathogens, vectors and plant diseases, approaches to control. Academic, New York, pp 45–56

Cohen S, Antignus Y (1994) Tomato yellow leaf curl virus, a whitefly-borne geminivirus of tomatoes. In: Harris KS (ed) Advances in disease vector research, vol 10. Springer-Verlag, New York, pp 259–288

Cohen S, Berlinger MJ (1986) Transmission and cultural control of whitefly-borne viruses. Agric Ecosyst Environ 17:89–97

Cohen S, Melamed-Madjar V (1978) Prevention by soil mulching of the spread of tomato yellow leaf curl virus transmitted by *Bemisia tabaci* (Gennadius) (Hemiptera: Aleyrodidae) in Israel. Bull Entomol Res 68:465–470

Coombe PE (1982) Visual behavior of the greenhouse whitefly, *Trialeurodes vaporariorum*. Physiol Entomol 7:243–251

Costa H, Brown JK (1991) Variation in biological characteristics and esterase patterns among populations of *Bemisia tabaci* and the association of one population with silver leaf symptom induction. Entomol Exp Appl 61:211–219

Costa HS, Brown JK, Sivasupramaniam S, Bird J (1993) Regional distribution, insecticide resistance, and reciprocal crosses between the 'A' and 'B' biotypes of *Bemisia tabaci*. Insect Sci Appl 14:255–266

Csizinszky AA, Schuster DJ, Kring JB (1995) Color mulches influence yield and insect pest populations in tomatoes. J Am Soc Hortic Sci 120:778–784

De Barro PJ, Trueman JWH, Frohlich DR (2005) Bemisia argentifolii is a race of *B. tabaci* (Hemiptera: Aleyrodidae): the molecular genetic differentiation of *B. tabaci* populations around the world. Bull Entomol Res 95:193–203

De Barro PJ, Liu SS, Boykin LM, Dinsdale A (2011) *Bemisia tabaci*: a statement of species status. Annu Rev Entomol 56:1–19

Dobson HE (1994) Floral volatiles in insect biology. In: Bernays EA (ed) Insect–plant interactions, vol 5. CRC Press, Boca Raton, FL, pp 47–81

Fargette D, Konate G, Fauquet C, Muller E, Peterschmitt M, Thresh JM (2006) Molecular ecology and emergence of tropical plant viruses. Annu Rev Phytopathol 44:235–260

Gillespe DR, Quiring D (1987) Yellow sticky traps for detecting and monitoring greenhouse whitefly (Homoptera: Aleyrodidae) adults on greenhouse tomato crops. J Econ Entomol 80:675–679

Gottlieb Y, Ghanim M, Chiel E, Gerling D, Portnoy V, Steinberg S, Tzuri G, Horowitz AR, Belausov E, Mozes-Daube N, Kontsedalov S, Gershon M, Gal S, Katzir N, Zchori-Fein E (2006) Identification and localization of a Rickettsia sp in *Bemisia tabaci* (Homoptera: Aleyrodidae). Appl Environ Microbiol 72:3646–3652

Guirao P, Beitia F, Cenis JL (1997) Biotype determination of Spanish populations of *Bemisia tabaci* (Hemiptera: Aleyrodidae). Bull Entomol Res 87:587–593

Hequet E, Henneberry TJ, Nichols RL (eds.) (2007) Sticky cotton: causes, effects, and prevention. USDA-ARS Technical Bulletin No 1915

Hilje L, Costa HS, Stansly PA (2001) Cultural practices for managing *Bemisia tabaci* and associated viral diseases. Crop Prot 20:801–812

Hogenhout SA, Ammar ED, Whitfield AE, Redinbaugh MG (2008) Insect vector interactions with persistently transmitted viruses. Annu Rev Phytopathol 46:327–359

Horowitz AR, Denholm I, Gorman K, Cenis JL, Kontsedalov S, Ishaaya I (2003) Biotype Q of *Bemisia tabaci* identified in Israel. Phytoparasitica 31:94–98

Horowitz AR, Kontsedalov S, Khasdan V, Ishaaya I (2005) Biotypes B and Q of Bemisia tabaci and their relevance to neonicotinoid and pyriproxyfen resistance. Arch Insect Biochem Physiol 58:216–225

Horowitz AR, Denholm I, Morin S (2007) Resistance to insecticides in the TYLCV vector, *Bemisia tabaci*. In: Czosnek H (ed) Tomato yellow leaf curl virus disease. Springer, Dordrecht, pp 305–325

Jones DR (2003) Plant viruses transmitted by whiteflies. Eur J Plant Pathol 109:195–219

Kontsedalov S, Zchori-Fein E, Chiel E, Gottlieb Y, Inbar M, Ghanim M (2008) The presence of *Rickettsia* is associated with increased susceptibility of *Bemisia tabaci* (Homoptera: Aleyrodidae) to insecticides. Pest Manag Sci 64:789–792

Lapidot M, Friedmann M (2002) Breeding for resistance to whitefly-transmitted geminiviruses. Ann Appl Biol 140:109–127

Macdowall FDH (1972) Phototactic action spectrum for whitefly and the question of colour vision. Can Entomol 104:299–307

Matteson N, Terry LI (1992) Response to colour by male and female *Frankliniella occidentalis*. Entomol Exp Appl 63:187–201

Moericke V (1955) U8 on the lifestyle of the winged leaf louse (Aphidina) with special consideration of behavior in the country. Z Angew Entomol 37:29–91

Mound L (1962) Studies on the olfaction colour sensitivity of *Bemisia tabaci* Genn. Aleyrodidae. Entomol Exp Appl 5(2):99–104

Mound LA, Halsey SH (1978) Whitefly of the world: a systematic catalogue of the Aleyrodidae (Homoptera) with host plant and natural enemy data. British Museum (Natural History), Chichester

Natwick ET, Durazo A III (1985) Polyester covers protect vegetables from whiteflies and virus disease. Calif Agric 39:21–22

Natwick E, Laemmlen FF (1993) Protection from phytophagous insects and virus vectors in honeydew melons using row covers. Flo Entomol 76:120. https://doi.org/10.2307/3496020

Nitzany FE, Geisenberg H, Koch B (1964) Tests for the protection of cucumbers from a whitefly-borne virus. Phytopathology 54:1059–1061

Oliveira MRV, Henneberry TJ, Anderson P (2001) History, current status, and collaborative research projects for *Bemisia tabaci*. Crop Prot 20:709–723

Orozco-Santos M, Perez-Zamora O, Lopez-Arriaga M (1995) Floating row cover and transparent mulch to reduce insect population, virus diseases and increase yield in cantaloupe. Fla Entomol 78:493–501

Palumbo JC, Horowitz AR, Prabhaker N (2001) Insecticidal control and resistance management for *Bemisia tabaci*. Crop Prot 20:739–765

Perring TM (2001) The *Bemisia tabaci* species complex. Crop Prot 20:725–737

Prokopy RJ, Owens ED (1983) Visual detection of plants by herbivorous insects. Annu Rev Entomol 28:337–364

Sánchez-Campos S, Navas-Castillo J, Camero R, Saria C, Díaz JA, Moriones E (1999) Displacement of Tomato yellow leaf curl virus (TYLCV)-Sr by TYLCV-is in tomato epidemics in Spain. Phytopathology 89:1038–1043

Stansly PA, Naranjo SE (eds) (2010) Bemisia: bionomics and management of a global pest. Springer, Dordrecht

Stern VM, Smith RF, van den Bosch R, Hagen KS (1959) The integrated control concept. Hilgardia 29:81–101

Terry LI (1997) Host selection, communication and reproductive behaviour. In: Lewis T (ed) Thrips as crop pests. CAB International, New York, pp 65–118

Thresh JM, Cooter RJ (2005) Strategies for controlling cassava mosaic virus disease in Africa. Plant Pathol 54:587–614

Vaishampayan SM, Kogan M, Waldbauer GP, Wooley JT (1975a) Spectral specific responses in the visual behaviour of the greenhouse whitefly, *Trialeurodes 6aporariorum* (Homoptera: Aleurodidae). Entomol Exp Appl 18:344–356

Vaishampayan SM, Waldbauer GP, Kogan M (1975b) Visual and olfactory responses in orientation to plants by the greenhouse whitefly, *Trialeurodes vaporariorum* (Homoptera: Aleurodidae). Entomol Exp Appl 18:412–422

Webb SE, Linda SB (1992) Evaluation of spun-bounded polyethylene row covers as a method of excluding insects and viruses affecting fall-grown squash in Florida. J Econ Entomol 85:2344–2352

Xu J, De Barro PJ, Liu SS (2010) Reproductive incompatibility among genetic groups of *Bemisia tabaci* supports the proposition that the whitefly is a cryptic species complex. Bull Entomol Res 100:359–366

The manufacturer's authorised representative in the EU is Springer Nature Customer Service Centre GmbH, Europaplatz 3, 69115 Heidelberg, Germany. If you have any concerns regarding our products, please contact ProductSafety@springernature.com

Printed and bound by CPI Group (UK) Ltd, Croydon, CR0 4YY
01/08/2026
02183523-0001